The Pacific Tsunami "Galu Afi"

The story of the greatest natural disaster Samoa has ever known

Written by *Lani Wendt Young*
Edited by *Hans Joachim Keil*

The printing of this book was funded by the Australian Aid Program.

ISBN 978 0 615 39840 2

Research for a book of this nature is a mammoth task and was carried out with the help of the following people –

Interview Assistant and Fieldwork translator: Lagitoa So'oalo
Additional assistance: Vanessa Niuwenhuizen
Coordinator of American Samoa interviews: Tereza Keil Steffany
Additional assistance: Francis Keil
Translation and transcription of interviews: Noa and Roberta Tiatia
New Zealand interviews and transcription: Carmel Peteru
Additional assistance: Diane Cederman
Tongan translation: Karoline Sopu
Tsunami science consultant: Dr Jose Borrero (NZ/USA)
Tsunami Psycho-social trauma consultant: Elena Peteru

Dedication

This book is dedicated to the memory of all those who lost their lives in the Pacific Tsunami on the 29th of September, 2009.

Table of Contents

Foreword

Tui Atua Tupua Tamasese Ta'isi Tupuola Tufuga Efi
Head of State of Samoa

On September 29, 2009, just before 7am, I was in my car on my way to Vailele for my morning walk. When we came to a river bank I felt the earth tremor. I saw nearby fences and trees shaking and a whole lot of leaves falling. I remember feeling a little disoriented. This book took me right back to those moments of disorientation. As I read I became engrossed in what is a most informative, most sensitively articulated and beautifully woven story. Not just about the impact of the tsunami on victims and helpers, but perhaps more tellingly, about its impact on us as a people, as a society and as a nation.

Albert Wendt has said: "We are what we remember". And, as Lani cites, Rudyard Kipling, once said: "If history were taught in the form of stories, it would never be forgotten". Stories can help us to remember history. They can help us to discern through emotion, wit, analysis and poetry what matters to us most and why; what holds, moves and defines us. They allow us to sift through and remember the good and bad; to learn and move on.

When we tell and retell stories about ourselves, our families, villages and countries we learn about ourselves; we remember and celebrate history, heritage and identity. When we tell and retell of the tragedies that befall us we tell of our strengths and weaknesses, likes and dislikes, achievements and failures, prejudices, biases, and all else that makes us human.

Stories of tragedies such as those captured in this book will be treasured and retold for many years to come. For this book it will not be just because it records with colour and feeling a significant moment in Pacific history, but more importantly because it captures our hearts and reminds us that what really matters in life is simple and enduring: what really matters is the love of family.

This book is about survival. Page by page it lovingly shares the voices and experiences of those who survived, those who did not and those who helped in the recovery. Page by page their personalities and relationships; family, community and friendship values and bonds come alive. And, in all these pages we are drawn into the comedy and trauma of their survival and are left to pause at the wonder of their resilience, strength, hope and healing. It does the soul good to be able to laugh, reflect and cry together.

Lani and Hans wanted to write a book that was not merely a documentary which retold facts but an interconnected series of tales carefully woven together to form a very special ie toga of voices and experiences, one that would capture the drama, lessons and feelings of the moment, keep us warm in remembrance and be cherished by all. They wanted to show through honest prose how the tragedy of September 29 was able to bring forth – despite the odds and in spite of pain and callousness – the best in us. They have succeeded.

One of the most salient points of the book is the need for national, village, family and individual preparedness. For Pacific island nations, surrounded by miles of ocean and close to the deep Tongan trench, we would do well to remember the message of disaster management experts and scientists, i.e. tsunami waves in our area are more than likely to follow earthquakes so we must look for the natural signs and be prepared to act rather than wait for outside confirmation.

Samoan oral traditions tell of stormy waves that swept fishermen and navigators to distant lands, sometimes carried by sharks and turtles. As a boy these stories not only captivated my memory for their danger and adventure but also gave me information about our sea and its waves.

Our Samoan oral traditions recognise the sea as kin, respecting its power and generosity. In Samoan stories of creation *galu* (wave or waves inside the reef) and *peau* (wave or waves in the open sea) are recorded as original ancestors. These creation stories note many kinds of waves, including tsunami waves.

O galulolo ma galu fatioo
O galu tau ma galu fefatia'i
O le 'au'au peau ma le sologa peau
Na ona faafua ae le fati.
O le peau lolo ma le peau ta'oto
O le peau malie ma le peau lagatonu
O peau a lili'a ma peau laasia
O peau asisifo mai gaga'e.
O le peau lagavaa ma le peau tagata
Ma le peau tautala o lona soa
o le 'au'au taa

The tsunami waves (inside the reef) and the waves which break and surf
The aggressive waves and the waves which interweave
The forming waves and the line of waves about to roll
They signal intent rather than breaking
The tsunami waves (in the open sea) and the waves in recline
The peaceful waves and the congenial waves
The fearful high waves and waves which reach the shore
The waves from the west and the east
The waves which lift the boats and envelops people
The waves which speak and the companion waves in support

These oral traditions evidence our intimate history, heritage and identity with the sea. They tell that the ills of the Pacific Ocean are our ills; its pain is our pain. When the ocean is in harmony with things around it, so too are we; when we are in harmony with our environment, so too will it be with us. It is uplifting to read of those survivors who have made peace with the sea.

Almost two hundred people died in the Pacific tsunami of September 29 and many, many more were deeply affected through loss of loved ones, property or livelihoods. Through this book the authors and the people whose stories we trace have given us more than a record for the memory; they have given us a personal gift of love. For that we will always be grateful.

Our gratitude extends to the Samoan, American Samoan, Tongan and overseas Governments, relief workers and tourists who without a second thought provided help in our time of need. Samoans and non-Samoans from all over the world rallied together. The response was that of kin. The love and support to Tonga, American Samoa and Samoa that poured in from all over the world demonstrates the best in our shared humanity. All those affected will remember for some time the power of this tsunami. We will remember this one not just for the paradoxes of its rage, but more poignantly for its rallying, reaffirming and cleansing of the ideals of family. The essence of family is its ability to come together in times of need.

Through this book we will remember the personal acts of love and kindness shown by so many. We salute each and every one of you. For me your acts and gifts grip my heart and give substance to the Samoan saying: *O le e lave i tiga, o le ivi, le toto, ma le aano*: he who rallies in my hour of need is my kin.

God bless.

The 29 September 2009 tsunami will always be remembered for the tragic loss of so many innocent lives in Samoa, American Samoa and Tonga.

It will also be remembered for the incredible bravery of those who survived, and of those who dropped everything to assist in the days and months that followed.

The Pacific Tsunami "Galu Afi" captures forever poignant memories of the day the earth moved and the sea roared. The day that changed forever the survivors' compact with the sea. From those who fled to those who did not have the opportunity and from those who answered the call to help, to those who needed every assistance. Embedded within these harrowing accounts are lessons learnt the hard way.

While Samoans suffered in their thousands, we too were touched. The ocean did not discriminate. Five Australians perished and countless visitors found refuge with Samoans who cared for strangers as they would their own. Bonds were formed in the aftermath of the tsunami that transcend borders and will never be broken.

From the powerful stories within this book, there are some key messages we must all heed in the Pacific: we can never be too prepared for a natural disaster; and we should head inland and to higher ground following an earthquake.

In the wake of the tsunami, Australia provided A$12 million to help the Samoan Government. This money was used to minimise distress and to sustain life and dignity. This assistance reflects the warmth and closeness of our enduring friendship with the Samoan people and our determination to help Samoa rebuild.

Australia is proud to support the publication of *The Pacific Tsunami "Galu Afi"* with funds from the Australian Aid Program.

We hope our support will remind Samoans that Australia stands ready to assist our Pacific neighbours whenever and wherever there is a need.

Matt Anderson
Australian High Commissioner to Samoa

Introduction

During our lifetime, most of us remember events that have special meaning or which affected the course of history. Where we were at and precisely what we were doing when we heard the news that President Kennedy had been shot in Dallas, Texas. When we were awoken in the early hours of the morning of September 11th and watched in horror as we witnessed on television the work of terrorists, the destruction of the Twin Towers, the World Trade Centre in New York City.

It's the same thing for many of us with 29/09, when Samoa was awoken early on that Tuesday morning by an earthquake that seemed to last forever, and then news started to come in, of the deaths and destruction at Aleipata and other villages along the south coast of Upolu by a tsunami, the deadliest in Samoa's history.

I was awoken in the middle of the night at a hotel in Pnom Penh, Cambodia, by a call from my daughter in Salt Lake City to inform me that there had been a massive earthquake in Samoa and that my wife Celine and our son Nathan were not harmed. She called me back a couple of hours later that a tsunami had hit Samoa, our family were alright but there had been many casualties, mainly in the Aleipata area.

On CNN that morning, pictures were coming through of the devastation that had hit our country. I cancelled my sightseeing trip to the famous temple in Angor Wat and tried to get a flight home that day. None was available so I returned to Samoa on my regular scheduled flight arriving Thursday evening.

Saturday afternoon I was able to drive to Aleipata and observe the devastation. It was unbelievable. Whole villages were destroyed. There was hardly anything standing and volunteers were still searching through the debris looking for bodies. It had a tremendous effect on me and I'm sure on everyone else who had driven out there.

An earthquake. A tsunami. Reports said 5 to 10, maybe 15 minutes in between. Why were so many people caught? Why didn't they take action and move inland immediately? We see the messages on television almost every week. Evacuate as quickly as possible.

We continued our drive all the way to Poutasi and to Sinalei Resort. I wanted to see my good friend Tuatagaloa Joe Annandale who had lost his dear wife Tui. Driving back I kept thinking, there has to be a record made of this. Someone has to write a book. Someone capable and very sharp. Someone who is a good writer and of strong character.

During that whole week while Samoa buried their dead and farewelled their loved ones, I kept asking myself, who will write a book about the tsunami? It had to be done as quickly as possible. Not a year from then when memories dimmed and survivors forgot their unpleasant experiences. I approached writer and English teacher Lani Young with my plan. Interview the survivors, medical teams, volunteers, anyone involved and get their stories. She accepted the task and we both agreed that any profits from the book sales would be given to assist victims of the tsunami.

We set a deadline for the printing and launching of the book. One year from the day of the catastrophe of 29/09. We endeavored to meet that deadline.

I would like to thank everyone who assisted in the formulation and production of this book. We realize some of the events may have been difficult to describe and relive for you. We are mindful of how sad this would have been. Your co-operation is truly appreciated. This is your book. This is about you. Thank you.

I remember the tsunami of 1960. It was an earthquake in Chile and warning was given for Samoa. I was a student at the time and it was discussed in class the next day. It was called a tidal wave then, or *galulolo.* The term tsunami was never used. We heard that the sea had gone out, people went to the beach to look and then picked up the stranded fish. It wasn't a huge wave. There were no casualties and very little damage.

The tsunami in 2009 was entirely different. Everyone felt the earthquake and the waves arrived within minutes. People were not prepared.

We hope this tragic event will never happen again. But we can never be sure. If not in our generation then maybe in another. We pray they will be better prepared. We hope this book will make them ready in case that event ever occurs again.

I know that reading this book will deeply affect many of you.

Lani has met the challenge and we wish to congratulate her for her dedication, courage, fortitude and perseverance in producing an outstanding record of this disaster and the stories of those involved.

Congratulations Lani.

Thank you.

Hans Joachim Keil.
2010

Acknowledgements

In October 2009, Hans Joachim Keil had a hope. A thought. A picture in his mind of a book. A record. That would speak with the voices of those who had lived through the 2009 Pacific Tsunami and tell of those who had died. A record that would speak with the voices of those who had worked to rescue, heal and rebuild. I was asked to take that hope, that thought – and give it substance. And so the journey began.

The material for this book was gathered from hours of taped conversations with tsunami survivors as well as with emergency, health and aid workers, from personal written accounts in journals and blogs, from filmed tsunami footage, from newspapers, magazines and televised interviews, from science reports and surveys, from the internet websites of many different organizations, from government reports and from a wide range of photographs. However, this book is certainly not offered as a comprehensive 'all-knowing' account of the 29/09 disaster as the stories contained here are only a fragment of people's experiences. I call on families and communities in Samoa, American Samoa and Tonga, to continue to seek out the stories of the tsunami. Listen, record, preserve and learn.

My journey began in the Aleipata and Falealili districts in Samoa. I went first with Lagi So'oalo to those whom the ocean had battered. Lying in hospitals, camped in tents, gathered in rough shelters in the mountain bush or sitting beside the new graves of their loved ones – survivors everywhere paused in their recovery and rebuilding to share their stories. I travelled to American Samoa. With gracious hospitality, survivors in both countries welcomed us into what was left of their homes and continually offered us the finest of whatever they had. With strength and courage, they relived the nightmares of that day in September 2009. With humility and faith, they entrusted us with their sorrows. And then they farewelled us with the fragile hope that somehow, their words would speak to you of all they have endured. To all those in Samoa and American Samoa that shared their stories with us, thank you.

I looked over the Pacific Ocean for those who had been on the holiday of a lifetime and then found themselves fighting for their lives. Again, many responded with grace and a willingness to help with Samoa's recovery in whatever manner they could. They wrote of survival, loss and heartache. They wrote with gratitude of miraculous rescues and compassionate service as Samoans cared for them in those difficult days after 29/09 when they were left with 'nothing but their lives' in a strange land. To survivors from

New Zealand, Australia, Germany and the United Kingdom that shared their stories, thank you.

I turned to those who had worked in the emergency response effort, the recovery and rebuilding. There were so many who dedicated incredible amounts of time and effort to see Samoa through this disaster – both local and those from other nations. Thank you for pausing in your ongoing work to share your stories. I remain humbled by the hard work and commitment of two people in particular – Commissioner Tony Hill of the Samoa FESA and Filomena Nelson of the DMO. I pay tribute to their service to Samoa during a most challenging time and am grateful for their contributions to this book.

The assistance of many Samoan government ministries was gratefully received during the research for this project, especially that of the Ministry of Health, Ministry of Natural Resources and Environment, Ministry of Police, Ministry of Agriculture and Fisheries, Ministry of Tourism and the Ministry of Works. The assistance of the Australian and New Zealand High Commissions in Samoa was invaluable as was that of the German Honorary Consul.

The chapter on Niuatoputapu was made possible with the help of Sakopo Lolohea, CEO for the Visitor's Bureau of the Kingdom of Tonga.
Local media were a great resource and I thank TV3 Samoa for access to their tsunami footage and Rudy Bartley for use of his tsunami video production. Many thanks also to the 'Samoa Observer' newspaper for their excellent tsunami coverage which I have drawn on extensively throughout the book.

This journey has not been an easy one. My personal testimony of God's love for us as His children has been strengthened as numerous survivors have testified of their faith and hope in the midst of great adversity. I am not a counsellor but in meeting with survivors – many of whom would speak for an hour or more about their experiences - I have had to 'mourn with those who mourn and comfort those who stand in need of comfort'. I have felt miserably inadequate for this task many times and wept many tears. Friends and family have given me their encouragement and support. Thank you. Throughout this project, my children have cheerfully endured a tired, short-tempered mother. My husband has given generously and unwaveringly of his strength and patience. Thank you Darren, Jade, Sade, Zion, Zach and Bella.

This book would not have been written if not for the following three people. My mother Marita Johnson who first showed me there was an amazing world of possibilities out there when one could read. My father Dr Felix Wendt who always believed that I was a 'good writer'. He waited many years to read the first draft of the first chapter of my very first book – and then ensured more would be written by continually telling me it was great (even if it wasn't!)

And finally, Hans Joachim Keil. Thank you for giving me the opportunity to live my dream - to labour over a book that really will be read. Thank you for entrusting me with your vision.

Here now, is the meagre work of my hands. Like an *ie toga* woven with patience and faith - I offer it humbly as comfort to those who mourn. I display it proudly for an honored tribute to those who served. I give it with my love and gratitude to those who carried me when the journey became too hard. And finally, like an *ie toga* of entreating apology - may the good intentions of this work be a covering for my remorse as I ask forgiveness of those who may find my words inadequate or be offended by their shortcomings.

Lani Wendt Young
July 2010

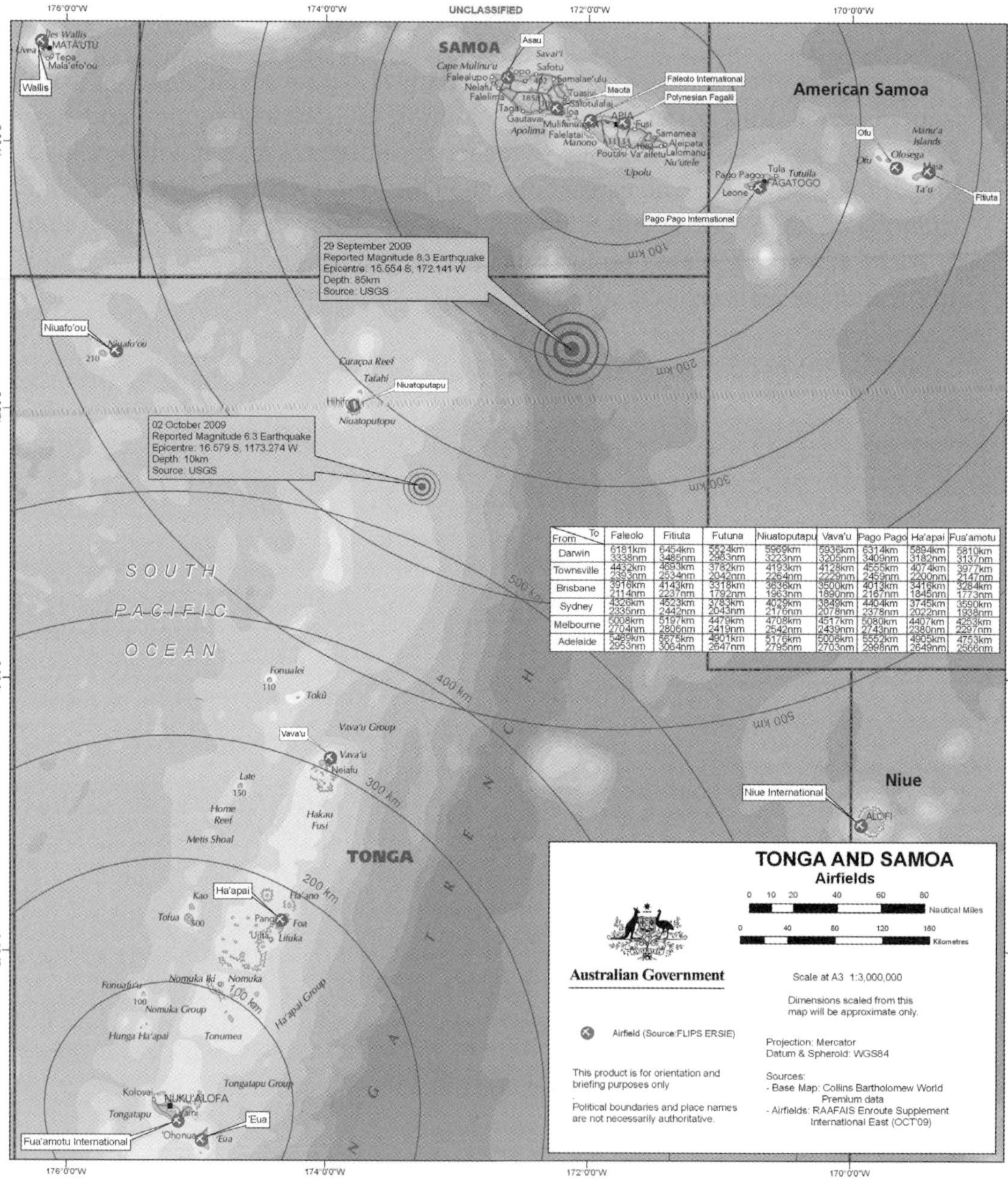

From \ To	Faleolo	Fitiuta	Futuna	Niuatoputapu	Vava'u	Pago Pago	Ha'apai	Fua'amotu
Darwin	6181km 3338nm	6454km 3485nm	5524km 2983nm	5969km 3223nm	5936km 3205nm	6314km 3409nm	5894km 3182nm	5810km 3137nm
Townsville	4432km 2393nm	4693km 2534nm	3782km 2042nm	4193km 2264nm	4128km 2229nm	4555km 2459nm	4074km 2200nm	3977km 2147nm
Brisbane	3916km 2114nm	4143km 2237nm	3318km 1792nm	3636km 1963nm	3500km 1890nm	4013km 2167nm	3416km 1845nm	3284km 1773nm
Sydney	4326km 2335nm	4523km 2442nm	3783km 2043nm	4029km 2176nm	3849km 2078nm	4404km 2378nm	3745km 2022nm	3590km 1938nm
Melbourne	5008km 2704nm	5197km 2806nm	4479km 2419nm	4708km 2542nm	4517km 2439nm	5080km 2743nm	4407km 2380nm	4253km 2297nm
Adelaide	5469km 2953nm	5675km 3064nm	4901km 2647nm	5176km 2795nm	5006km 2703nm	5552km 2998nm	4905km 2649nm	4753km 2566nm

1
The Waves of our Past

Standing sentinel high above the coastal strip of Saleapaga are the remains of an ancestral settlement. Stone platforms, star mounds hidden beneath forest foliage and ringed by rock walls. Nature has reclaimed her cliff top perch where once a people lived and loved, worked and played. From the cliffs one can look out over the vast expanse of ocean to the blue strip where earth meets heaven. From there one can well imagine that there was naught in the world but Samoa and the beyond. Beyond where the great god Tagaloa surveyed the workings of his children. Little wonder that the first ships carrying Western explorers appeared to have 'burst through the sky' *papalagi* literally exploding through the mantle of heaven. Although the French explorer Louis de Bougainville gave Samoa the name 'Navigator Islands', believing our ancestors to be experienced seafarers, we were in actuality at that point, definite homebodies. Our paradisiacal promised land had already been found. A lush collection of volcanic islands, rich and fertile – Savaii, Upolu, Tutuila, a smattering of smaller islands – each an emerald green jewel set in the shimmering topaz waters of the Pacific Ocean. To the south-west, Samoa's sister and royal collection of island gems, Tonga.

The oldest known site of human occupation in Samoa – at Mulifanua - dates to almost three thousand years ago. Vast earth and stone monuments can be found in various parts of the islands and include stone forts, pathways, agricultural terraces and walls – all pointing to a highly organized society from before the time of Christ. It was a society that preferred to settle inland as early European visitors consistently noted. The French explorer La Perouse wrote in 1787 that Samoans built their homes half way up the hills – because, as he assumed - it was cooler. He sailed along the northern coast of Upolu and described "a great plain covered with houses" some distance inland. Stone remnants of early settlement are littered throughout the inland country. It seems that for our ancestors, the ocean's roar was not something we wanted to fall asleep to. Yes, the sea was appreciated, but it was an appreciation tempered with a healthy dose of respect. We plied its depths for food and traversed its lengths to alternately visit and war with neighbouring islands but rarely did we settle along its shores. It seems our ancestors knew all too well the power of the sea, well enough to resist the lure of golden sands and the cool tropical breezes of the seaside. Then - we were people of the forest. Indeed, not until the Europeans came with their trading ships and their copra and cocoa plantations – requiring easy access to the ocean – did we begin building our villages on the coast. When the early missionary George Turner asked why Samoans did not live on the coast, we replied, "*Galu afi*". Wave of fire. Tongans have a similar descriptor "*peau kula*". Red wave.

What is a '*galu afi*'? The term conjures up an accompanying wave of images. A wave that burns. Rages. A wave that is unstoppable. Magnificent in its destructive power. A wave that consumes everything in its path and leaves the earth scorched. Once green land scarred brown. Skeletal trees reaching forlornly to the skies. A land and a people lamenting. It is difficult though to get any definitive answers. Some say '*galu afi*' refers to lava flow, similar to what occurred between 1905 to 1911 when the Matavanu volcano erupted, covering many miles of countryside and driving entire villages to flee before its fiery waves. Still others though will tell you that *galu afi* is the Samoan name given to what the Western world call, tsunami. People like, Oselani – a fisherman from Saleaumua village on Upolu – describing the wave that came on September 29th, 2009.

"I know this wave was a *'galu afi'*. It was different from any other big wave, any other '*galu lolo*'. The water was hot. It was so powerful. It destroyed everything and the people of my village who were killed – their skin was black like it was burned. And now, I feel in my chest where the water was, that it is not right. I can still feel a burning there, like I am sick."

Muaoa Fonoti from Vaovai village – "This wave is what you call a '*galu afi'* From what I saw, it was like how a fire works. When a fire spreads, it goes really fast. The speed of the wave coming was so quick like fire. I have seen '*galu lolo*'in the past – big waves that come – they were not deep or fast either. They just came gradually and there was no fish brought in like what happened this time. There were '*galu lolo*' back then but it wasn't the same, there weren't any tragedies like this time. This wave was different."

The tsunami killed Leua Vi'iga's child in Saleapaga village. "This wave was what you call a '*galu afi'*. It was a different kind of wave. It came to clear the houses…it was like a ravaged animal, a demon. It was so strong. Coming from the reef it stretched up so high and when it broke it came full force and couldn't be stopped and took everything in it's path. The other different thing was all the dead fish and the wave was hot…if you see the trees down there, they're all dry and dead where the path of the wave went. Everywhere the wave went it destroyed the vegetation…no wave brings a fire except for this one, this *galu afi*."

Yes there are those who will tell you that thanks to 29/09, they now know what our ancestors meant when they told the *palagi* newcomer about '*galu afi'*. A wave of fire. As the ocean heaves beyond its bounds, asserting its authority upon the land. *Galu afi*. Filling your lungs with the poison of fire water. Leaving a legacy of rot that will plague you for years to come. A hacking cough. A burning chest. Cuts and scrapes, seemingly insignificant in the face of the horror around you, becoming festering wounds. *Galu afi*. Tsunami.

Searching for historical tsunami though, is no easy task. Samoans and Tongans had no written language. In Tonga, oral tradition among the elders, tells of '*peau kula'* that can cover a whole island at one time and science is currently investigating monolithic boulders on various islands - some weighing more than 2,500 tonnes - believed to have been deposited by prehistoric tsunami. For Samoa, the earliest written record of a tsunami tells of an earthquake in Chile on November 7th, 1837 "and a sea wave started by it was felt at the Hawaiian islands; also at Tutuila in the Samoan group." Again in 1868, another shake in South America generated a tsunami that destroyed several settlements in Apia. The 1877 great Chilean tsunami that caused devastation throughout the Pacific, hit Apia with wave heights of four meters. In 1883 there was an earthquake and local tsunami reported by the captain of an American vessel in Apia harbor – "On the night of March 24th there were shocks of earthquakes…the east end of Savaii was visited by a tidal wave which swept away all houses within a quarter of a mile of beach for a distance of fifteen miles along the shore." In 1896, the US Consul-General reported on the strange behaviour of the sea as the ocean fell and then "great waves were seen outside the reef" and "considerable alarm was manifested over the unusual circumstance". Beginning in 1905 and carrying on intermittently through to 1911, the Matavanu volcano eruption generated a series of tsunamis "caused by avalanching material into the sea." One of the largest of these was in 1907 when a wave was generated "ten to twelve feet high" destroying houses, boats, copra sheds, "a 400 gallon tank of water was lifted bodily from its foundation and carried across the road." In 1917, an 8.4 magnitude earthquake in the Tongan trench generated a huge tsunami which hammered the Samoan islands.A detailed description of the event can be found in the June 30th edition of the 'Samoa Times' newspaper.

"Earthquake about 150 miles from Apia. Destructive tsunami on south coast…arrived in period of 18 minutes. On the Aleipata coast, the tidal wave is described as sweeping in, a white wall of foam fully 10ft in height. Although dead low water at the time, the advancing wall of water swept over high water mark and across the beach into adjacent native houses, carrying everything before it. In Lotofaga the wave swept right over the beach, flooding the houses and reaching out into the plantations at the rear. About two chains of solid cement wall, quite a foot thick and three feet high, was lifted bodily and carried away, pieces weighing over half a ton being shifted back for fully 30 feet…half the village submerged..At Palauli in Savaii, a bridge was washed away and native houses destroyed…In Tutuila a small tidal wave swept in through Pago destroying many Samoan houses…two large churches partly demolished. Many of the natives around the bay sought refuge in the mountains where they remained until morning."

From 1918 through to 1964, twenty-eight other tsunami are documented in the Carayannis 'Catalogue of Tsunamis in the Samoa Islands'. Most of these are described as 'minor trace' events only noted on tide gauges, the result of both near and distant earthquakes. Clearly, Samoa tsunamis are not as rare as many may think. But you would be hard pressed to find a Samoan today who can tell you about any of our historical tsunami. You ask, *did your grandparents ever tell you about tsunamis? 'Galu lolo'? 'Galu afi'?* Such questions are usually met with shrugs and puzzlement. No, our knowledge of such things comes now from television and the waves that have ravaged other nations. Tsunami are a nightmare that others endure, not here. Where once the sharing of oral history prevailed, now there are the wonders of mobile phones and instant satellite transmission. It is difficult to comprehend how such waves could possibly ever be forgotten. Cleansed from the collective lore of an entire people. But in the last one hundred years, it seems that the business of being busy with westernization has virtually erased that knowledge.

Now, we Samoans are a people of the coast. Gleaming tar seal roads encircle the islands and towering power poles hold hands up and down the shoreline. We build cement box houses on the sand where mangroves once breathed and adorn them with shiny television antennas. We go to the forest to plant our crops, clearing the bush to put up spindly *faleo'o* where we will rest from our labours. For those living in many coastal villages, the bush is not for living in. It's where the men go at the break of dawn. To weed. To plant. It's where we harvest taro from the black earth. And shoot pigeons and flying fox. The bush is not for people. It's for mosquitoes. And lonely spirits.

Now, we are a people of the coast. We build resorts that embrace the ocean. The tourist comes to scuba dive Samoa's blue depths, to delight in the multi hued bounty of coral reef. All of our major centres are coast side. Airports. Harbours. Townships. Apia. Salelaloga. Mulifanua. And another harbor and more recently, a slipway at Aleipata.

Go to Aleipata. Climb to the cliff tops of Saleapaga and Lalomanu that overlook the wild ocean. Stand on the stone remnants where once our people loved, lived, worked and played. Stand and look towards the faraway tectonic meet. Stand and imagine that long ago, when the earth rumbled and the waters swelled, a people could watch from the safety of their skyward home and marvel at the awesome power of '*galu afi*'. And warn their children to be ever watchful of the sea.

2
Mafui'e. The Earthquake

Legend tells that Mafui'e the earthquake god and keeper of fire, lives deep beneath the earths surface in the Underworld, Fu'ealoa. From time to time he shifts and stirs, shaking the earth in order to obtain food from mortals. The Asia-Pacific region or 'ring of fire' has always been a favourite playground of Mafui'e with more than thirty quakes of magnitude 7.0 or more recorded since the 1930's.

On the morning of September 29th, 2009, Mafui'e was in a rage. He reared his ugly head 190km off the shore of the Samoan islands where the Pacific plate and Tongan microplate meet in an uneasy subducting truce. The focus of his storming anger was only a scant 10 km underground while directly above rested more than six km of deep blue Pacific Ocean.

At 6:48am in Apia, the quake began like any other. The earth moved. Groaned. The corrugated roof clattered, waking us with its protests. At the National Aquatic Centre at Tuanaimato, more than two hundred tonnes of chlorinated water were heaved out of the pools. A similar thing was happening with Vaughan Simpson's swimming pool up the hill at Vailima, waves of water a foot high, sloshing back and forth and spilling out both ends, "like shaking a teacup." At 'Plantation House' design store at Alafua, tiles fell off the bathroom walls and crystal vases were thrown to the floor. The gold leafed statue of angel Moroni atop of the Mormon temple at Pesega swayed and lost its horn. At the seven storey Development Bank building in the heart of town, a crack seamed its way up through the wall. In a taxi with his children on their way to school, Vaomalo Kini, head of the National Hospital laboratory, worried about the hazardous chemicals lining the shelves at his work and the possibility of a fire. Along the narrow coastal road winding its way past Solosolo, mini-landslides dislodged from the cliff side, blocking the road.

On the south coast, in the village of Saleaumua, Lee Letiu's dog Missy started a panicked howling as she raced up and down the length of the lush gardens nestled beside a sleepy shore. Further along in the village of Malaela, tins of canned food fell from the shelves in Su'a Leota's store, sending his wife scurrying outside for safety.

At Lalomanu, Tufaiga Fatueseafioga was making tea for her mother when their television set smashed to the floor and her daughter called for her to get out of the house. Outside in the garden, they hugged each other - with cautious eyes on the swaying power lines beside their home.

Down the road at Litia Sini Beach Fales, New Zealanders Ngaire Woodward and Jill Barron, asleep on mattresses in their thatched roof hut, were abruptly shaken awake. "The whole earth was shaking and it was really going to town, we were rolling around…and having a bit of a giggle." At 6:50am Ngaire texted her daughter back home in Alexandra, *We are having a really big earthquake. Not a great place to be in a beachfront fale. Love Mum.*

Twelve year old Maxwell Wilson from New Zealand, was in the bathroom of the toilet block at Litia Sini Resort when the earthquake started. He thought someone had turned on a generator. His ten year old sister Poppy asleep in their fale, was woken by the rumbling. "In my sleep I thought it was a massaging machine but as I woke my Dad was yelling '*Get out of the fale'*. He grabbed me by my foot and dragged me outside. I realized we were in a big, low rumbling earthquake that went on and on."

Asleep at Taufua Beach Fales, Australian Claire Rowlands thought the tremor was a canine visitor. "It was quite gentle, like a dog scratching its ear underneath the fale. I came out to have a look but there was nothing."

At Saleapaga village, Peleiupu Susuga's assorted young charges were waiting patiently for their turn to have their hair braided for school. As the tremors rattled the roof and the wooden poles swayed, little Fa'amaile burst into tears and fearfully implored "We have to pray auntie." Their 85 year old grandmother, Fuailuma called from the next house, querulously demanding to know why the children were fussing.

At Faofao Beach Fales, fourteen year old Tia Ansell was on holiday from New Zealand, with fourteen of her extended family. The cousins had been thinking about going in search of breakfast when the earth shook. "We all went back inside our *fale* because some of the coconuts started falling off the trees. One fell on my *fale* (Samoan house) and it was really scary. The earthquake felt like a monster, it was stomping around and it was shaking, it went on for ages and it was like, *whoa!"*

In Lepa village, women were gathered at the communal pool by the roadside for the annual ritual of washing of *ie toga (*fine mats). Falling rocks from the steep cliff had them scattering for safety. One huge boulder landed right at the door of one woman's house and afterwards she called out to her neighbour, "thank God we have been saved from the earthquake!"

On holiday from Australia in her mum's village of Sa'anapu, Steph Fa'amatala was asleep with her cousins so soundly that the aunties had to use *salu's* to wake them – the fiery sting of the coconut broom sending them running outside.

At Maninoa Surf Camp, Tom Gogola from the UK had taken the fifteen minute boat ride out just past the shallow coral reef and never felt the shaking. But looking at the glassy, clean surf that broke at the four foot mark, he remembers thinking "this is paradise…27 degree waters, good clean surf and not a care in the world!"

Over at Coconuts Beach Resort, Siumu, Rachel Lynn out for her morning run, was startled to see the road dip and weave beneath her feet. Next door at Sinalei Reef Resort, the general manager Sose Annandale was walking the grounds and as the earth shook, she worriedly said to her son – "I think we're in trouble."

For most, the earthquake was interminable. It was the longest and strongest shake many had ever experienced. It felt far more consuming than its one and half minutes. Terrifying in and of itself. Panicked hearts beat faster. Pulses raced. Thoughts scrambled…what to do? Where to run? What to grab? Where are the children? Would the house hold strong? It was felt in much of American Samoa, Samoa, northern Tonga and as far away as Wallis and Futuna, and Niue. When the earth shakes for that long and with that much force, all kinds of doomsday images run through one's mind…giant cracks in the ground, smoke billowing from deep chasms that suddenly gape at our feet. So when the earth was finally still, there was widespread relief. Tense breath exhaled. Tightly clenched hands released. Silent prayers of thanks winging their way heavenward. Some hugged their young ones a little tighter, drowning the momentary 'what if' thoughts with a rush of grateful appreciation that all was well.

But was it? For a few, the ending of Mafui'e's tantrum brought no reprieve to the apprehension. For a few knew all too well what such a powerful earthquake signified. People like Tony Hill, Samoa's Commissioner of Fire and Emergency Services. Tony was a familiar television presence in Samoa as he spoke to the country about the dangers of earthquakes, fire and tsunami in a series of disaster awareness advertisements. A stocky man dressed in army fatigues wearing a green beret would instruct watchers solemnly to ' *walk, don't run to the nearest exit in an emergency… in an earthquake take refuge in a doorway, under a table…If you live beside the sea, run for higher ground…"*

A long time veteran of the department which he helped build from the ground up, Tony is a hands on man who can roll up his sleeves and fix a car engine, administer CPR, or read a story to his granddaughter. You can learn a lot about a man from his office, and Tony's speaks confidently of action. And work – get your hands dirty and sweat, work. There is little room to manoeuvre as it is piled high with equipment. There's a table made from a recycled bathroom stall door. Bags of first aid emergency response kits are stacked against the wall, threatening to totter over onto visitors. Behind him are piles of infant resuscitation devices which he will show you with all the excitement of a boy with a precious new toy. His thrill in the gear is understandable however when he explains that they are the only ones of their kind in the entire country. And that theirs has always been a department that had to 'beg and borrow' for their needs…fire trucks from Japan, rescue equipment through the British Embassy, another truck from Australia, their ambulance was a gift from SamoaTel, one of their service vans. The Faleata Fire station was built using materials from a condemned building that was donated to the Fire Service thanks to Tony's initiative. "A lot of this job is just common sense you know? See what needs to be done and get out there and do it."

For Tony, preparing the country for natural disasters was 'just common sense'. A response to global events – "we've got to be really realistic about what's going to happen within the whole Pacific region." His tsunami campaign hand in hand with the DMO, had many saying *that guy's lost it, he's gone in the wrong direction.* He concedes that he himself NEVER thought Samoa would be hit with a tsunami – but it would have been foolish not to prepare for the possibility.

When the quake hit, Tony was getting ready for work. The fury of the vibrations had him hastily throwing on his uniform. His consuming thought: tsunami. The final tremor had barely settled before he was in his truck, speeding to the meteorological office at Mulinu'u to await the Pacific Tsunami Warning Center (PTWC) report from Hawaii.

Several villages over, in Lotopa, thirty-five year old multi-tasking mum, Filomena Nelson was making sandwiches for school lunches. And harassing young ones to get dressed. And trying to get herself ready for work at the Disaster Management Office. As the house rocked to its very foundations and she gathered up her family to move outside, Filomena was already steeling herself for what she knew would be inevitable. "I knew. When that earthquake shook like that, that hard, I knew there would be a tsunami. And I knew there would be very little time for any warnings. If at all. The only thing I wasn't sure of was where exactly would get hit and how bad would it be."

As soon as she established her family was safe, she was on the phone to her staff. "We need to send out the warning. Now." She radioed to Commissioner Hill for a siren ride to the Met office, knowing that would be the only way possible she would get to town fast enough. From that moment on, as Principle Disaster Management Officer, Filomena Nelson became the most important person in the country. As she waved good bye to her children from the rescue truck, their mouths agape at the flashing red lights, could she know that her life would never be the same? They wouldn't see her again for three days. And then only for the briefest of moments over the next few months as she would dash home to steal a hug, a shower, an all too quick nap.

Filomena had started her public service as an accountant at the Ministry of Natural Resources and Environment(MNRE). She took her first step away from financial management when she was appointed Principal Ozone Officer in 2001 and then the Principal Disaster Management Officer in 2004.

The Samoan government approved a National Disaster Management Plan in 2007. A National Disaster Council was established, chaired by the Prime Minister and a National Disaster Committee made up of representatives from various government departments and other local bodies. There were more than twenty potential hazards for Samoa identified in the plan. Tsunami was number three on the list – after cyclones and volcanic eruption. Classed a 'rare' occurrence but one that would have major impact – "extensive injuries, a number of deaths, high level of building and infrastructure damage, major financial losses." According to the NDM Plan "a tsunami with a mean run-up of between 7 and 9 metres has a return period of between 50 and 100 years based on probability analysis of historical records. Significant damage to unprotected coastal areas will occur…loss of life expected especially for tsunami generated in the region, for example from the Tonga-Kermadec trench." Tsunami was high then on the DMO's list of priorities, but as many who work in the fields of prevention and preparedness will tell you, recruiting believers was no easy task. People aren't excited to put in time and effort or pay for things that haven't happened yet. Things that may *never* happen. More 'important' and more pressing needs take precedence in budgets. Things like health care and better schools. Filomena had a staff of three. They worked out of a rickety few cupboard like offices at the MNRE building. But they worked with dogged persistence to co-ordinate the necessary people.

A media awareness campaign was launched to educate people about the early warning signs of tsunami. Television ads told watchers about the danger of a strong earthquake. The water receding. The sound of the oncoming water. A handsome young actor named Peter Letiu, who also worked part time as a fireman, featured in the ads. Running away from the beach through falling forest.

At the close of 2008, the DMO began their education workshops in select rural villages. People were briefed about the hazard risks for their particular area. A disaster assessment team for each village was chosen and given special training. First Aid. Emergency rescue techniques. The aim of the workshops was to reduce the impact a possible natural disaster would have on the community. As Filomena explains,"We can't prevent a natural disaster like a cyclone or a tsunami, but there are things we can do to reduce the impact it will have on people and their families. *If* they are prepared and have the necessary knowledge."

There are some 342 villages in Samoa and the H1N1 virus crisis went a long way towards diverting the progress of the DMO workshops. Filomena and her staff were involved with shielding Samoa from the worst of the worldwide pandemic. By September 2009, the DMO had only covered five villages with their disaster awareness training. Aleipata district was scheduled to be covered in October.

The other main focus for the DMO tsunami campaign was putting a national early warning system in place. Previously the only way for alerting people about a hazard was via the radio and television – far from ideal in a country where power outages are a weekly occurrence. An alternative method of reaching Samoa's population of 182,000 thousand people had to be found.

Samoa has no seismic centre. We get all our earthquake data from the PTWC in Hawaii. Bulletins will only be sent to the Samoa meteorological office when there's an earthquake measuring 6.5 and above. The message is faxed. Emailed. EMWN through direct satellite hook-up. And texted to Filomena's cell phone. If the power is out and the internet is down, then the Met staff have a scalc of thresholds based on observations of physical damage that they will use to determine the threat. Rather than waiting for Hawaii – to tell them what's clearly obvious – the DMO will then go ahead and issue the alerts.

They are sent using SMS text message. There are two telecommunication companies in Samoa. Thanks to their extensive coverage and service affordability, just about every family in the country has a cell phone. Or two. The early warning system selected a little over a thousand people to receive the SMS alert. People like church ministers, school principals, village mayors, women's committee leaders, beach resort owners and certain government sector heads. There are two levels of alert. ***Tsunami watch*** does not require immediate evacuation. It means be on guard and spread the word. Be prepared to move if necessary. ***Tsunami warning*** means sound the alarm. Get out. Move to higher ground. Run for your lives.

When the warning is received, it then becomes the responsibility of these chosen thousand to initiate the alarm. Sirens are expensive commodities. There are only a handful of them in the country – at the fire station, airport, and at Sinalei Reef Resort. Hence the alarms are to go out using church bells and school bells. Some villages have whistles. Others use the *lali* drum. All boats in the harbor are to sound their horns. In a close knit community, the concept seems a good one. But it all rests of course, on the necessary information being received and the SMS being activated. In time.

Filomena would be the first one to tell you though, that even the very best early warning system can not be early enough in the case of a local tsunami which can hit within minutes of an earthquake. The value of an early warning system is greater in cases of distant tsunami, those generated by earthquakes in faraway places. Like Japan or South America. They take several hours to reach Samoa and allow adequate time for people to be warned. To get away. No, the best way for people to be safe, is to know the natural signs and act immediately.

Samoa tested its warning system during two countrywide tsunami drills, held in 2007 and 2008. Drills where many rolled their eyes and complained about the hassle of walking to selected evacuation points. Many hid in their offices and homes, waiting for the drills end. Police had to forcibly 'remind' some villages about their participation in the drills. One beach resort operator from Lalomanu remembers a policeman stopping their bus and ordering everyone to run for the hills. It was a scorching hot day and so the policeman was a very unpopular man as a result. The woman shrugs,"He said we were practicing for a tsunami. I didn't know what that was."

US Peace Corp volunteers Cale and Sara Reeves were secondary school teachers in Samoa during the October 29th, 2007 drill. Together with hundreds of others they took the four mile walk inland to their designated high ground. "We hoofed it up the mountain at 11:30am under the blazing sun for an hour or so, it was hot and sweaty and long and quite a work out."

The entire business community ground to a standstill as fire trucks screamed and police virtuously stopped traffic and told commuters to walk. Schoolchildren were delighted with the opportunity to get out of class. Parliament evacuated. The Central Bank emptied. But then had to dash back because someone forgot to turn off an electric jug. Vaosa Epa is the Secretary to Cabinet and a member of the National Disaster Council. A striking woman who moves with restrained elegance, she shakes her head as she recalls the general reaction of those who worked in the five storey block of government offices. "It was sort of like, *what are we doing this for?!* I think because it was quite unreal that Samoa would ever have one. So it was a disruption for most people in the building – they had to walk up the hill in the heat and then walk back down again and people were complaining and when they came back to the office, many of them would just get their bag and go home because they were all hot."

Earlier in 2009, there was even an opportunity to practice the evacuating – for real. On the 19th of March, a 7.6 magnitude earthquake hit 130 miles southeast of the Tongan capital Nukualofa and a tsunami warning was issued by the PTWC for surrounding islands. Including Samoa. Sara was teaching when they got the phonecall from the Peace Corp office at 8:50am. "Cale came into my classroom making the finger across the throat gesture that initially led me to believe that the power must be out at his school, but instead meant that I should stop teaching, drop everything and evacuate." They warned the vice-principal. "He told me that he already knew. I told him schools were supposed to be evacuating, but he didn't seem to be too concerned." The other staff showed little interest either. The two were the only ones from their coastside school to head inland on their bikes. They met other vehicles and some schoolchildren on the road headed in the same direction. Calmly and slowly. One family drove to the lowland grocery store first – "to buy snacks for the children and a newspaper so the morning wouldn't be wasted". Another only went to the evacuation point because their six year old daughter was frightened and wouldn't stop harassing them about the "terrible tsunamis' her teacher had taught them about.

In Tonga that morning, people didn't seem to be too bothered either. A police spokesman said residents didn't seem to be taking the warning seriously. He was quoted by a UK newspaper, "People are out on the roads, laughing at the warning. They are not moving from the coast even though there has been a strong warning of a tsunami."

Back in Samoa, the Reeves biked four miles back to school when they got the message that the warning had been cancelled. "Absolutely nothing had changed at school while I was gone and there was no reaction whatsoever to my sudden disappearance and bike ride to Faleata."

At Fagaloa Bay on the far eastern tip of Upolu, unusually high wave heights were recorded. The sea ran in to swish through several beach *fales*, but no houses were damaged. And nobody was hurt. Hardly anyone anywhere else even registered that the sea had risen. People shrugged. '*See, what's all the fuss about? Those palagi say tsunami when it's just a little galu lolo. A few big waves. It's nothing.*'

On the morning of September 29th were people remembering that day earlier in March? As DMO staff scrambled to contact the two telecommunication companies to send the SMS and as Filomena and Tony were flying through traffic at Vaimoso?

As Tony hit the sirens at approximately 7:10am and radioed his staff to begin the evacuation of all low lying areas? As the alarms went on at the fire stations in town, Faleata and at the airport?

In those next minutes, Filomena was merely a passenger as Tony drove down every side road with the siren blaring and the PA system shouting, telling everyone to get out and get up.

Tsunami. Evacuate.

29 September, 2009.

As the earth shook, we huddled together in the strongest room in the house. I held the children close and tried not to show the fear that cut raggedly through me. I wondered what I would do if the roof collapsed on top of us. At least... we are all together. We will die all together. It was strangely comforting.

3
The Evacuation

The Apia waterfront is a scenic stretch of seawall. The main government block of offices sits amidst green fields littered with palm and *fao* trees – their shade the perfect place for eating lunch. Or for a rowdy game of rugby in the late afternoons while walkers walk and runners run on their endless quest for good health. Sidestepping the odd stray dog or Vailima beer reveller. On a Sunday, the Mulivai Cathedral overflows with worshippers and women in white *puletasi* walk along the wall, their lacy hats flopping in the breeze. During the week, traffic halts at 8:45am as the Police Band marches the length of the waterfront, from a weary courthouse down to the paved front of the government building. There they play the national anthem and raise the Samoan flag. Once homage has been paid the traffic jumpstarts again and the capital of Apia is officially awake and ready for business. Blue uniformed boys hasten to morning classes at Marist Primary while next door, diplomatic staff at the Australian High Commission arrive to work in air conditioned comfort. Continue walking further and one passes the usual bistros and bars, shops and stalls. Taxi drivers lined outside Aggie Grey's Hotel will politely ask if you need a ride. A child will want you to buy her very last packet of matches, or a scarlet tropical print lavalava. Crossing the Vaisigano bridge, you will pause to marvel at the shimmer of white sails on fire as the sun plays among the yachts swaying in the Apia Marina. Sometimes you can catch children at play in the harbor waters, jumping gleefully from glistening wet rocks, pausing to wave to tourists.

On the morning of the 29th , no one can say exactly how long after the earthquake the grimy water in the harbour started to churn, like a washer on a super agitated cycle. Most people were busy going about their morning routines – now that a 'minor' upset like an unusually strong shake had passed. But workers on the top floor of the government building were astounded to see the ocean drop, like someone had pulled a plug at the bottom of the harbor. The water ran back and forth, white capped waves rushed over and under themselves. A line of black rocks just off the seawall emerged – one minute swirling water, the next exposed rock.

Everyone heard the sirens though. They shattered the morning air and were audible in all the surrounding suburbs. In front of the fire station on the main Beach Road, there was a commotion of activity as the roller doors lifted and the trucks pulled out with their lights flashing. Traffic came to a standstill and yellow jacketed police officers took to the streets, gesturing frantically for everyone to make way for the emergency vehicles.

The *Mahine Tiare* was docked in the Apia Marina on that morning. Owners Amanda and John Neal had just bought hot bread for breakfast from Farmer Joes when the earthquake had sent them running for the Marina, "eyes fixated on the waterfront watching for any signs of receding water." Their response to the sirens was immediate. "We had to make a quick decision – do we run or put out to sea? We noticed the water in the marina had started to move about and had quickly dropped four feet. Sea water was surging up and down, dropping only slightly lower after each surge." Several other boats put out to sea. "When we were entering the marina the previous day, we'd discovered the entrance is very shallow, we only had a depth of 1.7 feet under the keel in places and numerous large coral heads dotted the marina basin. We wondered if we would have enough depth to leave and what we would do if we got stuck on a coral head? We decided to run." Amanda told the crew to grab their passports and go. They ran three blocks along the waterfront before reaching the first road heading inland. Other yachties ran instead to Aggies Hotel "where staff welcomed them and sent them to the top floors of the five storey building. There they had a great view of the harbor going dry for several hundred yards out." Amanda and John joined the mass of people, cars and trucks heading uphill.

Initially, many people were vaguely bemused by the hullaballoo - not *another* drill?! Then curious. Video footage shows observers coming out of their shops and offices – walking *towards* the seawall for a better look at what all the fuss was about. Then, in response to police directives, people started to stroll. Up the roads towards the Palisi hills. Up stairs to the top levels of the high rise buildings. Still others kept on their course driving into town, unknowing or uncaring of the evacuation order. Tony's secretary, Peka Salapo and her policeman husband were in a car on the way to Matautu village. They tried to stop the traffic headed to the coast, but they were ignored since neither of them was in uniform. They soon gave up their fruitless task and hurried home to change before reporting to their respective offices.

As the sirens and police activity escalated, curiosity became bewilderment. Is this for real? And then the mood smelled like fear. A DJ on a local FM station was doing his breakfast show when he fielded a call from Falealili. *Oh my God, a huge wave just came and wiped out the village, it's unbelievable, oh my God, it just came in out of nowhere!* The caller cut out. The DJ sounded bemused and put on some jovial rap music. Listeners were confused. Crowds headed out of town on foot and in cars, started to pick up their pace. The exodus was no longer like an unexpected break from work – it took on a more serious air. Traffic jammed as impatient drivers took over the oncoming lanes and honked at each other to get out of the way. People began to look anxiously over their shoulders back at the harbor. Many boats anchored at the wharf, obeying the warning, turned on full power and headed out to open sea. Except for the Nafanua Patrol boat that was waiting for its Commanding Officer Petone Mauga. He was caught in the evacuating traffic and wouldn't reach his team for another hour. All over the Apia area, as the full import of the situation sank in, everywhere, people wanted to get their families and get to a safe place.

At seven in the morning, there were many people still at home in the throes of work and school preparation. Hearing sirens blaring and the radio rapping out orders – families found their disaster preparedness put to the test. How ready are YOU to exit your home safely? With children. And elderly? How calmly can you load everyone into a car at a moments notice, battling with thoughts that every ragged breath could be the last one you take before a wave sweeps over the trees? Or washes out your road? What if you have no vehicle? How able are you to transport your loved ones to safety in spite of how little they are, how old they are, or how heavy they may be? What if you are not at home with your family? Do you all know what to do and where to go? Where you will meet? What you will take with you?

People all over Samoa will tell you – even those who don't live in tsunami stricken zones that something happens when disaster strikes and you have moments to move and evacuate. Something happens to our attachment to so many things. Our houses. The new television that was our pride and joy in the living room. The furniture that took several pay cheques to buy. That white Sunday dress still in its plastic packaging, ready for your daughter's first *tauloto*. The dishes you wouldn't let the children use because they were too fragile. Even the beloved heirlooms from a grandmother – the only physical remnants of her love for you – suddenly all of that 'stuff' becomes just that. Just stuff.

As one woman wrote "Manuia is a teacher who had been at home when the quake hit. She was sitting in their *faleo'o* braiding her daughter's hair. When the shaking started, she rushed into the yard with her daughter in her arms. Her husband had also run from the main house, holding their other daughter. The family stood in the rock scattered yard, house on one side, *faleo'o* on the other and they knew in that moment that they held everything that was important in the world in their arms already. Without stopping to gather a single possession, they ran to the car and went inland."

Sixty four year old Amy Purcell at Malaela added, "I used to keep things for years, unwilling to let my kids use them. Nice sheets, towels, even clothes. I told my family we needed to save them for a special day. A special visitor. But then the tsunami came and took it all! Now I will teach my children and grandchildren not to keep things for years. They don't matter. They won't last. Worry instead about being kind and share with everyone because things of this world won't last."

If it was stressful for parents to evacuate their families, spare a thought for teachers all over the country who were responsible for evacuating milling *crowds* of children. Schools in Samoa start at eight and tardiness is uncommon. This meant that many of the country's children were either on their way to school or already in their classrooms. One minute they were preparing the day's lessons, the next, teachers were trying to simultaneously calm AND hurry along students as schools emptied out into the streets. Beth Hunt, the Principal of Ah Mu Academy at Lotopa, ran the 1 km distance from her house to the school to make sure all her pupils were evacuated.

"It was like swimming against the tide. I was the only one moving down the hill. Hundreds of people – old, young, uniformed students from neighbouring schools, parents pushing toddlers in strollers, big brothers packing little ones on their backs – all streamed up both sides of the road which was crowded with cars, buses, and trucks. All headed to the safety zone of the sports complex at Tuanaimato."

Matthew Leal is a Peace Corp volunteer teaching computer studies at Maluafou College, a secondary school in town. With their school bell adding to the evacuation chorus, Matthew took to the street with his students. "The uphill road was a one-way highway for cars and buses. There was a lot of confusion as to where we were supposed to go and where we could stop. Students asked me where we were going and I could only tell them we were going 'UP'. A couple of teachers also asked me and I told them 'I was following you!' Eventually, I set up camp with a bunch of senior students where we had plain sight of the ocean."

There were those who had to rush not away from the ocean – but towards it – as they moved to ensure the wheels for a disaster response were in motion. Lavasa Malua is the CEO for the Ministry of Natural Resources and Environment. As such, he is also the Chairman of the National Disaster Management Advisory Committee (DAC). Lavasa had been in his job for exactly twenty-four hours before the tsunami hit and Tuesday morning at 9am was to have been his first meeting with his staff. Instead, he found himself rushing down the road from his mountaintop home towards the coastside meteorological office. "People were running everywhere. When I got to the main Beach Road, the police were already calling people out to evacuate. I looked out at the sea and I saw the water was running out, it was flowing away from the shore like a stream and you could see all the rocks out there." Before starting at MNRE, Lavasa had worked in the private sector as a consultant – managing the project that was developing all the constitution and policy work for the Disaster Management Act – so he was familiar with the emergency response plan. "But since I had just started, I didn't have anybody's phone numbers yet, I didn't have the contacts, I wasn't sure where everyone was." He got a quick call from Filomena – *tsunami!* And the directions – the Disaster Management headquarters would move to the fire station at Faleata – away from Apia, away from the troubled sea.

Palantina Toelupe is the CEO for Health. She lives in the coastal village Vailoa, on the outskirts of Apia with her two sons. The younger son Vaitoa is Filomena Nelson's right hand at the DMO office. As the earthquake built in intensity and then shuddered to a still, Vaitoa yelled for his mum and brother to run. "Mum, there's going to be a tsunami. You've got about five minutes, you've got to run, you've got to go NOW!" Palantina ran but was bewildered to see Vaitoa accelerating in his car for the opposite direction. "Where are YOU going?" His reply clamped her with a cold fist of fear. "I have to go to Mulinu'u, to the Met office! I have to report, to get the computers up." She clearly remembers that awful moment. "It was one of the most horrible decisions any mother could make, knowing that you were supposed to run for your life and yet your son is going ***towards*** this supposed problem…It really killed me seeing him go. I ran and took his older brother with me and we were the first ones to get up to Faleata, that's how quickly we evacuated…but when we got up there, I stopped and said to myself, *what am I running for? What about my job? I've forgotten everything!* And so we turned around and came back home and when I got back, people were just starting to run – that's how early we ran. We got to the house, quickly changed and we went straight to the hospital. I got there and people were asking, *what's going to happen next?* And I said, *prepare – get ready for the emergency."*

Tautala Maoala is the Secretary-General for the Samoa Red Cross. She lives beside the Faleolo International Airport, a thirty minute drive from Apia. As she took to the coastal road to help with the evacuation – giving rides to the elderly she was on the phone with Red Cross volunteers all over the country as they reported to assigned evacuation points. "As I drove, I was monitoring what was happening on my left with the ocean…I could see the sea was receding, which we expected after an earthquake like that. I was worried but I knew we have '*auala galue'*, those good roads going inland in every village and I was very aware of what might happen. We have been taught that it only takes five to ten minutes for a tsunami to impact after these big earthquakes, but then after twenty, thirty minutes of driving, we didn't see anything so I came into the Red Cross office in town."

Dr Ben Matalavea is President of the Samoa Medical Association. A soft-spoken man with a calming, assured manner he juggles a private practice with mornings at the National Dialysis Unit as well as travelling often with the Manu Samoa rugby team as their doctor. He had an eight o'clock appointment with the Head of State that morning regarding an upcoming medical conference and was in a car trying to get to the official residence at Tuaefu when the roads jammed. "There were school kids all running towards the Malifa ground…and we were caught in the middle of it." They were eventually able to make the appointment but it was then cut short when the Head of State was interrupted. "While we were talking with him, the Acting Prime Minister arrived to talk with him and it was quite tense. I think we knew then that something drastic had happened." Their apprehension was confirmed when they received the phone call. From Dr Limbo Fiu, asking for all doctors to come in to the hospital. Immediately.

That morning, planes enroute to Samoa were abruptly diverted. At 7am a PolyBlue flight had taken off from Auckland International airport in New Zealand bound for Apia. On board was Dr Felix Wendt, director of LDS Humanitarian Services for Samoa and Tonga. The flight would be halfway to its destination when the pilot would receive the instruction to land in Fiji instead. A massive earthquake had hit Samoa and there were concerns about the safety of Faleolo Airport. The puzzled passengers would sit in Nadi airport for five hours, the worry for their families mounting as the clock ticked. A cell phone call to a spouse would only heighten the tension. *There's been a tsunami on the South coast. People are dead. We're not sure how many… we're not sure what's happening.*

For some, the evacuation proved to be a perilous thing. At a secondary school on the northern coast of Upolu – where no tsunami would strike - the principal loaded several students into his van to drive them inland. On the way, something went wrong. There was an accident and one of the students was killed. Several others were injured.

The Faleata First Responder Fire truck was racing along the south-western coast, warning low lying villages to evacuate. The truck rolled on a particularly troublesome patch of road. CFF Peter Letiu was killed and the other men badly wounded. Peter is the only Samoan fire-fighter to die in the line of duty.

New Zealanders Sarah Roberts and partner Richard Ingamells were on holiday in Savaii. Their frantic rush to get inland took its toll on a pregnant Sarah. Two days later, their son would be born a month early and require specialist care. Named 'Tamatoa', brave warrior, the struggling newborn would be airlifted to Starship hospital in Auckland. Four weeks later he would go home to join his brothers, still with a tracheotomy tube but valiantly living up to his name.

In Malaela village, fifty-six year old Simon Anae was visiting from Auckland, NZ. Two weeks earlier he had been in hospital after having a heart attack. The strain of evacuating was too much for the father of nine children and grandfather of two. He died later that afternoon in a National Hospital crowded with battered tsunami survivors.

Tony and Filomena reached Apia Park at Matautu to find the road jostling with people moving towards their safety point at Samoa College. Seeing several elderly struggling to run, Tony got on the PA to redirect them to the nearest high point. "Go to Apia Park – climb up the stadium and wait there at the highest seats. Go to the Park!"

He pulled over beside the John Williams building to give a similar directive, instructing crowds to forget the road evacuation and instead go upstairs for safety. He had just gotten back into the car when Filomena gave him the news. A wave had hit Aleipata. Stunned silent, Tony drove Filomena to her office, flying easily along the now quiet Beach Road. It was 7:26am.

"My train of thought was completely gone. I don't even know how I got to the DMO office. I just still couldn't believe that what was happening was real." Tony sent out the call for all rescue teams to respond to Aleipata. Then he started over the cross-island road, taking the longer route for fear of waves hitting the northern coast. All the training and preparation would now be called to the fore.

Behind him in Apia, the streets were drained. A lone policeman stood on the wall and watched the waters churning swirl. The white town clock stood sentry over an empty town.

Only a handful of people in town knew that a tsunami had already hit the southern coast. But there were many rumours and fearful stories clogging up the cell phone network. Vaomalo Kini's taxi driver got a call from a friend at Poutasi village. Yes there had been a big wave. He thought maybe three people were dead. Immediately Vaomalo started making mental calculations and a to-do list for whenever he finally got to the hospital…hmm, three dead, that would mean possibly up to fifteen people injured. There would be a lot of work to do, blood work, tests…He got on the line to call his staff to come in and assist with what would be an unusual workload. Vaomalo's lab handled up to fifty post mortems *a year*. In the hospital morgue there is cold storage space for five bodies. Three violent deaths in a single day was not only a tragedy – it was a potential administrative nightmare.

Lucia Hellesoe was in a car with her family heading for safety at Beverly Levi's house at Aleisa but her thoughts were with her elderly parents Jerry and Metita Schwalger at Malaela village, Aleipata. Again and again she tried calling her brother Jared to make sure they were evacuating. The landline just rang endlessly. Then broke to a dead tone of interrupted service. At Aunty Beverly's house, the evacuated extended family would chat over a breakfast of koko and hot bread. All the children would play on the slide set, delighted with the unexpected opportunity to laugh with their cousins. Lucia would sit crying with worry on the outside step.

As the road jammed with evacuating traffic, Vaosa Epa drove from her Siusega home to the government building in town. The Prime Minister Tuilaepa Sailele Malielegaoi was overseas and Vaosa needed to set in motion the wheels for the National Disaster Council. "I needed to get hold of the Acting PM immediately and call a meeting for the Council, So while everybody was heading uphill, there I was going downhill into town and the radio was talking about people dying and I was thinking to myself, *I must be really stupid!*"

Tony Hill continued over the cross island road, joined by the Assistant Commissioner Tipaula Laupu'e. Cars pulled over as the emergency truck flashed past them. People lining the street watched a fire truck, police cars, and then eventually Health department vans go by. Another fire truck flew by crowds hiking up the Aleisa road. There was urgency. There was confusion. There was fear. And a heavy sense of dreadful waiting. We all watched the sea. From the Palisi hills. From Vailima. From Aleisa. From grassy embankments where we pulled over, shushing crying children and appeasing hungry little ones with a stale bag of chips. We all watched the sun-dappled sea that lay like a piece of rumpled blue silk far below. We watched and we waited. And then the first phone calls started coming in to the 2AP radio station. About waves. And death. At Poutasi. Vaovai. At Siumu.

"The water came so fast and so strong. There are people dead here. We need help…"

"My house got washed away. The school is ruined. There are people missing…"

"Our village is here in the bush. We carried the old people. I think everyone is safe. We don't know what's left back there."

"There was a wave. It took the cars. The houses. I think only the church is standing..."

One cell phone provider gave $750,000 tala worth of call time to the nation. Three tala credit to every number so they could make calls. To family. Friends. To the radio. The announcer on the 2AP National radio station was Va'asiliega Iupati Laga'aia, a sports commentator. He fielded calls with calm detachment. As one explained it, "He was announcing the tsunami disaster as if it was a national event, like Independence celebrations or something. It came across very un-emotional, which in retrospect was a good thing. For us listening, it was very calming. He would say, *We've got a call from Falealili* and a lady says, *I've got three kids and they've died* and then he says, *okay we've got three kids dead, what else can you tell us?...* that kind of thing."

Late into the evening, Va'asiliega kept taking calls. And the nation listened.

29 September, 2009.

After the earthquake, I was afraid of a tsunami. That's the kind of person I am. Perpetually fretting about the possibilities. The maybes and what ifs. Always expecting the worst. Anticipating the nightmares so they don't catch me off guard. I said to my family – we should go, we should go now, what do you think? We turned on the radio. A song from the top ten countdown was playing. Jonah the announcer came on the air. Exclaiming about the shake. "Wow did you guys feel that? That was a big one..."

I was still standing there, waiting for the warnings, waiting for the confirmation that we should move. Someone called the radio station. Jonah answered him. It was a young man's voice. Panicked and screaming. Breathless. "Oh my God there's been this huge wave here, everything's been washed away, it took the cars, the houses, the people, oh my God, somebody help us, the people of Falealili have been hit, we've been hit!" The call cut out abruptly. Dead silence.

My eyes met my husband's. Fear threatened to overspill, threatened to frighten the children. Can this really be happening? Is this for real? What kind of sick joke is this crazy idiot playing? Jonah obviously felt the same scepticism. "O...kaaaaaay. Well thanks for that caller...hmmm..."

I didn't wait for any more. I didn't care if it was a prank. We moved. We rushed everyone to the car. We drove. I hated the time it took to lock the gate behind us. But the family was worried about robbers. Rushing to take advantage of our fear. We drove fast up the hill to Aleisa. And my every thought was screaming – can this really be happening? Is this for real? We turned to the government AM station on the radio. In time to hear the tsunami warning being announced. And more calls starting to come in. About waves. And running. And people dying.

The road uphill was starting to busy with evacuating cars. We saw friends and family rushing past. Faces of fear and worry. Faces of confusion.

The impossible had happened to Samoa. Tsunami.

4
Galu. The Wave

They tell you that before a tsunami comes, the water will often withdraw. One imagines a completely drained ocean floor, like Charlton Heston's parting of the Red Sea. With fish flapping piteously. Maybe a startled porpoise sitting on rocks here and there. The reef exposed in all its coral mystery.

And that day, the water *did* recede in many places. At Lalomanu. Saleapaga. And Poutasi. Vaovai. Only the reality was a bit removed from the movie version. It was mid-low tide that morning. So it wasn't immediately obvious when the water level dropped. And the drop didn't leave a drastically naked lagoon floor. Just a very shallow one. With some parts exposed. But big fish *were* caught off guard. As they fought against an ocean that pulled them mercilessly back. And left them high and dry in shallow puddles. For many people, the emergence of the reef was the most obvious sign that something was coming.

Moimoi Tumanuvao at Poutasi had just returned home from early morning mass with his children when the earthquake sent them outdoors. "We looked out towards the sea and there was no sea where the reef was and I said, *something's going to happen*."

To Lagi Tago the withdrawn sea at Vavau "was like a silken cloth that had been ironed, that's how flat it was."

Mike Whittaker of Auckland, New Zealand was holidaying at Lalomanu. They were at the resort office checking for tsunami warnings on the computer when they saw the ocean move."That was the scariest thing ever. Watching the water disappear over the reef, seeing the ocean empty. That was the worst part. Because then we knew that meant a big boy was coming…that's when we knew we had to haul ass. Then we were running so hard, I didn't have time to be scared anymore."

Mike and Sally Adye were guests at Taufua Beach Fales. They knew after the earthquake, that tsunami was a possibility. "[We] were semi-prepared. We had our sandals on and sat on the *fale* step watching the waves breaking on the reef. Time seemed to move so slowly but it would only have been ten minutes." They discussed the signs of a tsunami with people in their neighbouring *fale* but still were not concerned enough to act. "We sat and we watched and we waited…maybe a further minute or two at the most when changes started to happen. The waves stopped crashing. They started to retreat back from the reef. At this point, Mike shouted loudly – *That's it, let's go!"*

Also staying at Lalomanu were New Zealanders Mary Ann and Andy White. "We knew after the earthquake there would be a high chance of a tsunami. We looked out of our *fale*…then I saw the top of the reef and I knew something was wrong."

Named after her village, Saleapaga Sapelu was taking her daughter to school. Pausing to discuss the earthquake with a neighbour, she looked towards the sea. "It looked really different, it was like a bowl of water that was being shaken and just as I looked, the inside of the sea had gone right down and the sea disappeared and went right out and then it formed a big wave over the reef."

At Lepa village, seventy-eight year old Lotolua Niumatapele refused to believe the girl who called to her saying, "*Come look at the sea – the tide has gone right out and the sea has turned to mud!*" The sight of people running past, carrying their elderly on their backs finally brought her outside.

Surfers out by the reef were in the best position to see what was happening. American, Neil Lumsden was surfing at Maninoa. Two young boys were with him – his son Manoa and friend Kealoha Wallace - playing with Lego in the boat while they waited for him to "get a few early morning barrels to start my day." Unaware there had been an earthquake, Neil was puzzled to see the reef go dry. "The shelf was sticking up about fifteen feet out of the water. The normal channel turned into a puddle and my boat was nearly dry-docked sitting in a tiny pool of water. I knew this was serious." He struggled to get to the boat, fighting the current as the water began to surge back in.

Another surfer, further out, Tom Gogola, remembers seeing the reef and thinking "s@#* that's really shallow!" Within a minute, though, the surge began, filling the lagoon back up. "The sea was bubbling as if a ten foot plus wave had just detonated on the reef…millions of gallons of water were creating violent waterfalls. Surges carried on, one minute everything would be calm, the next there were the Niagara Falls like reefs unloading. When the reef was out of the sea there were all kinds of fish flapping and even turtles."

To some, the ocean didn't recede – the reefs rose up. Like a science fiction movie. At Poloa village in American Samoa, where the surf pounds a reef only a few meters from the shore, Simao Masoa was standing outside watching the sea "because I knew that tsunami follow earthquakes." She noticed the silence first. "Everything stopped, no waves, no noise. Everything went quiet but then there was a 'mmmm' sound..like a zoom sound, like when you listen to yourself humming, it was just like that. But it started from where the waves are and then everything dropped. I really thought the reef was rising like the earth was going to build a mountain or something, the reef was coming up out of the sea."

It would be misleading to say though, that the ocean will *ALWAYS* recede before a tsunami. At Coconuts Resort, the general manager Lina Bell told guests to be on the lookout for the water retreating but the ocean remained deceptively still and the first wave took everyone by surprise. One researcher who came to Samoa later with a tsunami survey team comments, "The sea does not always recede before a tsunami – the sea didn't recede in Sri Lanka before the 2004 event. When there is no receding of the sea, the first wave comes further inland instead."

The birds at Lalomanu knew something was coming. Leu Malaki was a cleaner at Lupesina Resort. The resort is nestled a short distance from the shore. In the pregnant silence after the earthquake ended, Leu saw the ocean recede. "Then, the birds started screeching, it was like a ringing in my ears. The birds were all crying and then they all took flight…I saw them take off from the trees. They must have known what was going to happen, but we didn't have any idea."

On the shore, you heard the wave before you saw it. At first it was a distant rumbling, like a storm brewing out to sea. Then it became a sharper roar like the deafening growl of a busy construction site. People ran out to look. Others followed asking, what is it? Is there a *misa*? A fight? The noise grew. It was a hundred jet planes taking off. A thousand bulldozers, gears stuck and grinding in first gear. It was like "war and machine guns."

You saw the wave, but it felt unreal. It was not the colour of the ocean. It was "black but the top of it was like smoke."

People stood transfixed, unable to process what they were looking at. Many were mesmerized. At Malaela, Leota stood and watched as the water smashed through the Aleipata wharf and still he did not run. Still he thought it would stop, turn and wash back out like any other wave. Not until the wave was almost upon him and his wife did he grab her hand and run.

You were unprepared for the sheer speed of the wave. It didn't seem possible that something so vast could move so swiftly. At Lepa, Sarai Fatialofa saw the sea "had gone." A boy yelled for her to start running. She went first to warn a neighbour, thinking she had time. But within seconds, the water was covering the road. "We saw the wave coming like fire, so we ran ."

At Matautu, Falealili, Tilasi Popese is still awed by the flow of the wave. "Five minutes after the earthquake hit, then the tsunami came very fast. If people were standing on the beach they would have gotten caught, it was just so quick." He shakes his head. "It doesn't matter if you're a good runner, you'd get caught. The wave was just too fast."

To some the wave was NOT a wave – it moved like a tornado. Tuluiga Utaulu described the tsunami at Lepa village thus, "This wave didn't move like other waves. It was spinning. The house was taken up, scraped up – the wave didn't hit it and break it into pieces. It was like seeing a wind, a twister, picking up a pole, picking up a line of bricks, picking up the cement of the house and taking it."

Maele Olosepu agrees. She was on the road to work at Litia Beach fales at Lalomanu village and saw the wave coming as she stood on the high road. "It was spinning as it came, picking everything up with it and roaring."

The sound of the wave is what most people remember most vividly. School teacher, Rebecca Hamilton from Gisborne, NZ was staying at Litia Sini Resort. Focused only on escape, Rebecca never looked back but the sound of the pursuing ocean was unforgettable. "The tsunami sounded like a train coming towards us as we climbed. As it got closer we could hear it smashing into *fales* and taking down everything in its path. The *fales* sounded like match sticks snapping. This was absolutely terrifying to hear, it was the sound of pure strength and force."

As the first wave came, it drove ahead of it, a torrent of air, a rushing wind that swept through the mountainside trees at steep coasts like Lalomanu and Saleapaga. Christopher Wilson of Auckland, NZ felt that raging storm of air as he climbed to safety with his children. "I heard the wave coming. It was a loud foaming sound that was interrupted with very loud sharp, shattering sounds…the sounds of trees and poles snapping. Then the wind hit us. It was the air that was being pushed in front of the wave. It was like a storm had hit the trees all around us."

People tried to make sense of what they were seeing. It was the ocean – but unlike any ocean they had ever seen. To most it had become a living thing. "A big black cloud that crawled forwards like a thief." "A beast that jumped up out of the sea and ran towards them…" "A wild animal that growled like a train." "A ravaged beast that was hungry for everything on the land". Six year old Europa of Saleapaga said it was a demon. Still others said that Satan himself was in the waters.

Many people remembered the warning signs from the DMO campaign or from lessons they had been taught at school. Enough to save their lives and the lives of others. School-aged children in particular, were quick to respond. Europa was walking to school at seven that morning with his cousins and together they climbed the steep hill behind Saleapaga without hesitating as the sea receded.

Many visitors staying at Litia Sini Resort at Lalomanu owe their lives to the actions of two schoolchildren. Ten year old Abby Wutzler of Wellington,

NZ was on holiday at the beach fale operation with her family. Her former teacher at Makara Model School, Kay Mudge, had used resources from the NZ Civil Defence and Emergency Management (CDEM) to teach her students about the tsunami warning signs. Abby was on the beach when she noticed the water start to go out. She ran the length of the beach yelling that a tsunami was coming, telling everyone to go to higher ground. "I ran up to the *fale* towards Mum screaming that the water was going out and a tsunami was coming. I explained to mum and dad…it took a bit of explaining. This meant we were not very fast and the water got to us. We were the last to escape, because the water was right behind us. We got as high as we could."

The Wilson family was also staying at Litia Sini. Twelve year old Maxwell was another quick-thinking youngster, who ran to warn others. Maxwell and his father Chris stood watching the ocean after the earthquake, "We watched the water…all of a sudden, the water went dead still, the reef turned into a backwards waterfall. My heart started racing and I was pretty sure a tsunami was coming. Dad yelled, *'Run for your life and don't look back!'* He ran to get my sister Poppy. Because no one else seemed to be aware that there could be a tsunami, I ran around the *fales* shouting, '*Tsunami – everyone run!'* I repeated my shouting nine times, determined to get people out of their *fales*. As soon as I saw people moving, I shot off towards the cliff."

At Poutasi Hospital, the Nursing Manager James Belford remembered previous drills carried out by the Health Department. After the earthquake, with a wary eye on the sea several hundred meters away, he had his staff evacuate the handful of patients a short way up the hill and over the main road. Everyone thought it was a joke and there was much laughter. "Eh – did you ever see a tsunami here?! Of course not…we know you are just finding excuses to make us get some exercise *a ea*?!" They scoffed at James as they strolled to wait under the shade of the mango tree by the roadside store, but a precise, efficient man, he refused to be budged from adhering to department protocol. There was no laughter when the water came rushing in to the village below them.

Often it was the elders who resisted the messages of the media campaign, shrugging the dangers off as "things that only happen in other countries." Tina Niusila of Saleapaga told her parents they should go to higher ground. "Like what they said on the TV. Nobody was listening to me so I just got my kids together and we walked to the high road with our bag." Sitting on the hill, Tina couldn't bear to even look at the wave that engulfed her village. "I shut my eyes, I was too scared to look at it. I could only hear its noise."

Notuse Semo's daughter called her from Fasito'o after the earthquake, telling her to go inland in case of a tsunami. "I said to her, don't worry. You are doing the Lord's work but you have no faith!" Two months after the wave killed 35 people in her village of Saleapaga, Notuse refuses to speak of what happened to her in the rush of water, or about the injuries that kept her in the hospital for weeks.

At Vavau, Lagi Tago's daughter also called her from town. Three times. Asking her mother to please go. "I said to her that nothing is going to happen and we are listening to the radio. I was saying they are waiting for further notice from Mulinuu. We will wait." It was her daughter's tears on the third call that got Lagi moving. Out of the house and up the steep slant to the main road. "It was about three minutes later that the waves came and broke our house. I know that if I didn't act on what my daughter was saying, I wouldn't be alive."

The earthquake felt to Tufaiga Fatuesiafioga "like the ground was going to open up." She told her mother they should get away from their house on the beach. "My mother said to me that nothing [like tsunami] ever happened in those days. I said to her that those days were different from these days." Tufaiga was separated from her mother and washed out to sea. A drifting log saved her. Her mother and grandbaby both drowned. Raw grief is a gaping wound in our conversation.

Tsunami. *Galu lolo. Galu afi.*

After the event, many teams of experts came in search of answers. One was the UNESCO-IOC International Tsunami Survey Team Samoa (ITST). Their objective was to "explore the nature of the tsunami and its impacts in Samoa…for the first time ever in a UNESCO-IOC assessment, we will attempt to explore the nature and linkages between the physical, social, economic and environmental systems in order to provide a more sophisticated understanding of the tsunami and its impacts – beyond a simple description of maximum run-up or total number of lives lost." This report is powerful testament of how man and nature are inextricably connected as it looked at things as varied as the local building code, whether or not villages had a well-conserved coral reef in front of their beachfront, whether the presence of mangroves had any impact on wave inundation and how traditional land ownership arrangements affected survivors' resilience and capacity to recover.

Other multiple reconnaissance teams travelled to Samoa throughout October and November 2009, bringing together many different people from all over the world. One of these was Dr Jose Borrero, an American oceanographer based in Raglan, New Zealand. He had been one of the first scientists on the ground in Indonesia after the 2004 Boxing Day Tsunami and to Papua New Guinea after the tsunami of 1998. The team that Dr Borrero was part of arrived in Samoa a week after 29/09. They travelled throughout the affected areas collecting data and carrying out interviews with witnesses and survivors.

Tsunamis are tricky things, virtually impossible to predict and difficult to study since the scientists have to work backwards to find the answers they need. Like studying the 'bull in a china shop' – ***after*** the bull has trampled the fragile dishes and run off...*Now,let's look at the mess and figure out how big the bull was? Why he smashed this plate and not that one? What made him turn this way and not that?* There can be many unexpected answers for scientists who go to study a tsunami. Bruce Jaffe from the United States, wrote of their expedition to the Samoa islands, "29/09 was the fifth tsunami studied by US Geological Survey(USGS) field teams in 15 years, and yet it presented many surprises...it was striking in several ways, including the variability of its impacts, the effects of coral reefs on inundation, the evidence of strong return flow, and the effectiveness of education in reducing deaths and injuries."

There are several different reports one can read on the Pacific tsunami that make a variety of conclusions but most of them agree on the tsunami arrival time – between eighteen and twenty-five minutes after the earthquake ceased. Talk to survivors though, and many will tell you it caught them anywhere between four and twenty minutes afterwards. Many were still reeling from the tremors, not even back in their homes.

Survivors say there were at least four waves. Memorable enough to be counted. Most people didn't have the time or presence of mind to take notes. Many agree the second wave was the biggest. A complex weave of factors determines how high, how powerful a tsunami will be, even how far inland it will go. Things like – whether or not there is a reef or offshore island in front of your village. A reef is your first protection against a tsunami. An offshore island will either reduce the impact OR amplify it. In one place, the Aleipata islands divided the wave, causing it to wrap and then meet with renewed fury before hitting the coast. In another, houses facing Nu'usafe'e island were somewhat sheltered, compared to those on either side of them.

If you lived by a river or stream, the tsunami caught you faster as it eagerly ran up the channel. Houses on both sides of the stream at Lepa were hammered hard. At Poutasi village, waves destroyed the bridge over the mouth of the river and easily swept in to hit the school that sat on a flat field. In the town of Leone in American Samoa, the wave reached inland for "an impressive 496 meters along the bed of the Leafu River."

The town of Pago Pago in American Samoa is located at the end of a four kilometre long bay which scientists described as a virtual "tsunami trap." A sheltered harbour that offered welcome protection from bad weather became a nightmare in a tsunami as yachts and other boats tried to escape. While the wave reached only three meters at the narrow entrance, it swept to a fearsome eight meters at the furthest inside reach of the bay, sweeping boats and shipping containers inland with deadly ease. Kat Pearson stood on the roof of the Evalani Motel and watched in amazement as a yacht sailed down the main road where it would eventually come to an abrupt halt in the walls of the LDS chapel.

Mangrove forest and other vegetation was a barrier that tried to slow the rush of the tsunami. The reach of their branches may seem like a vain attempt to grab hold of the ocean beast, like trying to catch running water in your outspread fingers. But in places where mangrove trees squatted over their muddy earth, the tsunami's attack was somewhat frustrated. At Vaovai, homes and gardens *behind* the breathing mangroves showed less damage than those without. A tangled thicket of beach hibiscus trees on the seaside of Sapo'e village helped shelter homes and slowed the wave, allowing people time to escape to the road. Utulaelae village, just next door, had cleared their coastal bush to build houses and for easier access to the beach. Their naked shoreline welcomed the tsunami with open arms and afterwards, shattered buildings sat dismally in the wet sunlight.

The terrain was of course another determining factor. Did you live on flat coastal plain that made walking to the plantation an easier task? Like at Malaela village? Then the wave had almost unlimited freedom to run and play, forcing everyone and their homes to play too – until it just plain old got tired and went back out. Wide flat coasts like at Malaela and Saleaumua also meant that people had much further to run before they could reach the safety of higher ground. Inland lagoons and water bodies "that wrap behind coastal communities exacerbated their vulnerability by cutting off evacuation." So at Malaeala for example, where mangrove swamp breathes behind the village, people had to run parallel to the sea for a distance before reaching one of the inland tracks that bridged over the water. A similar thing happened at Poutasi.

Did your house hug a narrow strip of land at the base of craggy cliffs? Like at Lalomanu? And Saleapaga? For you then, the wave slammed abruptly into rock wall and got angrier at the rejection. It threw a wild tantrum like a giant snake caught in a trap, whipping and twisting to get free. Churning your house, your car, your garden in a terrifying torrent. And if you weren't fast enough, strong or agile enough to climb up the cliff – then it whipped you along with it.

Simple questions then of – how deep was the water? How high was the wave? - will get you lots of different answers. Depending on who you're asking. Experts like Dr Borrero will tell you about *run-up* – the maximum vertical excursion of the water surface relative to sea level at the time of the tsunami's first arrival. *Flow depth* – the depth of the water as the wave comes ashore over the local ground. And *inundation* – the horizontal distance that 'got wet' from the tsunami. In low-lying flat areas, inundation distance will be greater. The ocean at Malaela came in for 400 meters inland because the village was at an elevation of only 2 meters above sea level. In areas with greater elevation, where hills and cliffs resist the ocean, inundation distance is less but run-up is high.

The maximum run-up of the Samoa tsunami was 14.5 meters at Lepa village and 11.4 meters at Lalomanu. "Tsunami amplitudes decreased rapidly westwards along the coast, reaching no more than 5 meters at Salamumu and 4.2 meters in Matafa'a. On the northern coast, there was 4.4 meters measured in the east but only 2 to 3 meters immediately to the west." On the big island of Savaii, the tsunami was angriest on the southern coast – running up 8 meters at the uninhabited Nu'u Black Sands beach. The village of Taga was hit with a 6 meter wave. On the islet of Manono, in the Apolima Strait separating Savaii from Upolu, the wave was 5 meters high.

A standard two-storey house is eight meters high. Set it in the Samoa tsunami at its worst hit points and you wouldn't even know it was there. But what does that mean for a regular person like you and me? I am five feet nine. It would take at least seven of me, standing on each others shoulders before I would be able to get my head above water at Lalomanu. What does that mean for a parent with a small child? Like thirty-five year old Maua Toa?

When the wave slammed into their village, Maua grabbed his seven year old son Aukuso and ran. They were smashed into a majestic-sized breadfruit tree. Maua held his son crushed between him and the tree trunk. They were underwater while the sea tried to rip them away. Maua pushed Aukuso up into the branches of the tree, helping him climb higher as the second wave engulfed them again. Higher than the roof of their Samoan *fale*. Sweeping over the edges of their neighbours two storey roof. Maua was exhausted. Debris continued to plough into him, weakening his hold on the trunk. But Aukuso was foremost in his mind.

"When I looked up the wave was on top of us. I couldn't breathe. The only thing I wanted was some air **so my son could breathe…**"

Maua shoved Aukuso further up the tree until he was safe. Crying above him in the final, smallest branches. Gulping mouthfuls of the sweet morning air. As Maua battled to keep his head out of the water, he thought about letting go. Giving in. But then Aukuso called to him,"Papa, be strong. If you die, I'm going to jump down so we will die together. *Fa'amalosi*!"

Maua hung on. A child's plea gave him the strength he needed. "I did what my son was saying. I ended up being saved because of what my son said, to be strong."

In Malaela village it was the ferocity of the wave rather than its height that survivors remember most vividly. Only a handful of houses were left standing when the water receded. Sixty four year old great grandmother Amy Purcell was running hand in hand with her grown son pulling her along, when the wave caught them.

"I didn't even feel it catch us. All that I knew was that I was diving under the wave. It was so strong that I didn't even know where I was. When the wave was calm, I was under the trees and the water was up to my neck. Our houses were destroyed. I heard many cries for help. My son told me to go inland because another wave might come and find me. He went back to look for my daughters and the children. It was very hard to go because there was a lot of rubbish everywhere and sharp things." Amy's great grandson Joseph was not found for another two days, underneath an upturned car. He was three years old. "He was my only great grandson. A smart boy. A good boy. Everyone spoiled him because he was the baby of the family."

Most often we imagine a tsunami like Hollywood depicts it. A massive wall of water that crashes down on terrified cities, shattering everything. But according to survivors, it is more insidious than that. It is an ocean that just keeps coming and wont stop. Evelyn Lili'o stood on the cliffside road above Fagasa village in American Samoa and watched the ocean empty in and out. "It was like watching a toilet bowl overflowing. The water just came up and up, spilling over everything and swirling."

At Lalomanu village the water chased people as they climbed the mountainside. One survivor who lived closest to where the road climbs the hill ran a short distance to safety before turning to look back at his village. Disbelief clouds his eyes as he tries to describe what he saw. "I looked and the wave came and covered our whole village. I couldn't see any houses. Nothing. It looked just like a calm sea. I thought when it went back out that then I would see our houses. But there was nothing left. The water went out and everything was broken. I don't understand how the water can do that."

It's not just the drowning depth of a tsunami that is the problem. It's the sheer power of millions of tonnes of displaced water. Strong enough to tumble cars. Deposit deep sea fishing boats in a tree. Reduce your brick house to rubble. Carve chunks out of your road. Being in a tsunami is not like suddenly being stuck in the deep end of an Olympic sized swimming pool.

TSUNAMI 29/09/2009

HIS HIGHNESS THE HEAD OF STATE - TUI ATUA TUPUA TAMASESE EFI WITH THE ACTING PRIME MINISTER, TUU'U LEOTA ANASII LEOTA DURING THEIR VISIT TO THE AREAS AFFECTED A FEW HOURS AFTER THE TSUNAMI HAD STRUCK.

The Head of State tours the disaster zone with the Acting Prime Minister on 29/09.

Lalomanu beach before 29/09.

Lalomanu beach after. *Photo - Australian Defence Force*

A beautiful day to be staying in a beach *fale* resort at Lalomanu.

Taufua Beach Fales before 29/09.

The wave wipes out Lalomanu. *Photo - Mike Adye*

The first wave pulls back leaving no structures standing at Lalomanu. *Photo - Mike Adye*

Aerial shot of Lalomanu village, 10 hrs after tsunami.
Photo - RNZAF Orion

Aerial shot of south coast villages, 10 hrs after.
Photo - RNZAF Orion

Aerial shot showing village still partially submerged 10 hrs later. *Photo - RNZAF Orion*

Aerial shot of remains of Lepa village. *Photo - RNZAF Orion*

Aerial shot of a battered Aleipata wharf. *Photo - RNZAF Orion*

One hour after the tsunami at Saleapaga village. *Photo - Bharat Chovan*

Lupesina Resort nestled in the foothills of Lalomanu, before 29/09. *Photo - Matthew Leal*

Lupesina Resort on 30/09. *Photo - Matthew Leal*

Matt Leal stands beside a church in Saleaumua, awaiting his turn to run in the Samoa Perimeter Relay, 28/08/09. *Photo - Matthew Leal*

Matt drives past the same spot as a Red Cross volunteer on 30/09. *Photo - Matthew Leal*

Pene T. of the DC Engineering Perimeter Relay team runs past a cement wall on 28/08/09. *Photo - Matthew Leal*

The decimated wall after the tsunami. *Photo - Matthew Leal*

Waiting for the next runner to arrive in a sleepy Lalomanu village on the morning of the Perimeter Relay. *Photo - Matthew Leal*

The day after the tsunami. You can still see the faint outline of the race road marking number 9 beside the relief worker in the yellow jacket. *Photo - Matthew Leal*

Wave damage along a section of road past Lalomanu. *Photo - Vaughan Simpson*

And trying to swim. It's the fact that the swimming pool is crammed full of debris. Like roofing iron that can cut you in pieces. Coral boulders torn from the reef. Timber beams that will break all your ribs. And you're in the middle of it. And something is shaking the pool. Like the way a mad dog shakes a kitten in its jaws. No, survivors will tell you that a tsunami is so much more than just a deep wave.

Talavou To'omalatai at Vailoa village had gone with his young son to feed the pigs but came running back when he saw the water receding. He met the wave coming and had to swim. It was a battle of wills as the sea resisted him with rubbish, rocks, furniture, and even floating cars. People called out that he was going the wrong way but he ignored them – his consuming thought was his family back at the house. His eighty-three year old father, his wife, their baby and two young children. Upon reaching the house, he could hear them screaming for help inside but stopped first to rescue his uncle that was rolling in the water, being hammered by debris. In the house, his father was also trying to keep his head above water as furniture washed back and forth. He had told his daughter-in-law Faiuga Talavou to leave and go inland when they first saw the wave coming but she had refused, "She cried and said that they would rather stay and die with me." The water was four feet high in the house. She tells her story,

"I kept asking the old man to leave but he wouldn't so we stayed with him. I took the children into the bedroom and tried to get them as high up as I could. I held the baby to my chest and made the other two kids stand on my shoulders. When the two waves came, we were underwater. I was afraid but I knew that at least if we died, we would die together."

The tsunami left their house standing. And Talavou's young family alive. They still struggle with the souvenir common for many other tsunami survivors - nasty lung infections. But they are alive and that precious fact outweighs all else.

Tsunami bring with them sea creatures of all kinds. And Samoa's *galu afi* was no exception. Survivors battled alongside sharks, dolphins, turtles, fish, stingrays, octopus and crabs. Leota from Malaela was caught with his wife in a dam-like build-up of debris behind their village which he credits with saving their lives as it sheltered them somewhat from the battering surges. Clinging to the wreckage they were taken aback to see the marine animals swimming alongside them.

"I was here – and right next to me there was a shark. Then a big stingray went past. So many creatures!" He shook his head in wonder. "You know, it reminded me of in the Bible where the scriptures say that in the last days, the lion will lie down with the lamb. That's what it was like. The shark was there and neither of us was afraid of each other. We were all just trying to do the same thing – stay alive."

Most of the marine life was left stranded when the water went home. Houses still standing were littered with fish. Some "as long as a man". Many provided the first meal for survivors on that first desolate night of the tsunami. A night when villages huddled together in the bush – wet, cold and hungry. Leota explained, "After the tsunami we sent the *taulealea* (young untitled men) to gather food. They brought baskets of fish and other animals – enough to feed everyone. We made a big umu and cooked them with taro and bananas from the plantation."

The Marine Protection Agency had been heavily involved with turtle conservation in the Safata and Aleipata districts and so of course they were anxious about the tsunami effect on the turtle populations. They documented at least fifty-one turtles had been trapped inland. Forty eight were successfully released back at sea while others died or were eaten. Two dolphins never made it back to the sea and were later buried by villagers cleaning their yards.

Up to a week after the tsunami, you were still stepping on small fish here and there. And the smell of wet rot was strong. From dead coral and decaying seaweed. And of course, other things.

What caused such waves?

In August 2010, a study led by the University of Utah Seismograph Stations announced that the Pacific Tsunami had been caused by not one - but three major earthquakes occurring within two minutes of one another. This is the "first known case of an earthquake occurring on the plate of the Earth's crust beneath the ocean and then triggering two major 'thrust' quakes in the subduction zone where the edge of one crustal plate is being forced below another." Co-author of this study, Keith Koper explained that scientists had been puzzled by seeming inconsistencies in the seismological data and the tsunami wave pattern. This led to the discovery that a first 7.9 magnitude quake had 'triggered' two more earthquakes with a combined energy release of an 8.0 magnitude. "It was essentially a great earthquake that was triggered, not some silly little aftershock."

This 'double jolt' generated a powerful tsunami. Several other ingredients added more power to the wave recipe. Like the depth of the actual earthquake. The closer to the earth's surface a quake is, the less energy is lost as it travels through the ground. Shallow earthquakes then, pack more punch. The epicentre of the 29/09 shake was only 10km under the sea floor.

Another feature that gave the Pacific tsunami even more power was the amount of water sitting on top of the epicentre. The site of the subducting plates is the Tongan Trench - a vast undersea canyon that "could literally swallow the US Grand Canyon and still have room for more." Located here is Horizon Deep, the second deepest spot on the planet. Scientists estimate that there was "well over three miles, maybe even four miles of water above the shaking ground. That means more water displaced and thus bigger waves. If there had been only one mile of water above the quake, the waves would have been about eleven feet smaller." The deeper the water, the faster a tsunami travels. Geophysicist Eric Geist explained, "This water was so deep that the tsunami could have been zipping along at 530mph. Usually a tsunami slows when it hits shallow water. Around the USA for example, the shallow continental shelf slows down waves dramatically. Samoa didn't have that protection until just before the tsunami reached the shore, Even then, it would still be coming at 30mph, so by the time you see a tsunami, it's usually too late to outrun it."

A key tool for alerting people that a tsunami has been generated is the series of DART buoys dotted throughout the world's oceans. Dr Borrero describes them as "pressure sensors at the bottom of the ocean, about 5000 meters deep. When a tsunami wave passes over the top it can sense it. Most wave gauges work by measuring water pressure – when a wave passes over and it gets deeper, it is registered as an increase in pressure, then as the trough passes over, the pressure at the sensor drops. By looking at the signals of up and down pressure you can see waves passing over a point. Tsunami waves are different from normal ocean swell waves because they have a very long period and very small amplitude. DARTs are positioned strategically across the oceans to most efficiently capture a tsunami signal from the most likely places and give the most advance warning to populated places." These sensors are always on, but after an earthquake has occurred close by, they are automatically put into a more sensitive mode. The greatest value of the DART system can be seen when a large tsunami has been created and could be threatening faraway places. Or not. "DARTs are also good for calling off unnecessary evacuations. Pre-DART, Hawaii was evacuated in 1994 after an earthquake in Kamchatka, which was really costly and embarrassing to the State. Several times since then, they have been able to say 'with certainty' that no destructive tsunami was on the way."

What DARTs are NOT good for are 'local' tsunamis "there is just not enough time to record the signal and process it before the wave hits. The system probably needs about fifteen minutes...In Samoa the wave was there in less than twenty minutes. So, just not enough time." Post-tsunami wave simulations graphically depict the speed of the 29/09 *galu afi* - the first surge hitting southern Upolu first, eighteen minutes after the earthquake, then southern Savaii and the southwest side of Tutuila in American Samoa, seven minutes later. One field survey report on the Samoa tsunami concluded, "This reinforces the concept that real-time tsunami mitigation in the near field must rely on self-evacuation, to be triggered by the affected populations themselves, upon feeling the earthquake or observing any anomalous behaviour in the level of the sea. Such warning must be individual, or organized at the strictly local level of a village or small town, and cannot depend on a centralized warning entity."

At 6:50 Samoa local time, the Duty Senior Standing Watch Officer at the West Coast and Alaska Tsunami Warning Center (WCATWC) was alerted to an earthquake by a seismic short period alarm and within two minutes he had an initial auto location of 15.3 degrees South and 171.6 degrees West. At the same time one of many tsunami detection buoys dotted throughout the worlds oceans – DART 51425 triggered back an alarm to the center. By 6:58 he had an initial reading of the earthquake, a magnitude 7.9. A magnitude that would be upgraded an hour later, to an 8.3 on the Richter scale. Reacting almost instantaneously he followed protocol and sent a message to the Pacific Tsunami Warning Centre based in Hawaii and to various other agencies via email. He followed up the email with a phone call to the PTWC and they issued an official tsunami warning at 7:04am for the Pacific islands. The entire process took a total of under sixteen minutes.

An earthquake off the islands of Samoa and Tonga. Alarmed and recorded in Alaska. Wave disruptions detected by a buoy sensor sitting in serene isolation in the middle of the Pacific Ocean. The dangers communicated to Hawaii. And then sent racing outward to the nations of the Pacific. Sixteen minutes. Thanks to the wonders of satellite technology and computer wizardry half the modern world knew there was an earthquake and possible tsunami in the Samoa region in only sixteen minutes.

But it was sixteen minutes too late. For the thousands who live on the Southern coasts of Upolu and the American island of Tutuila. For the nine hundred people living on the remote northern Tongan island of Niuatoputapu. It was too late. Because with the first violent shake of Mafui'e, waves began to resonate, gathering momentum as they raced across the ocean at over five hundred miles an hour. And directly in their path were the islands of Samoa, northern Tonga and American Samoa.

Too late.

5
Ou te fia ola. I want to live.

It was too late for the district of Aleipata on the Southern coast of Upolu.

The drive from Apia to Aleipata takes a confident driver about an hour. An hour and a half if you're the extra cautious kind who navigates the dips and curves up the Le Mafa Mountain Pass with hesitant nervousness. But the journey is well worth it. Children in ragged shorts will wave as you drive past and mothers sweeping up leaves with long handled coconut frond brooms will yell at them to get away from the road. You will pass roadside stalls hung with strings of dripping wet fish, rows of *niu*, bursts of orange papaya and sunshine fingers of ripe bananas. It seems everyone plays volleyball as there are riotous games occurring in every village. As you come over the rise of Tiavea, the coastline of Aleipata lies before you like the proverbial picture perfect postcard.

For those in search of Samoa at its untouched, idyllic best – then Aleipata district is the weekend destination of choice. Even locals head to Lalomanu and Saleapaga to retreat from the mad rush of dusty Apia. There is any number of places to stay. From the up market enclosed bungalows with their own beach front verandahs and a tropical breakfast buffet to the open *fales* where you can get a mat and use of a tap to wash off the salt from an all-day swim. On a clear day, the white sand stretches for miles and the azure waters beckon endlessly. Couples come from faraway lands to get married, pledging their undying love wearing garlands of pink and white frangipani and dancing barefoot on moonlit sand.

Saleaumua is the first village you come to along this flat southern end of the east coast, flowing into Ulutogia and Satitoa. It's a low lying coastal area that in many places was once mangrove swamp. The tar seal road runs parallel to the beach and houses are lined densely on both sides. As with all Samoan villages, families have a large 'guesthouse' in the front with a cluster of smaller dwellings in the back. A cookhouse, a bathroom, a sleeping house. Dogs squabble with chickens and pigs for scraps and for space under the house. As with all Samoan villages, there is colour and greenery everywhere, impossible to suppress. Red hibiscus bushes line the road interspersed with palms, clusters of grey barked frangipani trees. Leafy bunches of yellow play tic tac toe with white painted stones. In December, crimson flame trees are beacons of fire ablaze. There is a church every two minutes along the road. EFKS solemnity, Catholic grandeur, LDS uniformity.

There are rough roads – more like tracks that meander inland every so often – used by those going to their plantations, huge sections of bush where intrepid farmers would doggedly hack space to plant taro, bananas, yams. According to traditional communal land ownership, every family has their rights to a piece of land and can cement these claims with a bush knife and a lot of sweat. Increasingly though, many chose to pursue 'easier' methods of acquiring an income. The beach *fale* operations further down the road at Lalomanu provided employment for many of the young people. They went to work as cleaners, cooks, waiters, dancers and tour guides. Still others ran their own small businesses. Like Taupeau and Fiva who had a little countertop store next door to the church beside the sea, selling canned goods, beer and cigarettes, bread and chips. Their family lived in the two room house attached to the store – all fifteen of them, including six children under the age of ten and their 98 year old grandmother Suliane. Like the Solovi family of five children and ten adults who owned a taxi. The precious fares supplementing their little taro patch inland. Still others chose to rely on the lucrative business of remittances. *Meaalofa* that came from children, siblings, living in the lands of milk and honey far away.

Yes, in many ways, Saleaumua was like any other Samoan rural village. The *matai* council of chiefs met weekly to discuss village business. They handed out discipline and punishments as needed and made sure everyone kept their yards clean, the women wore long dresses, the youth didn't run wild, and nobody roamed the street during evening prayers. Sundays were for worship. Saturdays for worship preparations. Weekdays were for work. School. Games. Like most other Samoans, the people of Saleaumua were friendly and welcoming of visitors but somewhat suspicious of those who came to stay that didn't quite adhere to their communal code. People like Lee – a single woman living at the edge of the village who had painstakingly built a thriving garden nursery and wellness retreat. Yes, she was Samoan and yes, she even had several *matai* titles. But she didn't quite fit their model of a village dweller. Yet, no matter, life in Saleaumua carried on.

Until Tuesday the 29th of September when within a few short minutes, Saleaumua was no more. As one villager expressed it "Everything was going well and dandy until the day the earth shook."

It was a day that began like any other for twenty-eight year old Oselani who lived with his wife's family, he being the only breadwinner for his two children and the in-laws. He was a fisherman. As was his father before him. The sea spoke to Oselani. When the winds were rough and the tides were high, the waves would crash ominously on the reef and others would nervously stay close to the shore. But the rough swell would call to Oselani with the promise of a big catch and the white surf would only send an extra thrill of exhilaration as he paddled his frail wooden *paopao* out to the breakers. The sea was his friend and a benevolent employer, feisty at times but generous with her bounty. When income from his taxi driving job was sparse, Oselani always knew he could count on the sea to provide. A meal. A few extra *tala*. A fresh *malauli*, scales gleaming with newness would guarantee a smile from his stern mother in law.

On the morning of the 29th, Oselani was up at dawn with five other fishermen. They paddled their single hulled canoes away from Saleaumua, past Ulutogia and Mutiatele, past Satitoa, past the Aleipata wharf, and paused just beyond Namu'a island. The tide was low and they could see the ocean floor through the still waters. The fishing was slow and there was an easy, companionable chatter back and forth as the men called out to each other, while back on land the village began to bustle about with the day's preparations.

Just before seven, Oselani's boat started shaking, the oar placed lengthwise in the centre of the canoe clattering so abruptly that he hurried to reach and grab it before it fell into the sea. The others called out in consternation as they too noticed the movement of the ocean. But then the shaking stopped and they floated on a silken cloth once more. But Oselani was uneasy. He had never felt such strong tremors before; especially not ones that could make themselves known while he was on the water. Without knowing why, he felt for the first time, a sliver of fear about being out on the ocean. He asked the others if they should go back – in case there was something wrong? He thought about the television ads with the man in the funny hat. *What if there is a tsunami,* he called out? *Don't worry*, the others laughed back. *There aren't such things here in Samoa.* Oselani looked down into the water and couldn't make sense of what he saw. The ocean floor was now a churning mass of black, like boiling dirt clouds. His unease grew, but before he could act on it, the others screamed.

"*Galu!* Look at the wave! There's a big wave coming."

Rushing towards them was a rise in the ocean. A hill of blue. On the tiny island, Sara Ulberg, owner of Namu'a Beach *fales* was hosting a group of school children from St. Peter's in Palmerston North. With the ocean sweeping towards them, everyone scrambled up the hillside and stood awed by the sight of the wave. Sara watched in horror as "a beast jumped up out of the sea and ran towards them". The wave hit the island and divided in two, wrapping itself around the shore like a sinuous blanket of cobalt steel before meeting again in the front and continuing its onslaught towards the mainland.

"I thought we were all going to die. There was this massive wave that just sprung up so high and started rolling towards Lalomanu and Saleapaga while the wave on the right side of the island rolled onto Malaela and the other nearby villages. I felt so helpless watching the waves go to hit the villages, all I felt was that everyone would be killed."

The fishermen began paddling frantically back towards the land, towards where the wharf juts out at Satitoa. Oselani knew they would never make it to in time. His thoughts went to his wife and children and he hoped they could see the water coming. He stopped paddling and faced the hull of his canoe towards the wave with the hope that the craft would skim over the top of the water. As it drew nearer however, he realized that thought was a foolish one. The hill was immense. It was now a monstrous mountain of blue. He decided to change his tactics. He took a deep breath and dived into the sea, hoping that with his fisherman's increased lung capacity – he could wait it out and evade the wave as it powered over him. Another vain hope.

Oselani was swept along with the wave. He was but a wretched thing tossed in a vast washing machine on spin cycle. He couldn't swim, he couldn't breathe, he couldn't see anything. He was trapped in black water, buffeted on all sides. He knew he would die.

Thirty one year old Mamea Mikaele lived next door to the Catholic priest beside the grand old Saleaumua church. Known to all as 'Mika', he was a tour guide operator who worked on commission at Taufua Beach Fales. A keen entrepreneur, he also had his own business taking groups on walking tours of the cliff top hills of Lalomanu. Along the way he would regale the group with insights into Samoan life and culture, as well as explaining the different medicinal uses of native plants. He would show tourists the caves where Samoans would take refuge during times of cyclone and 'times of starving'. Vast caverns clefted into the hillside would shelter the villagers and one can still see the crops they would plant there to tide them through. It is doubtful whether many Samoans today could even tell you about them– the caves belong to bats now.

Back at Lalomanu, Mika would then give an *umu* making demonstration at the home of his wife Ave's family, speaking in faultless, clipped English. It would be a light hearted presentation with lots of group interaction as he invited tourists to have a go at scraping taro, husking coconuts and cracking them open. "Now see, to husk a coconut is quite a hard job, it takes a local man like me only a few seconds but for a *palagi*?! It will take you forty years to get it done..."

His tours were a popular highlight for the overseas visitor and provided a good income for his growing family. His son Iulai was five, streaked sandy brown hair and a cheeky glint in his eye – was his dad's shadow. Often Ave would stay the night at her mother's house in Lalomanu but always Iulai would insist on accompanying his father back to their Saleaumua home. At four years, Phillip was a shy dark eyed boy who adored his older brother. Together they would alternately pester and protect their little three year old sister Malo Maryanne. One year old baby Aleki was too small to merit much attention from them yet.

On tsunami day, Mika and his son Iulai were at home. The rest of the family had spent the night at Lalomanu. Iulai had just bathed and dressed when the earthquake hit. Mika took his son outside "in case the house couldn't stand". He still hadn't showered himself so once the shaking subsided he went back to get ready for work. About ten minutes had elapsed when he heard a rumbling, like the sound of a coming storm. Then voices shouted from the road.

"*Galu! Galu*! It's a wave, a wave!"

The church bell started ringing, a harsh clanging without rhythm. He ran to the road to have a look, just in time to see a wave coming from behind Namua island. It divided so that one wave came straight towards them while the other came at an angle, sweeping over the little wharf where a motorboat anchored that took people to Namua. The wave flounced the motorboat along with it like a piece of fluff.

Mika stood and studied the oncoming progress of the water "because I'm the sort of person that has to study things carefully so that I can explain them to others, that's what a tour guide does". The wave "was like when you roll a flax mat. I stood back and studied the water if it was going to affect anything and when I saw the way it was coming, I realized, *oh no this is the real thing*."

Mika grabbed Iulai and ran – over logs, bushes, pig fences and rocks. So did everyone else who could. The people of Saleaumua ran. Young and old alike – everyone was running to escape the water that wouldn't stop. People who hadn't run in years headed towards the inland. Many men had already gone to work in the plantations and the majority of school children were either on the road or already doing their early classroom duties. Left at home were the women, the babies, the elderly, the infirm. Mothers grabbed their little ones, their parents – and rushed to the inland bush. Faletaulupe half dragged her ninety-two year old mother Suliane to the car, roaring away in a spin of sand just in time. Junior Laki used a wheelbarrow to take his blind mother Aoina to safety. Iosefa Tagaleoa ran with his three year old daughter in his arms, looking back over his shoulder at the waves that "came sideways, making all the houses collapse." Teachers at the Saleaumua Primary school led the children several miles into the bush. At the secondary school, Maths teacher Pelesala Togafau was about to ring the bell for morning assembly when he heard the screams. "You could hear the rumbling of the tsunami and the trees and the houses falling but when I looked out to sea, I couldn't see a big wave coming, it looked only about three of four feet. But then the wave that came from the right, over the houses – it was the big one. We were so shocked. The kids were terrified. We all ran towards the back of the school and the race was on to get away. We told the students to drop everything and run."

Mika ran with Iulai about five hundred meters before the water washed past him. It was chest deep but most of its ferocity was spent by then and the two were wet but unhurt. They climbed a small rise and then turned to look back at their home village. The first wave was a misleading prelude to the waves that would follow. The second wave swept in, less than five minutes after the first. It was vastly more ferocious. It didn't come from the sea straight ahead but rather from the south-east, catching people off guard as it raced through houses and trees.

Second and third waves covered the entire village like a blanket leaving only rooftops and church towers peering free. The water receded and another two waves fringed over the seawall before retreating.

Although it didn't feel like it for those caught in its path, Saleaumua - as the last village on the south-eastern coast before the land begins to slope upwards into the hills - caught the tail end of the tsunami. As Lee Palupe explained "The water didn't really come crashing in – it was like a monster that crept in, a silent, gentle deadly rise of water. Gentle but deadly enough to move rocks and houses."

She should know. Lee's house was on the furthest end of Saleaumua, on reclaimed land. Her house looked out over the ocean. The earthquake had sent her scrambling for safety outdoors. Standing on her lawn calling to her distraught dog, the sight of the rising water coming towards her sent her running back into the house. She had always discussed with her staff, what they would do in case of an emergency, what they would grab, how they would evacuate but now that the emergency was at her doorstep she found herself lost in a stupor of thought. "I kind of lost my mind running back and forth in the house trying to decide what to grab. I finally decided I needed to put on some shorts and shoes. I had one leg in my shorts and I was looking around for my slippers when water lifted the front of the house and started coming through the windows."

Lee went from trying to decide what to save to trying to save herself. As the house began to tilt, awash from its foundations and the water level continued to rise around her, she knew the front door facing the road was her best chance to escape. The panic was a caged beast as she tried the door handle only to find that it was bolted from the outside. Something heavy fell on her foot, trapping her in place momentarily and she realized with a flash of clarity – that she could die there. Trapped in a crooked, drowning house. This lent strength to her efforts. Pulling her leg free, she began to kick in the wooden door, the water rushing her out through the jagged gap. Awash in the swirling waters, she slammed ribs first into the trunk of the *talie* tree that stood beside her front driveway.

"That tree saved me. I climbed up it and just hung on. I watched the water knock the second house off its foundations and I was so scared that it would smash into my tree and knock it over. I don't know what I would have done if that tree hadn't stood strong. It was all the debris everywhere that I was worried about. That was the dangerous stuff…that tree saved my life."

Lee wasn't the only one with reason to give thanks for a tree. Up and down the Aleipata and Falealili district coasts, survivors were scrambling up trees, hanging onto trunks as the waters threatened to suck them back out to sea, placing children on high branches out of reach, and holding onto floating banana stumps as life rafts. The UNESCO tsunami report estimated up to six hundred lives were saved because of a tree.

Iosefa Tagaleoa is a forty-one year old father of two children. From his house you can see the flowers on the altar in the Catholic church to the right and catch glimpses of the ocean across the road that runs in the front. Iosefa's mother was sick in the National Hospital in Apia and he was supposed to go visit her that morning. "But I didn't go because after the earthquake, I had a strange feeling." A few minutes later, Iosefa heard screaming, *Galu! Wave! The wave is coming!* "I looked out and the sea had gone but I couldn't see any wave coming. I picked up my little girl because I felt something bad was going to happen. My aunty was standing beside my five year old daughter. Then I saw the wave coming and a lot of the houses were collapsing from it. The wave sort of came sideways towards us, not from the direction of the sea. There was nothing else to do, except run. I just carried my daughter and ran. People were screaming, all I could think of was saving my children and so I ran with my aunty and my children. We got to safety at the back of the village, far inland."

After the tsunami, Iosefa swam back to his house. "I swam to collect our possessions that were thrown everywhere. A lot of people from other villages were coming and taking stuff that was still good. Five people from our village died and we couldn't find them all. We were searching until night time to find them. But I thank the Lord that He didn't take many more lives. He led many people to safety. I know that if the tsunami had come in the night then no-one in our village would have survived."

Iosefa's other aunt was one of those killed in the tsunami. That night, Iosefa's mother died. "They told her that her sister had drowned and her heart must have stopped at the shock of the news, I think she died from the shock of it."

Clustered in the bush with other villagers, Mika held his son tightly as people milled about, some crying, others talking excitedly and still others just stood in silence. He and Iulai had escaped with a single red Digicel bag. It contained a change of clothes for Iulai. Mika didn't know it then, but that represented the sum total of all their remaining worldly possessions. Oselani's wife Sina cried without a sound, knowing that the chances of her husband escaping the wave were slight. Her mother berated her. "He is a fisherman. He can go for a long time without air. He will be alright. They will all be alright. You will see." Sina hugged her son tightly and prayed her mother was right.

Everyone was agitated, wanting to return to their homes. To see what had become of their houses. To find their family members. Mika left Iulai with a cousin and made his way back to their house. Walking was no easy task. There were pockets of water – some waist deep - littered with razor sharp corrugated tin roofing. Fallen power poles. A sting ray flapped forlornly in a puddle. An eel – thick as a tyre slithered through the mud and coral debris.

Everywhere families were slowly returning to the shells of their homes. Cautiously. Leaving the children up in the bush with the elders, only the young men came back. The strong and nimble of foot. Those who could most likely make it if another wave came. There was an eerie silence over the village. No dogs barked as you stumbled through their territory. No chickens scattered. Someone cried out as they found their grandmother's body outside the church.

Slowly, the village came back to life. Limping and dejected. With a stunned sense of awe at the evidence of the oceans wrath. Water tanks squashed flat. A taxi thrown through the wall of a brick store. Giant *tamaligi* trees ripped out by the roots lay wet and miserable. The red roof of a guesthouse, a *fale tali malo* sat neatly on the ground like a pointed hat. A fishing boat sandwiched on top of a pickup truck. Everywhere there were oddities. A *palagi* style house a pile of dust and rubble. Beside it a traditional *fale* still standing proud. An orange two-seater sofa sat right side up in the middle of the road. Some child's school trophy rested in a breadfruit tree. Along with a wooden closet and a tricycle.

Mika's house was still standing. So were others. The wave had gone, leaving ditches of black water everywhere. And the leftovers of everyone's lives. But the sand was waist high inside. The kitchen house at the back was flooded with debris – dead fish, coral, mangled pieces of corrugated iron, a smashed television set. Mamea went to look for his wallet, left behind on the chest of drawers in his mad dash to safety. It contained the takings of the last two days worth of walking tours. Nothing. No drawers, no wallet. No surprise there. He found a fifty tala note tangled in the grass.

Methodically, numbly – Mika started cleaning up the mess in his house. Digging out sand with his bare hands. Throwing fish out into the yard. Anything to occupy his mind. As if by doing something, anything he could make things right again. And block out the shouts as others started combing the debris looking for their missing loved ones. Loved ones. Ave. Philip. Maryanne. Baby Aleki.

At about ten o'clock Mika gave up on the mess. With a shrug he explained – "I didn't want that house anymore." He was wet. And exhausted. His hands were cut from the corrugated metal. He decided to start walking to Lalomanu to see his family. The road had chunks carved out of it, but it was still passable. Power poles were down. People were crying. Boys walked past him waving silver fish in their hands. Another held a wriggling turtle. Mika slowly walked to the end of Saleaumua and to the next village, Mutiatele. He was vaguely comforted by the sight of houses still standing like his own. Piled high with sand and windows blown out to be sure – but still standing. *See? It's not that bad*, he thought. And he picked up his unwilling feet, eager to get to his children.

A white van passed Mika on the road. Filled to the brim with hospital personnel. One of them was Dr Ben Matalavea. "It was an unbelievable sight driving through those villages, it looked like when a bomb explodes in war torn towns… It was quite apparent starting from Saleaumua village that things were quite desperate and the magnitude of the situation was far beyond anything we had experienced before. We didn't see many people left in the villages. There were some youth collecting fish that were lying everywhere and others sort of sitting around, just looking at what was left. We were called to help one family along the way, but their person had already died and was there wrapped up. We had to explain to the families that we really had to get to the hospital to attend the injured ones, but for those already drowned, already dead, there would be help coming later. It was very difficult to get through to the hospital because there was no road in front of you, it was all debris. There was some police and the DMO already there and they helped us clear bits of the road so we could pass through." Help was coming to the tsunami zone.

Oselani can't tell you how long he was driven in the wave. Or how he survived. He can tell you that the water was hot. And that it raked him along sandpaper-like, flaking away pieces of his skin. He remembers the heavy objects that slammed into him, breaking ribs. He knows there was a satellite antenna on a rooftop that caught his clothing. And he has the jagged rip along his chest to remember it by. We know he inhaled a lot of water. Because his lungs will never be the same. Doctors warn that for years to come, he will probably still cough and choke on the rot left there. His journey in the wave is a nightmare that will be in every waking remembrance.

Somehow, he ended up in the branches of a *poumuli* tree. Beside a two storey house. In the village of Mutiatele. Which he clung to fervently as the wave tried to take him back with it, unwilling to let the fisherman stay. And when the water was at rest around him, he heard an old woman crying for help."*Ou te fia ola*. Save me. I want to live. I want to live."

She was clinging to some wood debris not far from his tree. He swam to her and took her to the roof of the two-storey house. Another call for help came from further afield in the mass of floating rubbish. A young woman. He swam again and brought her to the rooftop. The water level was slowly dropping and the roof seemed the safest place to be but Oselani needed to find his family.

Painstakingly – without shoes or clothes, Oselani half swam, half walked back to Saleaumua. He salvaged a piece of green cloth for a lavalava but there was no protection from the sharp coral and steel. Half-naked, cut and bleeding – he looked just like everyone else he met along the broken road. He went first to where his house used to be. There was nothing there. But he didn't care. All he wanted was his family. A neighbour scavenging through the rubble told him all the women and children had gone inland. He had renewed energy as he walked to the bush to hold his wife and children close.

All over the disaster zone, families reunited with their loved ones on September 29th were sweeping children into crushing embraces, holding spouses close. Standing in mosquito infested bush in wet, torn clothing, their homes and all their possessions swept away in a heartbeat – families held each other and cried, giving thanks that they were alive.

Mika saw many people weak with relief as they met up with their families. He was feeling positive about his walk to Lalomanu. About the fate of *his* family.

But then he came to Pu'e and Malaela. And there were no houses left. Only cement foundations with steel inserts gaping through where block walls had been ripped away. As if by a giant hand wiping a slate clean. A few people staggered through the rubble, calling out for the missing. The crumpled bodies of cars hugged skeletal trees. A shiny red pickup with silver hubcaps was half submerged in the mangrove swamp. A child's shoe floated in the muddy waters. There was a huge fishing boat sitting in a tree. The LDS church built to rigid US engineering codes, was still standing, but with walls and fences askew. The sun was high now and the sea, an indecent sparkling blue.

He reached Satitoa village. They had just pulled a body out of an upside down car. Another lay on the road. Beside the wharf, there was a coffin floating in the sea with a dead person in it. The embalmed skin matching the white silk lining. The body of a naked man was caught in the shore rocks. For a people who treat the dead with sacrosanct respect, Mika was stunned. "I couldn't believe it. The dead person was just lying there by the road and people walking by – he was like an empty bottle – nobody cared about it." Pickup trucks started passing him along the main road heading for town. Feet peered from underneath the lavalavas covering still forms in the back.

Mika kept walking, and as he surveyed the increasing devastation on either side of him, the fear within him grew. All he wanted now was to get to Lalomanu and see his family. Hold his daughter Maryanne. And see Aleki's eyes light up the way they always did when he saw his dad. There were cold shards of worry in his chest as he walked the once picture-perfect Aleipata road to Lalomanu.

6
Do you have hearts?

For the villages of Malaela, Ulutogia, Satitoa and Vailoa – it was too late.

After Saleaumua and Ulutogia, there is the village of Malaela, home to the Schwalgers. Both twenty-nine years old, Netta Slaven and Jared Schwalger had been married for a little over two years. They were parents to two year old Teancum. A gorgeous little boy with liquid black eyes and dimples. His grandfather would call him Elvis. And a daughter Abish, nicknamed Abby - born on the 10th of August 2008, was Jared's perfect Father's day present. Together they lived in Jared's childhood home on the Malaela shore with his elderly parents Jerry and Metita Schwalger. Jared was one of two children. He and older sister Lucia were a tight knit nuclear family that was somewhat unusual in the sprawling openness of extended family living in a Samoan village. Lucia and her husband lived in Apia and they would come to spend the weekends with the family, bringing their two sons. Metita loved having the house filled with her grandchildren and their energy. From the verandah of the comfortable two bedroom home, you could look out over the lagoon and Namu'a island danced just out of reach. The main road - focal point of all bustling activity – passed in front of their house. A *nonu* tree stood in the front yard. On Saturdays, Teancum and Abby would play under the shade of its sweeping branches while their parents would sell barbecue to the bustling stream of visitors making their weekly pilgrimage to the golden Aleipata shores. There was a boat parked in the yard. The kitchen looked out towards the back yard where the mangrove swamp struggled to survive in amidst the ever-growing cluster of houses. Both Jared and Netta had quit their real estate jobs in town when their children were born, deciding that the long commute everyday and the time away from their young family was too high a price to pay. As Netta explained,"Children are more important than a career. There's nothing more precious than children and being a mother was what I always wanted."

So instead, Jared worked the plantation. Fished. Netta and Metita would make food to sell at the local school canteen. "Even though we struggled, we had each other. We had the children. We had a home."

On tsunami day, Netta and Jared were asleep with the two children between them in the bed. Metita was preparing breakfast in the kitchen. Porridge for the little ones. The earthquake woke Netta first, then Abby and Jared. Teancum slept still. They went to the kitchen, taking Abby with them but the earthquake stopped before they had to decide if they needed to run outside. Abby was hungry so Netta started feeding her breakfast. Jerry turned on the radio to see what Apia was saying about the shake. Metita said worriedly, "I think we should go inland. To Pu'e. Just to be safe."

But the family told her not to fuss. They would wait and listen to the radio. Netta shakes her head sadly, "We didn't listen. You know how it is, your mother always tells you things and you don't want to listen. We always think they are overreacting. But a mother knows."

Jared went to the nearby store to see if they had heard anything. And as he did, the announcement came on the radio. Prepare for a tsunami. Netta can see every move they made in those next moments. As if ingrained on her heart.

"Jared came running from the shop. He was shouting that it was too late – the wave was already coming. I panicked. For my kids, I panicked. I ran to the room and I grabbed Teancum. He didn't want to get up, he kept saying – 'mummy *moe,* mummy sleep'. Abby was in her playcot. She had her hands up and she was saying, mummy, mummy. Jared was getting the car. I put Teancum in his seat and then he got quiet because he likes going for a ride. I picked up Abby, I got the bag of passports."

At the car with Abby still crying in her arms, Netta shouted for her in-laws. They ran outside and then Metita turned to run back inside as if she had forgotten something. "I was shouting, Mom the wave! The wave! She came running out and the last thing I saw was the wave hit the house and everything was broken into pieces. I turned to look at Jared and that's when the wave hit my face. I was hit out of the car. I was deep in the water and something was on top of me because when I tried to put my head up for air, I hit something."

Netta struggled vainly in the blackness and then after an airless eternity, she reached that point beyond panic. Beyond fear. When the survival instinct quits. Acceptance."I said to myself, if this is how God wants me to die, if this is Gods will, then why fight it. Just let it be. Just accept it. I let go and I sucked in a lot of water. I could feel my heart beating faster and faster. And in my head I was thinking about my kids and Jared. And his parents. And I was just hoping they would be okay."

The Schwalger's red pickup truck with shiny silver hubcaps was taken about a hundred meters away, coming to rest upside down in the swamp. Pinning Nettas right leg. She remembers regaining consciousness, shocked to find she was still at the back of the upturned car. Jared's father was slumped in the front of the vehicle. "The first thing I heard was Jared's voice calling me. The first thing I saw was his Dad. It was just him left in the car. When I looked around, I didn't recognize anything. It had all happened so fast. I couldn't see a house. I saw a boat in a tree. I saw a car wrapped around another tree… that's when I started calling for my kids, hoping to hear just a cry. I called their names, I called mum…"

Jared had been ripped out of the vehicle, receiving minor injuries to his legs. Of the children there was no sign. He swam back to the car and pulled his father out, attempting to revive him while Netta watched helplessly, half-submerged and trapped. "I saw that his father was all blue, his fingertips, his face. And when Jared turned him on his side, the only thing that came out of his mouth was water and blood. I was crying and I said to Jared, *it's too late, he's gone."*

There were people running past on the debris-strewn road beside the swamp. Jared called out to them for help. "*Help, please someone help my wife, please help us*! They just stood there and asked *whose family are you?"* Jared answered, *the Schwalgers*, and they just left. They just left us. But then there was another man, one of our neighbours called Niko. He swam over and helped Jared with his dad. But then we saw another wave coming and I felt sorry for Niko because his wife and children were back at his house. I said, *Niko go back, go back and save your family, we will be okay."*

With another wave sweeping towards them, Jared hugged Netta and his dad close. Once it had subsided, he used a piece of timber to try and free Netta's leg. That's when she passed out again, the pain too much to bear. "I don't know how long I was out for. But when I was out, I heard voices. I saw a bright light and I heard children's voices. They were laughing and they were playing. It was my children and I wanted to be with them. But then a voice whispered in my ear to wake up, that it wasn't my time yet. But I didn't want to."

Netta woke to hear Jared's plea, "*Netta, wake up please. I can't stay in this world without you. We've lost mum and dad, I don't know where the kids are and I don't want to lose you."* She states quietly, "that's when I tried to be strong for him, tried to come back for him."

Jared renewed his efforts to free Netta's leg. He appealed again to those passing by, "*Do you have hearts? Aren't you Samoan? Can't you see I'm trying to save my wife, she's fighting for her life and you're just standing there?"*

This time, they responded. Together, a crowd was able to dislodge the car enough to free Netta's leg and carry her to the roadside. Lying there, she watched while they carried out her father and mother in law. But she couldn't see Abby or Teancum.

"There was this voice in my head, saying that I had lost my kids but I refused to accept it. I was still hoping that they were somewhere out there and maybe Teancum found a branch that he held onto and maybe Abby – she got lucky somehow. And we would find them."

There are oddities that we remember, things that make us smile later. Netta was lying on the road when a passerby stopped to look, calling out to his friends, "He stood there with a fish in one hand and a tin of *pisupo* (corned beef) in the other and he just bent down and said – *hey look, this is Jared's wife, poor thing, you know, the one who looks like an Indian!* And in my head, I was thinking – you're so rude to stand there with a fish and a pisupo and say that and not even try to help us. You're lucky that I am too weak to get up and tell you off!"

A passing truck stopped to help. Strangers carried Netta to the truck and laid the bodies of her in-laws next to her in the back. Netta was losing a lot of blood and drifting in and out of consciousness. She was afraid in a strange truck with strange voices and looking vainly around for Jared. She heard someone say, *that poor man is still looking for his kids*. As the truck drove away, she prayed to God to please help Jared find her children.

The truck went first to Lalomanu Hospital, but it soon stopped because the road in that direction was impassable. They turned back and started heading for town via the Le Mafa Pass. Somewhere along the route, Netta was transferred to a police car. She remembers her overwhelming relief when carried out of the truck.

"The driver of the truck was going so fast and my body was like a yoyo in the back because of all the bumps. It felt like he didn't even slow down, he went straight over the bumps and my body flew up and went back down and I was saying to myself – if I don't die from the tsunami, I will die at the back of this truck!"

It was a nightmarish journey to the National Hospital, with policeman trying to keep her conscious, "shake her, slap her, *make her talk, make her talk, don't let her sleep because she might go to sleep and not come back..."*

But it was a nightmare that did not end. Netta Schwalger from Malaela village was among the the first survivors to reach the hospital. Her in-laws, Jerry and Metita were in the first truck from Aleipata bringing bodies to the morgue. But because Netta's injuries 'didn't look too bad' she was not seen by a doctor until the following day. When she would be incoherent with raging fever as infection from her wound set in.

Shortly after she arrived at the hospital, Netta saw Jared walk into the ward. But when he saw her, he turned away as if unwilling to look at her. A moment later, she understood why. In his arms he held their daughter Abby.

"I was still hoping you know? Still I had that hope that she would be alive. Jared came to me, he never took his eyes off me and I whispered, she's alive? Abby's alive?" Jared gently placed Abby in Netta's arms.

"No, Netta, she's gone."

Netta screamed. She held her daughter tightly and spoke to her. *Abby, I am so sorry for not running faster. I'm sorry I panicked and I wasn't thinking straight. If you see Nana out there, tell her that we are so sorry for not listening to her in the first place. Nana was right. She's always right. Abby, if you see Teancum, please tell him we're looking for him. And we love him. We love both of you so much.*

As Netta clung to her daughter, she tried to bargain with God. "People around me had tears in their eyes, but I know that their feelings were not really the same as mine. God had taken Abby and I prayed to Him, I said, *God please, you've taken Abby, please let me have Teancum. I promise I'll go to church all the time and take my son with me, I promise I'll behave, I'll do everything and anything you ask of me, just spare me one child, please."*

Jared couldn't stay with Netta that afternoon. His parents were in the morgue. His daughter had been found. His wife was going into shock. But Jared needed to go back to Malaela and find their son Teancum. "He had cuts on his arms and legs and I asked him to go see a doctor, but he said he wouldn't see one until he found our son."

By now Jared's sister Lucia had found them at the hospital - after going to Malaela and finding nothing left where their village had once stood. She had refused to accept the neighbors tale of her parents' death but now, as she stood outside a morgue overflowing with stunned survivors, Lucia could deny it no longer. Her parents and her niece were dead. And her brother needed to go back and keep looking for his son. Jared left the National Hospital with Lucia and her husband Goske. He would make the long drive out to Aleipata to continue searching. Netta was taken to Acute 7. Another patient would kindly use her cell phone to call Netta's family to tell them she was alive.

"She called my brother Christian and he was on his way back from our house at Malaela. He didn't know if we were alive. She told him I was here but that both our kids were dead and Christian just cried on the phone."

Christian called Netta's best friend Yvonne Mariner. She was sitting in the hospital chapel beside the morgue, weeping. She had come looking for news of her friend and had seen Jerry and Metita's bodies when they were brought in. She had seen baby Abby. And she had been going through the other bodies looking for Netta and Teancum.

The morgue was a busy place to be on September 29th. And in the days afterward. A busy heartrending place. Vaomalo's staff were struggling with their own waves. Of dead. Of unidentified and unidentifiable. Of grieving family. Of impatient and angry family. Of people bewildered by the 'procedures' that had to be followed. Much like the first truck Netta has been jostled about in – the injured were being brought to the hospital in the same cars as the dead. Trucks would reverse. The living would be taken to one entrance where reams of medical staff awaited them. The dead were taken to the other entrance. Where there was not enough room for them. The hospital chapel was converted into an initial holding space. People were laid out in rows on the floor, covered in sheets. Paramedic Steve Williams recalls the poignant image of five little bodies laid on a single hospital bed. Snug and tight.

In those first hours on that day the living were there too, crowded into every spare breath. Anxious not to let their newly dead out of their sight. They sat on the cement floor, lining the narrow corridors. They sat on the tar seal drive in the noonday sun. There was little sound. There were few tears. "People were just sitting there like they were waiting for a bus or something. But the look in their eyes, it was a dead look. Stunned. Shocked."

In the midst of their rush, Vaomalo thought to call the Reverend Oka Fauolo. Could he please come? The Reverend came. Standing outside the chapel that was too full of the dead to accommodate the living, Reverend Oka read from the Bible and led the huge crowd of grieving and stunned people in hymns. Samoans sing. When they worship. When they exult. When they mourn. Hymns of praise. Hymns of love.

The gloved hands of forensic staff carried out their grave task to the resonating sound of voices lifted to the heavens in song.

Jared searched for three days before he found Teancum. They were days that gave no sleep.

"Every time we tried to sleep, the only thing I saw was Teancum. He was running around in the mangroves and he was calling me "*mummy , where are you?"* I kept seeing him in the mangroves and so did Jared. Teancum was calling him, *"daddy, where are you?"*

There is comfort, however cold, in having a body. To touch. To look at. To confirm in that deepest recess where hope lingers – that your child is really dead. And if you believe in that 'something better' called heaven – to know that your child has gone there. They aren't trapped somewhere, suffering. Waiting for you to find them. Or wandering lost and alone in a dark mangrove swamp.

Netta didn't look at her son's decomposing body when he was brought in. She was spared that memory. Jared identified his son. A week after the tsunami, Teancum and Abby were buried together. The next day, Jared's parents were laid to rest in neighbouring graves.

Over the next month Netta would have four operations on her leg. She will be forever grateful to the medical team from Australia and NZ - the plastic surgeons who did the skin grafts that saved her leg from amputation. "My first thought when I saw my leg was – oh no the doctors are going to cut if off because the dead part was so huge and I didn't know how they were going to be able to stitch it up again. But the two doctors were so good, Dr Peter from Australia and Dr Andrew from New Zealand – they would joke and laugh with me, reassure me that my leg was going to be alright. I told them if I died in their operation then I would haunt them forever!"

Netta laughs. She says it's better than crying all the time. But for Jared, the memory of bystanders unwilling to risk their lives to help him save his family is a hard thing. It chokes him, weighing heavily on a heart already worn ragged with grief. "Can you imagine what it feels like? People you thought they were your friends, that they had known you all their lives and they would help you…it's like your best friend doing something that would hurt you the most. That's what it's like…"

A son who had to pull both his parents bodies from a tsunami. A father who had to search for days to confront the sun worn remains of his son. And bury both his babies. A husband who had to lose his fragile grip on self-control and rage before medical staff would put his wife's name on the urgent operating list. A husband who then slept on the floor by her hospital bed for four weeks while they battled to save her leg. While scavengers stole the hubcaps off his red truck upturned in a congealing swamp. And took every scattered remnant of their worldly possessions.

Yes, Jared is all these things. Is it any wonder then, that there is raw anger? Bitterness? Netta softly explains, "It's like if someone just came and stole everything from us, it's just a snap, just like that and everything that you knew that was precious to you just disappeared and there's a huge hole in your heart…it took your family, it took your house, it took everything but most importantly it took your life, your happiness, everything."

Where Jared struggles to forget those who wouldn't help, Vaelupe Olosepu's family will always remember those who did. Saving the lives of their three grandchildren during the mad dash for safety.

Vaelupe is a silver-haired man with an authoritative air. Two weeks after the tsunami, he sits in a large, solid *fale* built with his own hands in the wet forest. There are tents from the Red Cross at the rear for the extended family that spills over. He has already started clearing the bush and construction has begun on another dwelling for his daughters and their children.

When the tsunami came, the Olosepu family were hosting two young missionaries from their church – Wade Miller from the US and Heimuli Lelea from Tonga. Sixty year old Vaelupe had taken the car to work at the inland plantation. His grown daughters had left for work at the Litia Sini Beach Resort at Lalomanu. His wife was at home caring for the grandchildren with the help of various other cousins and extended family. Not long after the earthquake, Vaelupe received a phone call from his daughter Maele. Walking to work, she was on the hill before the Lalomanu resorts when she saw the wave coming. *You better go home and get the family. There's a galu lolo coming!"*

Back at the Olosepu home, Heimuli Lelea had finished showering in the outside tap when he heard the shouts, *Galu,Galu, Wave, Wave!* Looking towards the sea, he could see a large brown wave moving swiftly towards the village. He ran to get Wade, shouting,"We have to go…we have to run. Look out to the sea!"

Inside the house, Wade was writing in his journal. He had heard the shouts but because of his shaky grasp of the Samoan language, he had only thought them puzzling rather than a concern. One glance out to sea though, had him sprinting behind Heimuli as he ran towards higher ground. The two men were among the last to leave the village. They could see many other people in the distance, climbing up trees and yelling back at them to hurry, to climb as well.

Malaela village is on a flat plain. The hills are far away. Running for your life, pursued by the sea - a hungry beast that roared and churned behind you – would you have had ears to hear the fearful crying of a small child? Would you have had eyes to notice two children huddled together? And if you had, would you then have stopped to take them with you?

Wade and Heimuli did. They paused. They grabbed the children in their arms and kept going to the nearest big tree. Wade, the more agile of the two, was up the tree first. Heimuli heaved the children up one by one, before heftily pulling himself up to safety. A woman ran by heavy and laborious, carrying a baby in her arms. Seeing the men, she stopped and begged them to take the infant. Heimuli climbed back down to grab the baby while the woman continued on to another tree to try and pull herself up. The missionaries kept careful hold of the children as they tensed for the wave impact.

"We were scared that the wave would knock the tree over and take us out to sea. The kids were crying and we were just hanging on."

The two men prayed aloud for deliverance as the waves rushed over the land, sweeping its hoard of debris with it. Their tree stood strong. When it seemed as if the waters had stilled, men called to them from further inland. *Come, bring the children, its safe here on higher ground.* Wade and Heimuli carried the children on their shoulders through the five feet deep water, joining an exodus of shaken survivors as they wended their way to the inland plantations.

Back at the village, Vaelupe brought his car to a stand still on an impassable road. His village was unrecognizable. "I saw the devastation of God's work. There was not one house standing in our village, everything was destroyed and even household items and material goods."

He left the car and tried to make his way through the flood waters to where his house used to be. "I got there and there was no house, nothing. I didn't know where my family was." His worry - for the children in particular - was a coldness on that sun filled morning as he surveyed the destruction. But a phonecall to his wife brought reprieve. They were safe. All of them. And thanks to the missionaries, even the smallest among them were unhurt.

Several weeks later, two of Vaelupe's grandchildren sleep under a mosquito net, at peace in a steaming hot afternoon. The other, a shy five year old girl, regards the camera with solemn eyes. There are tears in Mrs Olosepu's voice as she thanks God for the two men who stopped to save her grandchildren.

Sally Taiva'asele is seventy-one years old. She is the matriarch of her family in Malaela – her three children, their spouses, and her ten grandchildren all lived together in an assorted cluster of brick houses and open Samoan *fales*. She is a beautiful woman, long fair hair coiled in a bun, powdery white skin and with an easy gracefulness about her. She has lived in Malaela all her life. There were earthquakes during her childhood that were big enough to leave an impact on her memory. And she can recall other waves coming into her village. But nothing like what happened on September 29th. "Yes, I remember, there were other *galu* (waves) in the past, but they weren't big, people just stood and watched them. They didn't reach the houses, those waves ended at the road."

Sally is not in the best of health. She has diabetes and tires easily. She is alive today because of her daughter Leata who ran with her and refused to leave her. When the earthquake happened, the two women were the only ones at the Taiva'asele home. The children had already left for school. One son-in-law was away in New Zealand. Another had gone to the plantation. Leata was making breakfast for her mother when they heard the sound of the wave. "The wave made a huge noise. I didn't see it, but I heard it, it was like machine guns going off in war."

Leata ran outside to look. "The wave was something so different…it looked like glass with a fierce animal inside it, it was so big, it looked even bigger than the Namua island."

People were running and screaming. Sally and Leata ran along the front road that lines parallel to the ocean, making their way towards the rough road that led to the plantations.

"I didn't grab my bag or anything near me, I just crawled outside the *fale* and then we ran. I had no shoes on, we just ran and I felt I couldn't run anymore because my feet were sore from running on the sharp rocks, but my daughter kept telling me to keep going, the wave is coming, its coming."

At the plantation road, the two women turned inland, passing several houses with the roar of the glass animal at their backs. Sally's strength failed her then. She stopped and told Leata that she could go no further. Leata pulled her to stand beside a knot of trees. The wave would only have been a heartbeat away, but that moment, cowering beside a tree, with the air consumed by the crackling roar of the water as it ripped through Malaela homes – seemed an eternity to Sally.

"I could hear it coming…I said to my daughter, there's nothing more we can do, this is in God's hands. If it is His will that we die, then we will die. What can we do? That wave was a killer coming for us. When it reached us it hit me in the back and it was such a powerful force, it was so strong, I couldn't do anything. I felt my leg hit the trees and it got stuck there and my lavalava got ripped away. Leata was beside me, she was crying and holding onto my top, trying to keep me from going with the wave."

Sally grabbed onto the nearest tree. Unseen sharp things slashed through her arm and something heavy crushed into her leg, pinning it to the tree. Fallen logs and other large debris began to build up all around the trees, trapping the two women behind a wall of wreckage. They were able to keep their heads above water and once the levels began to drop, Leata renewed her calls for help, someone, anyone, please help them. Sally was in a lot of pain, her arm dangled uselessly now and she couldn't tell how badly injured her leg was, stuck as it was still, in a mass of house remnants. Finally, men came to pull them free. They had to use bush knives to reach them. "They chopped down the trees and the *fue* vines next to us – that was the only thing I could see, the *fue* and boards with nails all around us. Then they took us to dry ground and that's when I saw how bad my leg was…"

Leata had the usual tsunami cuts and eventual bothersome chest infection – but Sally was in hospital for ten weeks. Her arm will never work properly again and half of her foot was amputated. Neighbours tell us amazedly, *you know Sally has only three toes left!* There is wonder that anyone could survive such an ordeal. Sally sits in a tsunami relief house, newly built by her church and tries not to remember what happened to her that day. "I don't think about it, I try to forget about it. There was another woman in the hospital with me who was really affected by the tsunami, she is not right in her mind now. She would scream and swear at all the doctors and they couldn't do much to help fix her mind. But for me, I try not to think too much about it, the only thing I can't forget is running from the tsunami…"

Sally knows her family will never forget though. She motions to the cookhouse where her daughter is making the food for the evening meal. The daughter who stayed by her side as the glass beast rushed to devour them. "I know my kids can't forget this tsunami. They think about the pain I went through and how I survived this tragedy. They worry about another natural disaster like this coming again."

There is a beast that crashes through forty-two year old Sapua Toilolo's dreams. It clambers up out of the morning ocean, over the seawall, and the coconut trees tremble before its towering height. It is the colour of night and the sea spray at its head, rises like a cloud of smoke. It is a beast that killed Sapua Toilolo's neighbour - her cousin Fa'apopo Toeuli and six others in their village. It is a beast that has driven the Toilolo family to settle in the plantation bush. It is a beast that troubles Sapua's sleep and makes her five children wake up early every morning "because if it had come at a time when we were all asleep, we would all have died, so I say to my children, remember what happened, remember what came that day, don't ever sleep in late, we need to get up early…"

On the day the beast came to Malaela, it was only Sapua and her two sons at home. Her husband was away overseas and the other children had already left for school. Pe'epe'e her eldest boy was preparing to go to the plantation when they heard a ruckus of sound from their neighbour's house which was closer to the ocean road. "We were sitting there to have our breakfast when we heard people screaming, I thought they were arguing down at Levao's house because there were people running around and yelling and there we were just sitting in our house unaware of anything. I looked out and saw an old lady running past with some kids. She was only wearing a towel, like she just came from her shower."

Pe'epe'e ran down to see what was happening, then ran back even faster. "He said to me to *run there's a big galu, a wave coming.*" Sapua's little son took off immediately up the road beside their house that led inland but Sapua wanted to see for herself what was coming. "I didn't run at that time, I looked out to make sure my son was away and then I looked out to the wharf and the *galu* was coming. It was very high and I knew that our houses might get affected because it was coming over the seawall but I thought maybe that was as far as it would go, so I just started walking back on the inland road."

Sapua still wasn't too bothered, still she thought the wave would be nothing more than a vaguely annoying surge. Then she turned her head to the left and saw it. The beast. Another wave that was rolling in over the land, travelling almost parallel to the ocean. "It was different to what was coming from the sea, it was much bigger and moving very fast." Sapua was torn with indecision. Back in the house was a sizeable amount of money, the children's school fees that she had been planning to go and pay that day. She ran back and had reached the verandah when her son shouted out to her.

"*Run! You'll die if you go back in the house...run!*"

Sapua listened to her son. She left the money and began running with him up the inland road. They met other groups of people clustered a short distance in. "They were just standing there, they couldn't run anymore and they thought maybe they had gone far enough, but I told them, *run, be strong, let's keep going!* I looked back and I could see the wave coming and taking roofing iron off the houses like it was paper. So we all kept running."

The wave caught some of those in her group as they ran. "The wave hit Lele's house and it was still coming and it got some of us, that's where I lost Fa'apopo and some others that were running with us as well. That's what spurred me on. We crossed the bridge that goes over the mangrove stream and got up to higher ground. We stood there looking back, we saw trees being uprooted right in front of us because of the strength of the water. I saw people in other trees trying to pull up their kids to safety."

Sapua cannot forget what she saw. "The wave that came across the land was like an animal coming up with high fins on each side, I looked at it and it scared me. I said to all those that I was running with *oh my gosh, we're all going to die from what's coming*...because I saw the coconut trees were lower than the wave. It was like a wild animal with huge fins, but it was pitch black and the mist or sea spray that was coming out of it was also black, like the smoke from a *umu*, it was brownish-black."

That night the Toilolo family slept in the bush. "Once the fear had left us I went back down to the shore to see if there was anything to salvage from our house, but it was too late, everything useful was already taken by the scroungers and they even took all the big fish as well."

Sapua's little son is still unwilling to return to the coast. "When the tsunami was over, I tried so hard to get him to come with me to have a look at the aftermath but he said *no, no let's not go down there, we'll die down there*. I kept telling him there was no more tsunami but he said no. Finally, a leader from our church took him for a ride to the beach in a car and he could see that nothing bad was going to happen. I don't want him to have that fear."

Months later though, sitting in her plantation house, Sapua herself still has those fears. Of the wild animal with huge fins that breathes smoke. "We go down to the village to get water and just clean up a bit and then come back here. I still get scared…when I hear machinery making noises because I think the *galu* is coming back and I have to get up and look and then I want to hurry and get away from there."

How many of us listen to the tales of our grandparents? Stories from their childhood. Times when they were growing up and life was (of course) harder, and slower, and they had so much less than we take for granted now. Do we listen – and remember? Do we listen – and learn? Or do we roll our eyes and sigh, politely hiding our yawns, but with our thoughts miles away?

Falimu Mautu's family at Satitoa village owes their lives to their mother, Aniva. Who long ago as a child, listened to her father tell stories. Among them a story of when *he* was a child. And there was an earthquake. And afterwards, the sea came. Not a mildly irritated sea, but a roaring demented one that ripped through their village. He told a wide-eyed Aniva, 'whenever there is a big *mafui'e*, then you can be sure, the sea will come. You must not wait. You must run. Quickly."

Aniva's daughter Falimu is twenty four years old. She was on the roadside waiting for the bus to town when the earthquake came. She went back into the house to check on her family, in time to hear her mother.

"My mother was saying the earthquake was too strong and we had to start running because the sea would come soon. She called out to my father who was cutting grass at the back of the house and my dad said, no there wouldn't be anything like that."

But Aniva was insistent. She screamed at the children who were preparing to leave for school. "No! No-one's going to school. We all have to run. Now!"

Frightened, the children followed their grandmother's directive and began running with her back towards where the distant hill began its slope upward. Falimu picked up her baby and followed them. Her father put down his bush knife and started after his wife, still shaking his head at the strangeness of women. Still unconvinced but unwilling to risk the inevitable tirade that comes when one ignores a wife. As she ran, holding her grandchildrens hands, Aniva continued to shout – calling out to the neighbours they passed, "Run! Run! We all have to run because the sea will come soon…"

They were midway up the hill when the sea came and overpowered their village. Hearing the wave, Aniva's husband turned back to try and get his mother, 109 year old Fa'aliu. His son went with him. The two men could not fight their way through the water in time. Both were injured. Falimu's brother was crushed under a broken cement wall and would later be taken to New Zealand for surgery to help him walk again. The family matriarch Fa'aliu was buried in Satitoa by the remains of the Mautu home. With her are the others in the extended family who died – forty-five year old Sefulu, seven year old Satelite, and a seven month old baby, David.

After the tsunami, Falimu is emphatic, "No, no one in our family is going to live near the sea again. We're all going inland because this isn't the last natural disaster. There will be more mafui'e one day." And for now – Falimu will tell *her* grandchildren stories.

In the village of Vailoa, the To'omalatai family owns the only gas station in this corner of the island. Beside it are a shop and a spacious two storey home. Next door lives more extended family. Inland is another expansive home where the patriarch of the family, seventy-five year old Toomalatai Lauvai lived with his wife Sesa, their son Ian and his wife and three year old daughter. That morning, Sesa left the house early to go open the shop beside the road while To'omalatai Lauvai listened to the radio. When the earthquake came, "I saw it as something that would cause death, it was so strong. I held on to my seat and faced the road. After the earthquake finished someone called the radio station to ask them why don't they send out a warning about a tsunami? The announcer did not reply. I heard a rumbling like a wave about to break. I looked south towards the sea and I saw many children running to their homes, yelling out that the tsunami is coming. As I watched them run, I heard a rumbling like a big storm. It was very loud in my ears. I walked outside and looked towards our two-storey house by the sea. I wanted to see where my wife was so I started walking towards the house and when I got there, I was hit by the wave. It was so strong that I was struck down and carried with it. It carried me close to my brother's house. Then the wave came back again and I was rolling around in the water. My right leg was caught in a tree that had drifted in the wave. So I held on to a branch until I was able to touch the ground and then I crawled to my brother's house. I had to climb over things to get inside because of all the trees and rubbish that had gone inside the house. I was very surprised to see my brother inside there. He cannot walk properly, he has to use walking sticks because he's been sick for awhile. He was sitting there in a chair that was rocking to and fro in the water and all the windows of the house were broken. He said to me, *why have you come here? You should have stayed at your own house, it would have been safer.* So we both sat in the wet house, facing each other for awhile until we got the news that my wife had died."

To'omalatai's wife had been at the big house with her son Ian when the wave came. He was making breakfast for his parents and did not see the wave until it was too late. "I could see people panicking and running but I didn't know why. I saw people pointing out to the sea, they were saying that a wave is coming. But I wasn't paying attention and then it was like the wave was boiling over and I got such a fright. It wasn't even thirty seconds and the wave hit us where we were standing at the back of the house by the kitchen. My wife was carrying our daughter and I caught hold of my mother. The wave hit and destroyed everything. We all went together out to the sea, I grabbed a tree branch and hung on to it with my mother while my wife and daughter were taken far away. We were floating in the sea and I was trying to pull us in because I could see we were drifting further and further out, so I tried to pull us more towards the land. Another wave came and that's when my mother slipped from my grip, the wave took her. I ended up stuck on the seawall but my mother had gone. My wife and daughter made it to the roadside near the church. People from the village came to help. We looked for my mother and we found her at the back of our house. I wasn't badly injured, but my mother was dead."

Sesa was the only person killed at Vailoa village. She was seventy years old and the mother of ten children. Ian's little girl is named after her grandmother. He holds her gently with huge hands and cries in the afternoon sun as carpenters hammer and saw, rebuilding the petrol station.

Melody and her husband Lesa don't live right beside the ocean at Satitoa village. They have a little wooden house further inland and up a small rise where the beach starts to embrace the bush. But Lesa's family did. His parents lived in a solid European house with their other sons and daughters. Beside them was another big brick house and store belonging to Lesa's aunty. Standing in Melody's front room, you can look out down the road and see concrete foundations – the leftovers of the Lesa family's homes.

Melody came a long way to marry Lesa. She was born in Quezon City, in the Phillipines. But she's lived in Samoa for eight years now and her Samoan language ripples fluently as she talks to co-workers and patients. Melody is a nurse at Lalomanu hospital and she was on night duty the Monday before the tsunami. Which means she was still at work when the earthquake hit. Lesa – like most farmers had already left for a sweltering day in the plantation far up in the bush. Which means their two children were at Lesa's house being cared for by his mother who loved to babysit. Lesa felt the earthquake as he walked with the other men that new morning. They hadn't gone too far up the plantation road. Shortly after the shaking settled, they heard something unusual.

"We heard a rumbling noise and it was just different from anything we had ever heard before. Then we could hear people screaming, *galu lolo, galu lolo*! The sea is rising, the deep wave is coming!"

The men started running back to the village, coming to an abrupt halt as the rough road ended in swirling water. The sea had covered Satitoa, reaching all the way to Lesa's house bringing with it wreckage it had ripped out along the way. His front yard was now a mass of fallen trees and broken houses. A blue rental car was completely flattened underneath a fishing boat. And some distance ahead on what was left of the road, Lesa saw the roof of his father's car – stuck between two crushing trees. And almost completely covered with water. Lesa ran, and when the water got too deep, he swam, feeling his way through the water that was filled with jagged things. Lesa couldn't make out who the occupants of the car were because it was filled with brown water. He tried the door handle, feeling blindly underwater, but the door wouldn't open. Frantically, he smashed at the windows with the handle of his bush knife. Again and again until the glass shattered and the thick water rushed out.

There were nine people trapped inside. His parents. A brother. A sister-in-law. A sister and her husband. His sister's baby. And Lesa's two children. Three year old Quezon. And his four year old daughter Merilynne. All of them were still and silent. Heavy and unresponsive as he worked furiously to pull them out through the window, one by one. Searching for a safe place to set each one before returning to his desperate task. He pulled out all the adults and the older children. But he couldn't see the babies.

"He was looking everywhere inside the water for the babies. He couldn't find them in the car and he was thinking maybe they didn't bring the babies with them? Did they forget them behind? But then he was feeling in the water around the back seat and he felt something like flesh. And so he grabbed and it was the two babies and he raised them up and brought them out and he couldn't believe it because they were both still breathing. The last ones out of the water and they were still alive."

Three others were not so fortunate. Lesa's mother. His brother-in-law. And Melody's son, named after the city where she and Lesa had met, little Quezon.

"They drowned. They had no wounds on them like other people because they were protected from everything in the car. They drowned. It was too long and too late for them."

A little over ten minutes drive away, Melody was waiting out the end of her shift. At the Lalomanu District Hospital.

The hospital is located on a hill past the primary school. It sits behind a chain link fence. There are wide open grass spaces on either side of it. And a cement driveway. It breathes newness. Cleanness. It has a six bed ward, two treatment rooms and a resuscitation room. There are flowers in the garden – tended by the nurses who use their breaks to weed. There is a window where you check in. And wooden slat benches to sit and wait for a doctor. And colourful posters to look at. Telling you to breastfeed. And brush your teeth. After eating five servings of fruit and vegetables. And not to forget wise family planning.

Melody was looking forward to the end of her shift, when the earthquake rattled the hospital roof and sent the nurses running outside. She was used to strong earthquakes in the Phillipines – and had grown up with "almost daily earthquake drills at school that we would get sick of." But this magnitude was a first for her time in Samoa and she was as unnerved as the other staff. It didn't take long for the first wave of injured and dead to arrive at the small district hospital.

"About half an hour and we had our first victims. That's when we first knew there had been a tsunami, when people came running here holding their dead ones."

There were three nurses at the hospital at 7:30 am. It would take another three and a half hours for medical help to arrive from town. And in the meantime, the people of Aleipata were coming to bring damaged and dead ones. The morning passed in a blur for Melody as together with the other staff, she struggled to cope with the influx.

"There were people lying in the corridor – some of them dead and covered with a sheet. Some of them alive and waiting for help. There were people carrying children, babies, following us around asking us to please look and see if we could do anything to make them wake up. We didn't have enough beds, so we just put the injured in here on the floor."

Even in all the chaos, there are some images from that morning that stand out for Melody. "There was one woman. A woman I know well from Lalomanu. She always came in for her antenatal checkups. She knows me very well. Her baby died. She brought him here but there was nothing we could do, he was blue and his mouth and nose were full of sand and dirt. She came here and was calling for me, because she knows my name. And she didn't want to believe that her baby was dead. And she sat over here and kept talking to her baby, and she was trying to breastfeed him. And I couldn't do anything for her."

Avaia Lautusi is a nurse from Lalomanu. It was her day off and she was on a bus to Apia when the tsunami hit. She got a phone call from a nurse at the Lalomanu hospital. Crying. Telling her there had been a *galu lolo*. And there were too many injured people. And six dead. And could she please come right away to help? Immediately Avaia's thoughts went to her husband in their coastside home. She tried calling him, but there was no answer. She then called the National Hospital to make sure they knew about the situation at Aleipata. Meanwhile the bus driver had turned back, many passengers anxiously wanting to return to check on their families.

"Everyone on the bus that morning was scared for their families. They kept standing up and asking the driver to go faster. I was trying to calm the people down and the driver too. I was worried, maybe we won't die in a tsunami but we will die on the road when this bus goes too fast like this!"

It was an interminable wait for Avaia as they drove to Saleaumua village. There was still no answer from her husband's phone. The bus could not pass through to Ulutogia village because of the debris, so everyone alighted and made their way on foot. She cried tears of relief when her sister from New Zealand called to tell her that she had spoken to Avaia's husband and he was fine. He was on the beach helping to search for their neighbour's children. It took Avaia an hour to walk through the disaster zone to the hospital. It was 10:30am. She remembers counting thirteen dead when she arrived. An hour later, the first white van pulled up from the National Hospital.

Dr Ben Matalavea arrived to a hospital with no running water and dead power supply. Where crowds of bewildered survivors clustered – outside under trees, in the open halls, along the cement driveway. Where nurses who had lost family of their own, cried as they treated the wounded. As the senior doctor, Dr Ben took charge. "There were about sixteen bodies already there when I arrived…our first priority was to stabilize the very sick ones, triaging and then going to work. We were worried about how we would evacuate such large numbers of the wounded when we only had two ambulances. Wc sent the ambulances full and it would be more than three hours turnaround for them to come back, so we couldn't wait. Without power or water, it was a priority to stabilize people and get them out of there to the National Hospital in town." The decision was made to use trucks to move the wounded, which required taking the deceased off first. "I am so grateful to our nurses, doctors, the ambulance drivers, even the grounds people, everyone came to help… they are the ones who put the wheels in motion for the evacuation process. We had to tell them to take the bodies off the trucks so we could use them to transport the injured. First we put down mattresses and mats and anything else we could find and then carry them on the trucks. The dyna truck could fit about ten people lying down and helpers were needed too – to sit beside them holding the IV drip..whatever space was left on the truck we placed the injured who could sit upright. There was also a bus that came to bring health workers. We folded down all the chairs and tried to fit people in there. It was quite difficult manoeuvring in the restricted space – there was a *palagi* woman with a broken leg and she was quite good, she was very tolerant of our moving her around with her IV placed between the seats. Our people were very co-operative – if we saw some whose injuries could wait then we would ask them to go off the truck and allow another injured person on, they were really good and patient."

Dr Ben remembers dealing with the 'walking wounded' – those injured who did not require hospitalization. "I would treat them and then say *just go home and come back the next day for another wound dressing* and they would answer *well, we haven't got a home to go to now*! Obviously we quickly learnt that were no homes for these people to go, a lot of them were transported to town or to other areas by their extended families. That's something that our helpers from overseas were quickly impressed with about our Samoan way of life – we will always find a relative to stay with!"

Even a medical professional as experienced as Dr Ben was shaken by the magnitude of the suffering. "We were in a situation that we'd never been in before and there was just shock, the quiet shock where people look at you – but they're not seeing you, they're staring, just stunned. There were a few couples that were just sitting there without a word who had lost their children…I didn't know what to say to these people – because you just know that whatever you're going to say or do is not going to change the situation, and just being in that crowd where we all shared in those feelings…as a doctor, you had to change your tactics, the usual thing that we normally do when we treat people is ask *hello, how are you, why are you here, how can I help you* but on this day it was nothing like that. I couldn't ask any questions because I didn't know any questions…you know the child is lost and you can't find the child and they are so distraught, what can you do? These couples have lost their children and there is so much destruction and sadness but you can't allow yourself to feel anything. Not then. We had to just keep moving from one person to the next and focus on the work. The sight of those three couples without children, for me the shock came later when I thought about it and I still think about it and it always brings tears to my eyes when I remember them."

Another doctor sent to Lalomanu Hospital was thirty-one year old Teriki Puni. "It was shocking to see how severe the disaster was – you really felt helpless with the amount of people coming and bringing the dead and asking if you could do something…many of them just broke down and cried, they knew that there was nothing to be done but I think they just wanted to be told, they wanted that reassurance that nothing could be done. *Can you help my baby?* And you saw this baby was just completely purple and you open up their mouth and there's just debris and everything, it was just covered in rubbish. I knew they were past helping. It was quite heartbreaking just to feel like you're not able to do anything. But for those who were severely injured but still alive, you could do something and just try and stabilize them before you rushed them off to National."

Coping with injuries the like of which had not been encountered before, presented challenges. "Initially the majority of injuries were lacerations – deep, dirty lacerations, like down to the bone, cuts and bruises – superficial ones and very deep ones but pretty much from head to toe, like people had been in a washing machine being tossed and turned. We found that a lot of injuries, it would have been better to just leave them open because there was so much dirt that we couldn't get out. Post-infection was a big problem we found out later as well, but things really became more helpful when we got foreign health workers coming in and helping, especially those with experience in these types of situations."

Dr Teriki spoke of the struggle to remain detached enough so one could do the job at hand. "You had a choice, yes you can be emotional about things, but it's not very helpful when the patient sees you crying…it was difficult though. There were some of the staff members that were literally very emotional, that needed to find a quiet little spot to get themselves together, some time to cope with what had happened."

Melody needed that 'quiet little spot'. At midday, Melody got a phone call from Lesa. Using his cell phone with three tala free credit. Telling her that Quezon was dead. And they were taking him and the others into town to National. A month later, Melody shrugs. She is quiet and matter-of-fact. "I kept working. There was too much to do here, too many people who needed help. And I had to keep going. Later on in the afternoon when there was space in a van going back to town, then I went with them to the hospital."

Back at Lalomanu Hospital, everyone kept working. And the people of Aleipata kept coming. The hospital was like a beacon, an inevitable destination. For all those with lost ones. For all those in search of hope.

Mika was almost to Lalomanu when a girl he knew from his work met him on the road. She told him Ave's mother was dead.

"She didn't tell me in a way to let it down easy for me, she just came straight up to me and said it. 'I think Malo is dead.' I heard her voice, but I didn't care about it. I didn't want to hear her. I asked her where. She said they are at the hospital, everyone is gone to the hospital. I kept walking."

Mika wasn't going to turn up the inland road to the hospital. He was still determined to get to his family in Lalomanu. He was still clinging to the surety that he would find them there at the house. But then he met another co-worker, the cook at Tafua Beach Fales."She said to me – I think two of your children are missing. I didn't think. I just felt pain inside. And she told me, everyone was at the hospital and so I went there."

What did he see? On this day, the hospital drowns in misery. Quietly. Misery mingled with pockets of hope. Cars reverse slowly to the side so they can gently carry the injured. Laying them on mats, ripped sheets. Ready to drive the long, jostling route back to National. Over potholes and speed bumps. Weeks afterward, drivers will need counselling as they talk about driving over Le Mafa, slowly, slowly with their precious cargo. And having someone die in their car. Which would be haunted with their ghosts now.

The media have arrived. Local video producer Rudy Bartley is there with his camera which will shoot the first images of devastation. Images that will eventually beam to international networks. Wet people stand at every entrance. Waiting. For their injured to be seen. For their dead to be taken to town. Some with their own raw tsunami reminders. There are rows of people carrying children. Walk the open hall and you can feel the precious edge between life and death. A little girl stares blankly at the wall, held by a grandmother. She has cuts to her face. And her back is purple with bruises. She doesn't care about the camera. Just stares at you. Is she here? The grandmother is talking to the women around her. Telling them about what happened. Stopping every so often to cover her face with her free hand, to cry. And to softly rub the little girls back.

Another man in a blue shirt holds a naked toddler. The boy has sand crusted eyelids. And flies buzz around the scratches on his arms and legs. He wants to sleep. Eyes closed but letting out a wail now and again as sleep slips away from him. The man speaks to the camera, "The wave came and we were still asleep in the house. We found this one on the sand, down by the sea. After. Just lying there. Alive." He shakes his head and turns away to let the tears come.

A mother with a bruised face sits patiently in the hall. Beside the pink flowers in the grass. In her arms is a sleeping chubby baby. Bursting with goodness. Twenty-six year old Iota Shen Yi and her son Matthew. "I held him in one hand and swam with the other. I kept trying to hold him above the water. I just wanted him to have air." Iota's other child, three year old Fa'atamali'i was taken by the wave and Iota's husband is out looking for her body.

At the end of the walkway, the bodies are laid on the ground. Covered with floral bedsheets. A large woman sits beside one of them. Patting the shape tenderly. And shushing the fussing baby in her arm.

There is a couple that sits on the floor of the treatment room. They had been at the main restaurant fale of their beach resort when the tsunami came. Their child was asleep in the beachside bungalow. Awash in the cool morning breeze. They had tried to rush back to their bungalow but the wave was too strong. The child was taken. The man, wearing only a lavalava, is in shock. He cannot speak, he sits and shakes uncontrollably, rattling the IV tubing that hangs suspended from a ceiling hook. In his eyes there is a look of desperation, of guilt. Mirrored in the eyes of his wife sitting beside him.

There is a tour bus parked in the front of the hospital. A *palagi* woman tiredly tells the camera crew, "We're just waiting. We just want to get to town to the embassy. We've got nothing."

There is a thin old woman wearing a red sweatshirt and pyjama pants. Short gray hair. She is unhurt. But agitated. Asking everyone who passes, *have you seen my suitcase? I need my suitcase, please can you help me find my suitcase?*

Another two *palagi* men lean against the gate in the shade. Brent Cederman and Andrew White from Raglan, New Zealand. One speaks for the other, "His wife has died. We don't want to talk to anyone right now."

All these things greeted Mika as he walked in the gate. He looked and saw Ave sitting on the cement edge. Not far from the dead people."I asked her, *who is it?* And she said, *It's my mother, she died.* I asked her where the children were and that's when she started to cry and then I knew what happened. She told me they hadn't been found. Only Phillip."

Mika asked his crying wife, "What happened?" And slowly, she told him.

October, 2009.

Today the children gave the program in church. Dressed in their cleanest, sparkling best, they sang songs about God's love and carefully said their parts. Scriptures and talks. Little sunbeams lisped, 'I am a child of God/And he has sent me here/Has given me an earthly home/with parents kind and dear.'

Watching my children up on the stand on Children's Sunday, always makes me cry.But today was different. Today, my eyes were on the couple who had edged into the back of the chapel a little late. A young man with dark hair and eyes carefully pushing his wife in a wheelchair. His every movement spoke of protectiveness. She was thin. The chair was too big for her huddled frame. Her leg was bandaged, there were healing cuts on her face. I recognized the couple – and as I did so, my heart sank. No, this was the worst possible day they could have chosen to come to church. No. Not today.

I watched them furtively. The children sang and the woman's eyes filled with tears. Her husband gripped her hand in his. A little boy grinned mischievously as he stumbled over his words then laughed aloud triumphantly when he was done. And the man started to cry. His shoulders shook quietly as he covered his face with his hands. It was the woman's turn to comfort him.

I wished someone could have warned them. Told them to stay away on this day of all days. Other parents all around them were smiling as they gazed upon their little ones. Children singing, wriggling under a teacher's watchful eye. Children smiling and waving when they caught sight of their mum and dad in the audience.

It did not seem right to be so happy when another was suffering so much. It did not seem right to rejoice in one's children when another's had been ripped from them. At the end of the service, we talked to the young couple, but words seemed so inadequate. I'm sorry, I wanted to say, I'm sorry that while I have my children, yours are gone. I'm sorry that they can't be here today. I'm sorry that happiness is even an option for us.

I'm sorry.

7
I am tired.

The night before the tsunami had been a busy one for Ave Mikaele. Mika went to spend the night in Saleaumua, taking the oldest, Iulai. Ave stayed to be with her mother, sixty-nine year old Malo Laufiso. She had not been well and so Ave would often sleep at her family home at Lalomanu to help with the chores. Ave had the other three children. Philip, Malo Maryanne and baby Aleki. Like most Samoan families they lived communally with extended family. Ave's sister, thirty-two year old Tua was there. Her brother Vitale was there with his wife and children. Another sister, her husband and two sons. Uncles and aunties. Cousins and their children. A finely woven web of family ties, worn but beautiful, like a fine mat.

On Monday night, Ave went for a ride to Apia to take her sister in law who was due to have a baby. She got back to Lalomanu at about two in the morning. The children were already long in bed. But shortly after Ave had gone to sleep, her three year old daughter Maryanne woke her up.

"She said, *mum wake up and lets have a prayer*…I was tired but she wouldn't sleep, so we said a prayer." Mother and daughter bowed their heads in prayer as the sea breeze rustled the white folds of the mosquito net draped above them. "After our prayer, we went to sleep and the next morning the earthquake hit and my daughter was saying goodbye."

When the earthquake started, Ave was making breakfast for her mother. Malo was sewing in the front house. Ave turned on the radio to see if there were any reports about what was happening, but the radio said they too had no information yet. She was troubled, uneasy. She woke the children and told her mother maybe they should do something? Go somewhere?

"My mother said – why should we worry, it's just a tremble and nothing is going to happen. She just thought that tsunamis were from other countries and not here. I walked back to my small house with the children. I didn't feel good, because the earthquake was too strong. When I got back to my house I heard a grumbling sound and the school children on the road were screaming that a tsunami was coming."

Ave is a lean, strong and fit woman. She is used to working in their vegetable garden, clearing land and carrying buckets of water from the faraway tap. She can husk coconuts, dig taro and cut grass with a bush knife. But Ave had three children. And only two hands.

"I couldn't do anything. One child I put on my back, another child I carried, and the other child was holding onto my skirt. We started towards the mountain, but it was too late. The mountain was far away and the wave had come. I was very scared not only for my children but for my mother who was at the front house and also a lot of children were still at the front. I felt the first wave hit me and my children. I went to grab my children and I felt something heavy on top of me and I saw that it was the wave. It grabbed us and thrashed us about – my children were lost. I couldn't save them because the wave was too strong."

Ave and her children were swept towards the mountain and trapped in the backwash. "The wave couldn't go straight because of the mountain – maybe if it was something that went straight like at Saleaumua, we all might have lived, but it couldn't, the wave was going straight up and I think that is why the children were taken."

Ave remembers two waves. The second was strong enough to pull her a ways out to sea where she was caught in floating rubbish. "When I realized the wave had gone, I felt that I was safe because I was stable and I was in the sea. I lay there in the rubbish for awhile. I was so tired and I thought I would rest and get some strength before I could swim towards the shore before another wave came. I prayed and thanked God for letting me live. Then I swam and when I reached the land, I sat there and cried and called for my children and my mother."

Her brothers found her there and tried to take her up the road to higher land for safety. They told her there were warnings from Apia that there would be other waves coming. But Ave would not go. "I wasn't worried about another wave. All I wanted to do was go and find my children and my mother and if I found them I would thank God, but if I went to find them and another wave came then so be it – I will die with my children and my mother."

After the tsunami, Ave sat alone in the wasteland that was once her village. She estimates it was about two hours after the first wave before she could start making her way through the devastation. "All the people of our village were already gone, they were all up the mountain. I just sat there and waited. When I saw the water die down, that's when I started to swim back to where our house was. In some parts I couldn't stand because the water was too high and there were other parts with trees where I could stand. I swam to find my mother and I was calling out for my children. I was swimming close to where my brother's house used to be and I heard someone calling my name. It was my brother's daughter who had run with my mother. I went over to her and she was safe and alive."

Five year old Sama sat on a fallen pole. Crying but unhurt. Beside her lying face down in a pool of water was her grandmother Malo. "I tried to lift my mother's head but she was dead. I lifted her and put her on the pole next to Sama."Ave then heard a sound she will never forget. " I heard someone screaming and it was the voice of my son and he was calling my name and I called out. I called Philip and he answered and said mummy."

Four year old Philip was entangled in a small tree that lay over a pool of brown water. She had to call many times to find him. Calling out with the wretched hope that only a mother who has lost her children can understand.

"I couldn't see him, all I could hear was his voice, calling for me. I went and looked for the voice, and the voice still came and it was my son. I kept getting closer to him and on my fifth call, I finally saw him. All I could see were his hands, hanging onto this tree that was about the size of his arms, and his leg was trapped in the tree as well."

Philip had been caught there for at least four hours. Ave cannot bear to think what would have happened to her son if she had not found him then. It was a hard thing to reach get him out because he was suspended over water that came up to her throat.

"I know that if I had been any later, he would have died because he was holding on for his life…he would have drowned if his arms got too tired and he let go."

Ave carried both children to safety, walking down the road wet and bedraggled. Her brothers met her there, walking up the Lalomanu road. "I told them to get the children to safety and then come back with me to and get the old lady…They were crying and so shocked that the children were alive. They took Philip and Sama while I went back to look for my other two children…"

Ave continued to search for her family. While others searched for their families. Until her brothers came to carry her mother's body in a lavalava, struggling over mounds of garbage to the road and up the hill inland to the Lalomanu Hospital. By then it was past noon and the sun was high. And there was still no Maryanne or baby Aleki.

Ave's cousin is married to Nofo Letau. On tsunami day she was six months pregnant. They have three sons. After the earthquake, Nofo gave the youngest child, PK to her husband so she could clean the house. Her oldest son Junior left to walk to school.

"As I was cleaning my house I could hear rumbling like something was going to happen. I called to my family to have a look out to sea and look what was happening. It got louder so I came out of the house to look. Just as I stepped out on the road, the wave hit the coconut trees, after the trees it came onto the road and then to our house."

Nofo yelled for everyone to run. Turning, she grabbed the nearest child. A toddler – her one year old niece. The child's mother Miriama had already left for work. Nofo's husband held their two boys – PK and Komiti, in his arms. He could do nothing to help Nofo as she struggled laboriously to make it to the hill.

"As we were running, I'd fall and then get up because my husband kept telling me to keep going in case I got washed away. He said run and save yourself - because I can't save you."

The water caught up with Nofo and dragged her under. She couldn't hold on to the child as she was tumbled about in the wave. She blacked out. Then the water caught her husband too, ripping Komiti from his arms, but unable to wrestle away the baby. For a short while, Nofo's husband was underwater but he held baby PK above his head, refusing to let the sea take him.

Nofo regained consciousness in a tree. That she clung to gratefully until the water level dropped enough for her to move uphill to join the others. She was relieved to find her husband and all three of the children alive. Junior had run for the hills as soon as he saw the wave. Then stood there crying as he watched the sea wipe out his home and sweep his family away. Komiti was found – also safe in a tree. And baby PK was still nestled in his father's arms. But Miriama's daughter was lost. They would find her the next day buried under rubble. Along with her two year old brother Jason. And further down the road, still wearing her school uniform would be Miriama's oldest, eight year old Nifo.

Eight weeks after the tsunami, Nofo and the children are staying with Samoa Victim Support in their refuge house at Motootua. They moved there after she was discharged from the tsunami ward at the hospital. Her husband is building them a house inland at Lalomanu. But Nofo's sons won't go with him. Every time he tries to take them to see their new home, they cry. "They are too scared." Nofo explains. "They think another wave will come."

You know that feeling you get when you've forgotten something but you don't quite know what it is? You've left the house, you're on your way to work – and you have that niggling feeling that you're not quite complete?

Rita Romeo of Lalomanu had that feeling on September 29th. As she chased her children up the steep embankment at the back of their beach fale operation. As she tried to tell her guests to follow them to the mountain. As she went back to scream at tourists who were holding tightly to the posts of her dining *fale*, telling them – "Don't be stupid, those posts won't save you! You have to run up the mountain. Run!"

As she watched the girl who works at the ice cream shop on the beach, turn back and run to her house as if she had forgotten something, ignoring Rita's shouts. "Come back! Can't you see the wave is here? Come back. We have to run."

As she saw the girl's house collapse in the wave with a loud banging sound.

As she scrambled up through rocks and bushes. Pulling on roots and branches to get up the steep mountainside. Turning to grab an older *palagi* woman by the arms, scraping her bare legs raw as she dragged her up and over rocks.

Even as she gasped for breath at the top of the hillside, watching the ocean consume their village, their home, the twenty open beach fales they rented out to visitors for a day, a night, a weekend. Yes, even then, Rita had that terrible feeling that she had forgotten something. Or someone.

Rita has eight children. The youngest still toddles over to nurse. At the close of day, as sunset burned the evening sky, Rita would usually take a few moments to sit in the main *fale* with her child at her breast – a few moments of contentment as she surveyed the tourist operation that she and her husband had worked so many years to build. They were extending their business – a restaurant and a roadside store. A White Sunday opening was planned.

But on that Tuesday, as she stood with her family and guests on the mountainside, Rita realized with horror what she had forgotten. Her six year old son Livi. Asleep in the bedroom when the wave came. "After the wave finished I felt sadness. As I looked out over the villages there were no houses. All I said was '*oh Lord what have we done to deserve this?*' I felt so much grief when I heard people crying, wanting to live. I thought Livi was dead too. I couldn't do anything."

It didn't seem possible that anyone could survive the ocean beast. But Livi did. An hour after the tsunami, he was found underneath a pile of rubbish.

"He was found by a woman named Salome. She said she saw him crouched down under all the wood and roofing. He must have been trying to get up for breath but he couldn't because of the debris. Many people were under all that rubbish. Salome saw a body down there close to where she was standing and then she reached down to bring it out and he came up and it was my child and he was alive."

Livi's chest was cut and badly bruised. He had ingested a lot of seawater and would need some serious antibiotics for the infection. But two weeks later, he would be his usual mischievous self. Pestering the team of Australian army volunteers who came to clean their beach with amazing machines that could drive into the ocean, growling over coral on the sea floor.

Rita would watch him run over golden sand beside crystal water and shake her head in quiet awe. "We are thankful to the Lord for sparing us and our eight children. Especially my son Livi who was lost and then was found."

Lesa Foti Tuvale is seventy-five years old. He is not a man who lives a sedentary life or who overindulges on heavy fare. Two weeks after the tsunami, Lesa sits under a canvas awning with a bush knife beside him. His hands are worn with black earth. He is the father of six children – one boy and five girls. Lesa'a daughter Leueta was killed in the tsunami. Often when a child dies, we are plagued with questions. Were we good parents? What could I have done differently? Our grief is serrated with guilt and self-doubt. Lesa is no different. On the morning of the tsunami, Lesa had angrily punished his daughter Leueta before she went to work at Taufua Beach Resort.

"This is the only daughter of mine that is naughty. I'm always looking for a stick to smack her with because she was very difficult and wouldn't listen to her mother. This is the duty of the parent to discipline the children… but it's so hard. I don't know. I just pray to God and hope that I was doing the right thing. Who knows if I'm wrong or if my children are wrong – only God can judge."

Lesa lived on slightly higher terrain than his neighbours, and further back from the ocean. The first wave brought his neighbours with it, slamming them into his wire fence.

"The wave brought that old lady here, because their house is quite low and next to the beach. The lady was crying out *I want to live, help me*. I went to help her but she's quite large and I couldn't lift her over the fence. Some boys came and helped me to carry her up where the water didn't reach."

The next waves were bigger. They reached Lesa's house and destroyed it. Reaching for him hungrily as he ran up the hill to the road. "I didn't wait, I just ran…whatever things from my house I wanted to save, I just left them because there was no way I could save anything."

Lesa's other daughter was caught in the wave at Saleapaga. She ended up high in the branches of a tree where her husband eventually found her. Lesa was at Lalomanu Hospital when she was brought in,

"It was hard to look at her because she was doubled over on the bed and her eyes were staring at me. She said, she thought everyone was going to die in this wave. I told her that it was the grace of God that saved her."

This daughter would spend a month at National Hospital, her lungs clogged with sediment and seawater. But Lesa's other daughter, the naughty one Leueta, would be found late that same night.

"It was dark when I got word that she had died. And her body had been found. I felt my heart breaking and tears sprang to my eyes. I left everything at that time and just went and sat at the hospital where they were bringing people. I just wanted to see her body. They brought in a group of bodies and I saw my daughter amongst them and I just went and wept on her. It's hard to explain. It's right about the saying that death is like a knife going through your heart, you don't know how to move on."

Peniana Iuni is sixty-six years old. He is married to Soli and they have nine children. He has the wiry body of a plantation farmer while Soli struggles with ill health. That morning he was cutting his grandson's hair in front of the *fale*.

"I couldn't cut his hair properly because the ground was shaking so much. I told my wife to come outside because the house was like it was suspended in the air from the shaking."

When the ground stopped moving, the family continued preparing for the day. The grandson went down to the beach in front of their home to wash his hair. Peniana gave his daughter twenty tala to buy some bread for breakfast from the store several houses down. When the tsunami came, she would drop everything in her dash back to the house as she screamed for her parents to run. Months later, Peniana would still shake his head disapprovingly,

"You know, she dropped the shopping AND the change from the shopping – she dropped it all on the road in front of someone else's house…. there was twelve tala change and she dropped it! What a waste. We never found it…"

Peniana's house was not far from the hill but his neighbour's wire fence made running from the wave difficult.

"I couldn't get over the fence, it was too high. The water came. It caught me and broke the fence – the wave shattered the fence poles and I fell face down. But I was able then to run and climb further up the hill."

Soli's son saved her. He grabbed the back of her shirt and never let go. Not even as the waves washed them back, slamming them into the houses uphill behind them.

No one from Peniana's family died that day. But the man who used to walk to the plantation every morning with his bush knife, work all day and then return carrying a basket of taro and coconuts on his shoulder, now sits in the shade of his tarpaulin at eleven in the morning and shrugs.

"My family has been going everyday to clean up and they are starting to rebuild. But...not me. My strength is not good at the moment. I am tired."

That afternoon on September 29th, sitting outside the Lalomanu hospital amidst tears and dirt and blood, Ave was tired too. As she waited for a car to take her dead mother to National in town. As she shook her head numbly when a nurse asked her if she wanted them to clean her cuts that seeped dirty red. As her heart, mind and soul screamed silently for her two missing babies. Yes, Ave was tired.

But now Mika was here. And surely together, they would find the children. Wouldn't they?

Mika and Ave spent the remainder of the afternoon searching for Malo Maryanne and Aleki. It sounds easier than it was. Looking for two small people in an area that you know like the back of your hand. Your birthplace. Your home where your family have lived for generations. But the Lalomanu that greeted the waking sun that morning was not the same place that watched the evening come at the close of that long afternoon. As one explained it – "there were no familiar landmarks there anymore. The only thing left that I recognized was the speed bump on the road. That's it. The only way I knew where exactly we were."

It didn't help that the section behind Ave's family houses was somewhat submerged. Mika describes that afternoon,

"That whole area, the water was up to my neck and the water was dirty so you couldn't see anything. Corrugated iron, wood, trees, rocks, nails – so you would step and try and balance and then you would try to take another step. And we could hardly see anything. That whole afternoon, we were looking for our children and we didn't succeed. I was wet from morning until night, wet and hungry. There was food lying all over the place but I was too disgusted to eat it…except for packets of twisties. Because you can tell if they're fresh or not. If you press the packet and its crunchy then you know it's a good one, but if it's soft then you know the water got inside. So all I ate that day was some twisties."

Night brought darkness. And darkness brought a halt to the searching. Mika's sister came from town to offer them shelter. An exhausted Mikaele family made the long drive to Apia. Without Malo Maryanne and Aleki.

8
A magical place in the world.

More than 100,000 visitors come to Samoa every year. The bulk of them are Samoan nationals returning home for holidays, funerals, church dedications and chiefly *matai* titles. Others come for business, sports and from other Pacific nations to attend regional meetings. Many are American Samoans hopping back and forth, keeping their family connections strong. About thirty thousand of these visitors are bona fide 'tourists.' They come for the beaches. The weather. The green environment. The welcoming friendliness of the people. To experience a traditional culture still vibrant in spite of the ever sweeping onslaught of Westernization. In Samoa there are no military coups and the most dangerous wildlife is a wriggly centipede. Or the one footloose mongoose roaming in Satitoa village, a stowaway on a container from Fiji. In Samoa you can walk the seawall at evening and not have to worry about muggers or child abductors. You can sleep in an open hut on the beach and only pesky mosquitoes will pester your sleep. Yes, there are many things in Samoa that call to the overseas visitor.

Mike and Sally Adye from Gisborne, NZ first discovered Samoa in 2003 with their two sons. "We had a brilliant holiday and stayed at many different places in Samoa. Together we spent hours snorkelling and relaxing and then in the evenings we were delighted by the hospitality of the Samoan people. We knew we had found a magical place in the world."

Rebecca Hamilton and Harriett Sherratt, both from Gisborne, were looking for a holiday destination in 2009. "We decided on Samoa…neither of us had been and it looked absolutely stunning in the travel promotions. We had also heard from a number of people that the Samoan people are truly wonderful, caring and very hospitable people. Being school teachers we also wanted to take part in some school life with local children in a village setting."

Sarah Trafford and eight month old son Jai, from Gisborne, NZ, came to Samoa in August 2009 for three weeks. To go on holiday with a baby when you're a solo mother can be a scary thing, fraught with stress. Who will help you with a crying infant? How will you ever get to do anything relaxing when you have a baby on your hip 24-7? "I chose [Samoa] because I had heard so many good things about the people there and how great they are with children. Our whole trip, I felt like everyone was looking out for us, even on the plane on the way to Samoa, my Samoan neighbours on the plane beside me all took turns cuddling Jai and making him laugh on the flight, then one of them actually gave us a lift into Apia and dropped us at a good place to stay."

The Cedermans are from Raglan on coastal New Zealand. They own the Raglan Sunset Motel together with Andrew and Mary Ann White. The Cedermans first came to holiday in Samoa in 2002, spending time in Upolu and Savaii. "When we got home we both agreed that had been the most relaxing holiday that we've ever had, we just thought it was absolutely fantastic. Because normally we're just full on and when we travel we go up to Asia or Europe and you come home more tired than when you left. Samoa was just so relaxing, so lovely that we always thought we must go back."

The Ansell family from Auckland, NZ has a long history with Samoa. Graham spent three years living in Samoa as a child back in 1973. He returned with his wife Diann and children in 2005 and they enjoyed the visit so much that they brought their extended family with them for a holiday in September 2009. "We didn't want to go to other places in the Pacific where it's just like going to Auckland with warmer weather…because all the Aucklanders are there. In Samoa, you don't stay in one place. You get around and look at different places. It's just such a neat place, the people are fantastic and the foods good. A lasting memory I have is driving behind one of the wooden Samoan buses and seeing five BIG Samoan people sitting across the back seats. It's just pretty magical there really."

Yes, over the years, Samoa has cast its spell over many, creating a growing tourist industry. Many tourists are repeat visitors who form close bonds with their local friends. They cease being tourists and instead become family. This is particularly evident with the small beach accommodation operations at Aleipata where an extended family will host visitors in traditional style wooden huts. Shared facilities and communal dining adds to the casual, closeness of the experience.

Taufua Beach Fales at Lalomanu Beach is one of the larger of these and arguably the most popular. Before 29/09, they were hosting between 100 and 300 visitors a day, many of them group excursions from schools overseas. The Taufua accommodation *fales* were right on the beach, shaded by scattered palm and *tamaligi* trees. The *fales* looked out over the sea. They were built in two lines so that one hut sat between the gap of the other – ensuring every *fale* shared in the ocean breeze and every guest could wake up and look straight out to the surf that laps the reef. This arrangement also meant however, that there was no clear view of the ocean for any of those who lived across the road. There, sitting at the base of the mountain was where the villagers lived in assorted western style houses mingled with traditional

Samoan open *fales*. In front of the houses were their gardens – hibiscus and ginger bushes coaxed from the sandy soil. At the rear were the rougher style dwellings – the cookhouses, bathroom blocks and pig pens – many of them made with sheets of rusty corrugated iron and held together with not much more than good intentions.

The business was started back in 1970 by the patriarch of the family, Taufua Leifi. His daughter Faafetai (better known as 'Tai') explained, "It started out as *fales* that our family used for resting on Sundays after lunch, we'd go to the beach and sleep there. Then we started having tourists come in asking if they could rent the *fales* for a day or two and we'd allow them." And so it began, modestly, with a few huts on the white sand and over the years it grew. In 1999, Taufua Snr handed over the business to Tai and her husband Sili Apelu. Initially there was a temptation to completely change the look of the place, to do away with the beach *fales* and build a more Western style resort. "But we thought it would be very un-Samoan to do that. We decided to maintain the beach *fales* but to at least make them more comfortable for the tourists. Over ten years, we have built on slowly, adding new *fales* with the availability of funds…that has been a blessing in disguise because it has been a time for us to learn and to nurture our workers and that is where our strength is. I think what makes this place special is the atmosphere that we created. Guests always feel at home as if this is their home as well. We make them feel very welcome and we see that in the repeat guests. Many of them return, its like friends and family coming back. Of course, Lalomanu has its own beauty, like the sand and the clear blue lagoon and all those things, but service – quality service and the attitude towards the tourists is what makes it special."

Sara Trafford delighted in that family atmosphere. "What really stood out for me was how great everyone was with my son – from the ladies in the office, to the young men waiting tables – everyone helped to look after him and gave him loads of cuddles and hugs. They would often come and whisk him away from me for a walk so I could eat my dinner, go snorkelling or just have an hour by myself, it was wonderful! Even the night watchman would go and sit outside my fale in the evenings after I put Jai to sleep and then run to the restaurant to get me when he woke up for a feed. When I would get up in the night to soothe Jai, the night watchman would shine his torch in to make sure we were both okay." Tai's sister-in-law Sina Ale helped in the business. She had three little ones of her own, the youngest - Etimani Jnr, was the same age as Jai and the two babies would bathe in a plastic tub – one rosy-cheeked peaches and cream, the other sugary cinnamon and raisin cheerfulness.

Tina De Suza from Auckland, New Zealand calls Sili and Tai, 'dad and mum'. She first visited Samoa twenty years ago and "fell in love with it immediately." Her and husband Greg were married on the beach at Taufua Beach Fales in 2007. "We have been to Samoa many times, for periods of one to two months…during our longest stay we lived and worked at Taufua Beach Fales, helping the owners in exchange for board. The Taufua family welcomed us into their home and family…we have seen most of the kids grow up and a lot of the staff members too. Sili's daughter Lumepa is as close to me as my own blood sister."

There have been others that come to stay at Taufua under similar arrangements, that are welcomed as family, given a place to stay and who then pitch in to help. Like Frederico Fernando who arrived midway through 2009. "He was a chef, he came and offered to teach our staff how to do pasta. So I agreed that he can stay and work for his upkeep here. He stayed with our family."

What happened to Taufua Beach Fales on 29/09? The night before the tsunami, the Taufua children were having their White Sunday practice. Every second Sunday in October, Samoa honours its children with a day devoted to them. It is a day when the youngest take the place of the pastor and churches are overflowing as proud elders listen and learn from their children. Young ones practice for weeks on end to memorize passages of scripture and rehearse often elaborate Bible story skits. It is a day when children wear new white clothes. Satin and lace. Ribbons and bows. Frills and flounces. Crisp shirts and lavalavas. It is a day for feasting. When the children have the first servings of corned beef, chop suey and mutton flaps. Boxes of neon pink ice cream and sheets of chocolate cake sticky with spirals of cream frosting. On that Monday night, the Taufua's were putting the finishing touches on their Sunday program. "I was with my sisters, all my nieces and nephews, we were rehearsing our White Sunday skits and we sang the Peace Chapel song, the one that mentions mountains falling and the ocean covering the earth…so we only had a couple of hours sleep because of the late rehearsal."

Tai is a habitual early riser. Her father would have everyone up at 4am for family prayers – regardless of any late night practices. "Dad expected everyone at that time and I'm the one that has to wake all the grandchildren from their houses to attend…so on that day every child in this household was up and the prayers actually started right on the dot. All our children, my nieces, nephews, sisters attended that morning. After the prayers we went about our morning chores to get the guests breakfast ready and the children prepared for school." As was her usual custom, Tai was chatting with guests in the dining *fale* as the first ones began wandering in for breakfast. By six that morning she was sitting on the restaurant deck with sisters, Petria and Rebecca Martin from Matamata, New Zealand. It was another breathtakingly beautiful day at Lalomanu beach and one could never tire of the view as sunrise burned the sky and solemn blue ocean lightened to a frolic of glittering gold. From worshipful ode to light-hearted rhyming verse.

When the earth shook, Tai's husband Sili was asleep in their home just behind the resorts dining fale. The tremors had him worried enough to quickly dress and get the Landcruiser truck that was parked between the *fales* and drive it out beside the road. He turned the car radio on to check for warnings. The car clock read 7:05. Sili looked out to the ocean "as I walked across the road I was looking at the sea and there was water in the lagoon. I went into the office to check the internet for a tsunami warning. My daughter Lumepa was already on the computer. I told her to get up and let me check. I sat down and typed in the USC website address. As soon as I finished typing I looked up and there was no water in the lagoon. I knew a tsunami was here."

On the restaurant deck, Tai saw the ocean move. "I saw the ocean shrink back so fast, that was my signal. I got up and started screaming…the Martin sisters ran back to their *fale* and I think that was why the wave caught them." She started running up the length of the beach, shouting to the guests to *get up, run, there's a tsunami, run!* Many were still sleeping and Tai was frantic with the delayed responses of many. "I was practically throwing stones on top of the *fales*."

Tai's thirteen year old son Mike ran to the far side of the beach where the extra guests were accommodated in spare *fales*. "He ran there to the overflow *fales* – we use them when we are fully booked and need extra space – he knew many of our guests were staying in those. He ran across there yelling at people to run and luckily he remembered a track up the hill that's behind there, it's not as steep a section as the mountain behind the main *fales*. So he took some people and they all ran up that track to make it up alive."

After screaming the full length of the resort, Tai then turned back to where her own family was. "It happened so fast that all we could do was to get people to the hills and then I ran back to my sisters and my brother and his family. I could see it was too late to take my father to the car at the front of the *fales*…and my baby was with my babysitter, one of my relatives."

From the office, Sili saw several waves fast approaching from different directions. "I saw a wave forming just in front of Sieni's Beach Resort – which is the accommodation next to us on our left. I could see two waves actually, one coming in front of the other, building up quickly. I called to Tai to run and I ran myself. But when I got on to the main road, then I looked right and there was a wave already on the road, it was actually coming down the road. So it must have hit the upper part of the village and now it was coming on the road and bulldozing it's way towards me."

With waves rippling towards him on two sides, Sili ran across the road. He saw his four year old nephew standing by himself, bewildered. "I picked him up and started to run towards the mountain but as I turned the corner of our house, I saw that the water had already gone before me, and it was coming towards me from the mountain. I don't know how…maybe because our home had two doors opposite each other and the water went through the first door and spilled out the back?"

Now water comes at him from *every* direction and Sili holds in his arms a frightened little boy. "I didn't run then. I just reversed, walking backwards and put my back against the wall of a house. In a split second the house was gone and we were in the wave. We were taken up to the foot of the mountain."

Sili fought to keep hold of his nephew as the waves met and became one, churning them along with houses and cars. He was pinned underwater in a mangled mass of timber. Still in his arms was the little boy. Sili tried to break loose but it was impossible. In those panicked moments as the pressure for air became unbearable, Sili had to make a choice. He was trapped, but his nephew was not. "I couldn't move. There was this big thing against my back and I suspect it was a tree and maybe parts of our house. Something was holding my leg so I couldn't move. When I looked up the water was about two meters above me and I didn't know what to do with my nephew. When I thought he may have a chance, I just released him, I pushed him up towards the light."

In the next instant, another wave hit. It was strong enough to dislodge the debris that held Sili and take him in a brutal swirl of madness further along the mountainside. "The wave kind of pulled me up and I was spinning in the water, going anti-clockwise. I remember turning every way and trying to grab on to things. I held on to something, I don't know what it was and luckily when I put my foot down I was able to keep myself stationed there. When my head came up out of the water and I looked up, the first people I saw were our four year old daughter Ake and one of her aunties looking down at me." Sili's nephew was nowhere to be seen.

Sili was able to join those huddling on the mountainside. Little Ake was there because of a pasta chef. Frederico Fernando had been shaken from sleep by the sounds of people screaming. "He woke up and saw people running. The only thing he could do was run as well and he was following the pigs up the mountain. When he looked down, he saw Ake on the roof of a beach *fale*, gasping for air. She was standing there by herself, somehow. So he came down quickly and luckily the *fale* was very close to the foot of the mountain where the water was so he grabbed her and took her up. That's how she was saved."

What of the other Taufua children? Tai and Sili's eleven month old baby, Sili Jnr? Their granddaughters? The nieces and nephews who Tai would call for family prayer every morning? The children that she had been practicing the White Sunday program with the night before? '*mountains fall and oceans will cover the earth…*

As soon as the water began its retreat, Sili yelled for the village boys to start searching for those caught below. He was injured but he made his way up the hillside, looking for Tai, looking for the children. "When I got to the top, I met Tai there and she was limping. She was cut badly, you could see the bone in her thigh but she was walking around trying to take care of people. I suppose it was shock or something, because we were both walking around on top and people were telling me, *hey Sili you're bleeding everywhere* but I didn't feel any pain and neither did Tai."

In the meantime, the Taufua family's personal losses were adding up. Fourteen people in total – including all three of Sina Ale's children - were killed in the tsunami, spanning four generations. Tai's sister Su'e and Sue's daughter Matala had gone to try and get the patriarch Taufua Leifi out of the house and all three never made it. Tai's cousin Si'imoa who lived with them and helped with the business. Lafatua, another cousin who cared for the babies. Tai's niece, Anita Nu'ualitia who looked after the ice-cream shop and cocktail bar. Tai's cousin Sieni who lived next door. Sili's nephew that he pushed up towards the light, towards air. Sili and Tai's baby. And their granddaughter Moanalei Long. All of them taken in a few raging moments.

Moanalei's father Jon Long, wrote of his daughter's death. "My nine year old daughter was trapped in a truck while trying to escape the tsunami with her mother Lumepa and her five year old sister Galo. Lumepa and Galo were outside the truck when it hit them and they were separated. Galo washed up amongst a number of bodies and walked up the hill and found her mother. Lumepa found a kayak and held on to it until a boy grabbed her from the top of a coconut tree and pulled her out of the water. The truck with Moanalei in it, was dragged 300 meters underwater, into the mountain and then back down into a whirlpool. She was trapped inside with no way of getting out. When I found it, the truck was upside down underwater. I retrieved her body from the truck when I arrived from Apia."

Tai and Sili were taken to Lalomanu hospital. Doctors said they needed to be transferred to town but the couple insisted on first taking care of their guests. "We went to the mountain house because we knew many people had been taken up there. Tai was giving out the instructions, getting people to move the timber, sweep the floors, put down cement boards so people could lie on them. There were so many injured people, some of them had no ears, they were bleeding all over the place so everybody was trying to help." It was with great relief that Sili saw a good family friend, Mika Maiava who worked for the United Nations. "Luckily Mika came, so it kind of took the burden off us. At least we knew somebody we trusted that we could leave behind to look after the guests." By two in the afternoon, Tai and Sili decided the situation was stable enough that they could go in to town. To have their wounds checked at the hospital. To buy food for the family and the guests. And to take the bodies of some of the recovered family to the morgue.

They stopped at MedCen Hospital as they went over the Cross island road. "It was filled with guests from Coconuts Beach Resort. One of the doctors tended to me when we got there but he couldn't work on Tai's wound because it was too big. He wanted another doctor to be there. We told him that we just wanted some quick treatment then we would go shopping and return to Lalomanu but they said that was impossible." Tai went into theatre at about six that evening and came out three hours later. The couple spent several weeks in hospital. As did many others of their extended family.

A week after the tsunami, there was a memorial service held at Moanalei's school, Vaiala Beach Primary. One family friend commented, "It was a beautiful service with all her classmates taking the microphone to tell their stories about Moanalei. Afterwards, Lumepa took the microphone and addressed the kids, many of whom were terribly upset. How she managed to talk to those kids and thank them for their heartfelt stories and words about Moanalei, I will never know. She turned it around and had the kids laughing and smiling before she put that microphone down. She is one of the strongest people I know."

Sili says of 29/09, "The loss of the children is the most painful thing. The elderly had lived their lives, but the young kids were so helpless, nobody could help them. Our nephews and nieces, even though they had their own parents they call us mum and dad because we look after some of them. They would get up early in the morning and come to our room, *mum, mum, hungry, fia ai.* The physical association with the kids, every morning, every evening, they would always be asking us for something…they would come looking for us. We're missing them, because even though we believe in God and we are sure they are in a better place, sometimes there's a feeling of emptiness, not being able to see them anymore."

Nine days before the 29/09 tsunami, Greg and Tina DeSuza flew back home to Auckland after a two week stay at Taufua Beach resort. "Of all the times we've holidayed and lived-in at Taufua, that particular two weeks was the most beautiful and memorable of all, perhaps because we were able to introduce two of our friends to the pure bliss and joy of living in a Samoan *fale* on the edge of a beautiful beach, combined with a return visit for our other two friends, both of whom had been at our Lalomanu wedding in 2007."

At 7:30am on 29/09, the DeSuzas got the first news of the earthquake and tsunami in Samoa. Tina immediately called Taufua Beach *fales,* knowing Lumepa was working in the office, hoping to get an update. The phone lines were down. Tina called Lumepa's sister Via who lived in Auckland. "By now it was ten am. Via had spoken to Lumepa and learned only that Lumepa's youngest daughter Galo was alive but injured. Lumepa was alive but Moanalei was missing. At that time we hoped that maybe Moanalei had been swept along to the next village and was with people there. We didn't know how bad things really were."

By that afternoon, the De Suza's had a clearer picture of 'how bad things were' and in the following days, there were many emotions. "I can't describe the total feeling of helplessness and hopelessness that we experienced. We couldn't get there, we couldn't do anything, we couldn't fix it. Our minds tortured us with the thoughts of what would have happened if we'd still been there. Tai had asked me to stay on for another week or so – what would have happened if I had? If we had been there, would we have run for the hills? Would we have gone with Fa'asega, Su'e and Matala to help get the old man out of the house? Would we have gone searching for the kids to take them up the hill? Would we have tried to get everyone in the truck? These thoughts are still with us three months on. They come in the dead of night and when you least expect them. There's no getting away from them. I feel relieved for us not having been at Lalomanu at that time and in the same breath, I feel guilty that we weren't there to help."

Greg and Tina came to Samoa a week later, bringing what help they could. Gifts and money donated by many generous friends – "through the generosity of many past guests, friends of the family and even from people who have never met them and never been to Samoa. The kindness of these people completely knocked me sideways." And photos of the family, of the Lalomanu that used to be. "I asked people to send any photos they had of the kids, of the family, of the *fales*. We printed over 400 photos and put them into albums. I printed off every single message we had received, totalling over 80 pages and put them into flip file books for the family to read, just so they would know how much people were thinking of them and praying for them."

Meeting with the family at the hospital was very emotional. "When we first saw them, we all broke down and cried. Tai was bereft at the loss of her children, her sister and father and other family members and for the losses sustained by the many tourists who had been staying at the beach fales. Tai also felt a huge responsibility for those tourists who had perished and this added immensely to her grief. It's one thing to grieve a death, but how on earth does anyone grieve for fourteen family members at the same time? Not to mention all their friends in the village who also sustained huge losses in their families. Nothing in life prepares you for the losses that occur due to a natural disaster such as this – nothing.

We gave them the photo albums we'd compiled and they started to cry before they even opened them to look. It was heartbreaking to watch as they talked to the people in the pictures and gently kissed the images. Months later I still can't think about that moment in time without feeling their pain once more.

The hardest though was seeing Sina who had lost all three of her children. She'd already had extensive skin grafts on both arms and was drugged up to ease the pain of the surgeries, but I could see that nothing would ease the pain of losing her children. There were no words to say, and we could only sit there and look at each other and cry."

Of the Taufua family's loss, Tai simply states, "We left our family and we came to save our guests first…as a result, fourteen of our family members passed away. It's because we just ignored them and came trying to save the guests. It is fortunate that we're still alive."

Another Taufua who is fortunate to be alive is a young man called Otele. And because of him, a young New Zealander is also still alive. Woken by the earthquake, Otele was across the road from the main restaurant *fale* when people first noticed the troubled ocean. Hearing the shouts of warning, Otele ran to the beach and began banging on doors and yelling for guests to run. The wave chased him as he ran through the trees at the base of the hillside. He was caught and thrown face first into a breadfruit tree, cracking his front teeth. Dazed, he climbed up the tree but the force of the next wave was too strong and the tree fell – with him underneath it. He was pinned underwater for a few moments but fought his way out of the tangled branches. He swam to the hillside and was making his way up to safety when he saw a *palagi* woman in the water some distance away. Otele went back into the water, swam to the young woman and pulled her to safety. The woman's name was Olivia Loeffen and she had been holidaying at Taufua Beach Resort with a group of friends from New Zealand. Olivia was badly injured. Otele carried her up the hillside on his back. He applied a tourniquet to her leg to stop the bleeding. Others helped him to bandage her cuts. Then, Otele ran to the Lalomanu Hospital, carrying Olivia in his arms.

There had been three others in Olivia's travel party – sisters from Matamata, New Zealand, Rebecca and Petria Martin, and their cousin Jodi McGlashan. Jodi was injured. The Martin sisters were killed. After the tsunami, their parents came from New Zealand to wait for their daughters to be recovered, to take their children home.

The Salavert-Wykes family from Australia, was holidaying at Taufua Beach Fales – Jorge Salavert, Trudie Wykes, twin four year old boys and their big sister, six year old Clea. Clea was one of the victims. The family released a statement to the media, "We are torn by the tragic loss of our beautiful daughter Clea. She was the happiest of miracles for us and a wonderful big sister to her brothers. She was a very special person and we miss her terribly. Her death is deeply felt by her family in Australia and in Spain. We offer our condolences to all who have lost loved ones in this tragedy."

British couple Gary and Gill Cunliffe who had emigrated to Nelson, New Zealand - were at Lalomanu with their two year old son Alfie. The family was caught in the wave and Alfie was lost. The UK 'Sunday Mirror' quoted Ulrich Moritz, another tourist who had been on the same beach as the Cunliffe's. "At first it was only Gill in the hospital. She started crying, *I've lost Alfie.* We thought he could be with Gary. But then they brought Gary in. He was in quite bad shape. That's when she realized that no one had Alfie."

The Cunliffes were airlifted to New Zealand. Alfie Cunliffe was listed as missing, presumed drowned.

Australians Claire Rowlands and Vivien Hodgins came to Samoa on a beach holiday. Both women were caught in the waves and Vivien was killed. Claire was badly injured. She spoke to the media from her hospital bed. Pale and weak, the interviewer had to strain to hear her story. "Everyone shouted, *run* because they noticed the sea had gone flat. And everybody started running. And some people didn't run fast enough. I didn't either. I hung on to a tree but the force took me off down to the bottom of the sea, scraped me along. You come back up and then it would go out again, then you get a little bit of time between the surges. You tried to climb up then it would hit you down again. The last time it hit me, then I can't remember anything after that because I couldn't see. A Samoan lady came and told me to get up and walk to a higher spot. I couldn't see. They brought a truck down, started picking people up. I had a gash in my ankle, but you don't think of things like that when you're in that situation…."

Vivien was missing for many hours and her daughter Carla back in Australia, had posted a plea for help to find her mother on the social networking site, Twitter. It was evening on tsunami day when the Hodgins family received confirmation of their mother's death. They told Australian media, "We had so much hope…We loved her dearly, very, very dearly…we are just glad she was somewhere she loved when she died. She was at the beach and she was happy."

On the Sunday before the tsunami, two couples arrived at Taufua Beach Fales from Raglan, New Zealand. Brent and Diane Cederman and Andrew and Mary Ann White. They were owners of the Raglan Sunset Motel and had come for a board meeting holiday. This was not the Cedermans first time to Samoa and they were looking forward to relaxing at Lalomanu although Diane did have her worries about how Mary Ann would take to the island style accommodation. "I had warned her that it wouldn't be like the resorts she had stayed at in Bali or anything, but after our first day, she absolutely loved it. At the restaurant everyone eats together at long tables and we got to meet so many new people and she was having a great time. She said, *This is so fun, it's just like a school camp!"*

On Monday, the couples went on a tour excursion with a man called Mika. He led the group on a hike up the steep hill backing onto his wife's home telling them about the rich history of the area. Then everyone was able to try their hand at making an *umu*, a traditional above-ground oven made using hot rocks and banana leaves. His demonstration was peppered with continual banter. *Do you all know how they crack open a coconut in Vanuatu? The grandmother will put the coconut on her head and a grandchild will shoot it with a gun…And in Tuvalu, they will throw coconuts off the cliff onto the rocks below to open them…But here in Samoa, we use a stone to hit it in the right spot and crack it open just like this!*

Mika's wife Ave helped show the women how to make *palusami*, little parcels of coconut cream wrapped in young taro leaves. Mary Ann in particular proved to be a natural at the cooking class. "Her *palusami* parcels didn't leak like mine – and I didn't quite master the art of fire lighting like she did either!" recalled Diane. Later that afternoon, Mary Ann spoke to her daughter Jo back in Raglan. "She was having a ball, cooking a Samoan meal the traditional way and was all hot and sweaty and looking forward to having a swim."

After dinner both couples retired to bed pleasantly tired by their day. The next morning Andrew and Mary Ann were awake in their *fale* when the earth rocked. "It wasn't that violent, more like a continuing tremor." Accustomed to life by the sea in the coastal town of Raglan, the Whites knew a tsunami was possibility. They looked out to sea, "I saw the top of the reef and I knew something was wrong." Andrew ran next door to warn the Cedermans before both he and Mary Ann started their dash for the hill. They ran across the road and through the closely clustered houses, ducking under clotheslines and weaving through gardens. "We could hear huge cracking sounds as the beach *fales* were exploding behind us."

The Whites found their path blocked by a corrugated iron fence just at the base of the hill. "I jumped the fence and managed to get up. Mary Ann fell in a hole and was pulling herself up when the wave reached us. She turned around and saw the thing in her face. It just swept up two houses and dumped them on her. For a nano-second I thought of trying to run back for her but it was too quick. I couldn't have saved her, I would have died with her."

Mary Ann was swept out with the wave. Her body was found later that day by another guest who had been on the same tour with Mika.

When the Cedermans ran from the wave, they met also met with a corrugated iron fence. "It was too high to jump over going straight at it so we kept running along, hoping to get to the end of it...but when I realized I couldn't see the end, we clambered over it. We were at the base of the hill then and we just went straight up on all fours...I remember digging into the dirt and grabbing at trees." Diane is grateful for the tour they had gone on with Mika the previous day. "Going up that hill the day before with Mika was kind of like a training, a preparation. We had an idea where to go."

The Cedermans ran barefoot. It seems an impossibility now. "I remember thinking at the time, *oh my God I can't run, I haven't got any shoes on!* There was just no time...but we did it."

Midway up the top of the cliff, Otele Taufua, overtook the Cedermans. He was carrying a badly injured woman over his shoulder. "It was just amazing, we're still running away and he's coming down and he passes us and goes back into the water, then he came back again with Olivia and takes off running, telling us to follow to get to the hospital."

At the top there was a harried father with two small crying children. Diane offered to help him and took the youngest, two year old Briana. Everyone made their way to the hospital. But there was still no respite. "A truck came along and people are yelling and screaming at us, *go higher, there's another one coming!* We didn't know where we were so we jumped on the back of this truck and it took us way up further inland." There was a large gathering of people there outside a church, locals and tourists. "The minister, Isaia Tiatia was there with his wife and there were a whole lot of kids all dressed up in their school uniforms. I was really thirsty, absolutely parched and I'm looking around at the grass and thinking *I'll lick the grass... what did Mika tell us about what to do if you're really thirsty?* The minister sent this little boy over with a bottle of water but I said, *I'm fine*, because I felt bad taking the children's water. But then this guy climbed up the coconut tree and chopped down all these coconuts, knocked the tops off and handed them round to everyone. So we had a nice drink. Then this girl came with round pancakes and another with ripe bananas, so we all had something to eat. It was like a church picnic."

The Cedermans still had the little girl, Brianna and her parents were somewhere back at the hospital "I told myself it would be alright because they knew the little girl was safe with us. The poor thing was thrilled to see the food. She grabbed a pancake and started eating it and when the bananas came round, she grabbed that and started eating it. I said to her *hey, you can't eat that you've got to finish your pancakes first!* And then I thought, *you fussy old bag, we've just had a tsunami and there's chaos here and you're worried about her manners!* So she crammed the whole pancake in her mouth so she could have the banana."

Diane and Brent couldn't see the Whites anywhere. Worried, they started walking back down the burning hot tar seal road towards the hospital. "Another tourist, Mike [Whittaker] came walking up the hill and he had made these shoes out of coconut husks and tied them on his feet and they were really cool. So we found us some coconut husks and ripped a shirt and bandaged our feet up so we could walk back down." It was a long hot walk as they shuffled along in their makeshift shoes. An old woman sitting under a tree by the side of the road, called to Diane to *come, sit and rest!* "I must have looked haggard at that stage, she put out a *lavalava* and I sat down and she was offering me water and food. Every time I went to get up she said, *no, no sit down, rest!* But she had no idea of what was happening down below, she told us she thought maybe a couple of people had been killed?"

The Cedermans eventually continued on to the Taufua house where they found Andrew. "But nobody knew where Mary Ann was. I said to Andrew, *we'll go and look for her.* And he said, *no* and that's when he told us that he'd seen her being swept away by the building." Shortly after, a truck pulled up with several bodies in it. Accompanied by New Zealander Joseph Walden. "He and his wife and younger son had run up together from the wave, but they didn't have their older son Josh with them and he thought Josh was lost, so he went back down to the beach to look for him and paddled out in a kayak and helped people, bringing them back in. He found Mary Ann, he just saw her foot looking out from under some rubble and he cleared it off and he recognized her straightaway. So they delivered her up to the house."

The men called to Andrew, *It's your wife, you better come and check it out.* "They had a cover over her and he just saw her arm sticking out and he didn't even need to look at her, he just knew straightaway." The trio went with the truck down to the Lalomanu Hospital. " It was sort of like a morgue, we got off the truck and it went round the back and that's when I thought, *well I better go and have a proper look* because Andrew didn't identify her*, I'll go and see where they're taking these bodies to and what's going on.* So I went round the back of the hospital and all the bodies were piled up there, I think there were about 25 of them. There were some police guys there, but quite frankly, useless. They didn't know anything about disasters, they weren't doing any tagging of the bodies or anything. I thought I needed to identify Mary Ann properly, to make sure it was her. There was a guy taking photographs of the bodies and he pulled the cover off and I could see it was her."

While Diane was standing outside the hospital, a man showed up on a motorbike and introduced himself, *I'm from the NZ embassy here, can I help you?* He was lovely. But then he asked me, *You don't have a pen and piece of paper on you do you?* I was like…*I don't think so!* there I was in an old pair of bike shorts, a cotton top, my coconut husk shoes, just a real mess… It was one of those funny things you remember …I didn't realize until much later that he had actually come to look for his family that were on holiday at Lalomanu and as soon as he heard about the tsunami, he had jumped on his bike and rushed out there to find them."

Andrew White and the Cedermans were transported to town late that afternoon. "It was quite bizarre the ride back to Apia, driving through the place and looking up at the trees. There was a car, completely smashed but the back windscreen wiper was still going furiously. There were these pockets of total devastation, areas totally flattened. One totally bizarre sight was this guy out cutting his grass using a weed eater, just cutting his grass calm as could be – and I thought, *does he not know?!* Everything before his house was destroyed, then there was his house and after his house, destruction, just in this little patch of paradise, nothing had changed, untouched. And he was cutting his grass!"

It was night when they reached Apia. The driver of the truck asked if anyone needed supplies from the pharmacy. "Some of us did because we'd lost medication so we pulled up to this pharmacy and we're a bunch of bedraggled people. The pharmacist, Matthew Mualia, comes out and he says, *I'm not going to charge you for the drugs.* Then he goes into the freezer and comes out with a case of beer for us…that was just so amazing because that was just so welcomed. We were parched and a cold beer was just, *oh wow!* It was just what everyone needed."

The Samoa Hotel Association had arranged for their group to stay at Sebrina's Lodge, a motel that was thankfully located high inland. The staff had cooked up a huge meal for everyone and there was a box of clothes to search through. "They had gathered like an op-shop box of clothes and I ended up with a t-shirt that had a big thing on the front that said, *Samoa Paradise* on it. So that's what I wore all the next day and on to the plane home."

The Cedermans spent the following day at the New Zealand embassy making the necessary arrangements to get home as quickly as possible. The wait was long and to fight the feeling of helplessness, Diane started talking to other survivors, gathering stories, names and contact details. "I thought we would all need to communicate with each other and be in contact, so I got people's emails and started putting together a list of people."

The Cedermans were on a flight home to Auckland that night. They were welcomed on the plane by Ross Vanderhelder – an Air NZ flight attendant from Raglan. He had two trolley bags prepared for them with a complete uniform ensemble. *It's freezing cold and raining in Auckland, you'll need some warm clothes and I've had the girls sort something out for you.* Diane was thrilled to change into some clean clothes. "When we arrived in Auckland, we were dressed like flight attendants. I had this very smart coat that they wear with a lovely scarf and everything and I was feeling pretty good, much better than I had before. Then I looked down at my feet, I was wearing some white rubber jandals that were much too big for me and I couldn't get the grime of my feet so they were all muddy and horrible. There I was with this very smart outfit and then you looked down and there were these dirty legs and rubber jandals!"

It was a flight they remember with gratitude. "All the Air NZ people were wonderful. They had put on a bigger aircraft and it was full, there were quite a few people that they had to wheelchair on the plane. And Ross was amazing, he made sure that we had beers, the beers kept coming. Our insurance company had organized someone to meet us at the airport and booked accommodation for us. They even had someone meet us and take us out through the back because there were media and tv crews waiting. When we got home, we were met by the Salvation Army and group counsellors and our kids were there waiting for us – anybody that could help had been mobilized for us. It was so good."

The White family had to wait twelve days before Mary Ann was returned home as there was some confusion with official identification. "There was another woman from Australia who looked a little similar to Mary Ann. I think because there hadn't been any proper tagging done and by the time the bodies got to the morgue they had been sitting outside in the sun for a long time and so they had to be exactly sure who it was. We were worried but they were just going through the procedures, making sure there was no way they would make a mistake."

It was a long wait, but they were days which Andrew White did not have to endure alone as the close-knit community of Raglan came together in a fellowship of mourning. "Everyday people were coming in to visit them, so you can imagine that time at the house, it was just one continuous crowd at Andrew's place. Then about 600 people came to her funeral, it was huge."

There are 2,700 people living in Raglan. And 790 people in Lalomanu. Two coastal 'villages' divided by an ocean, united in sorrow. Mary Ann's farewell service began with a karanga and mihi as local iwi welcomed her back. Her casket was adorned with a Maori cloak from Raglan's Poihakena Marae – and a tapa cloth from Samoa. *Live well, laugh often, love much.*

Mike and Sally Adye had spent three sun filled days relaxing at Taufua Beach Fales before the tsunami came. "The days passed slowly, leisurely, wonderfully…in the mornings we woke to the noises of the sea, the waves breaking on the reef, blue sky, a sandy beach and the thought of snorkelling in our minds. In the evenings we ate and chatted to others around us and would return to our *fale* feeling contented."

After the earthquake, the Adye's were cautious about the possibility of a tsunami. Mike was dressed to go swimming but they both put on their sandals and sat on the *fale* steps to watch the ocean. The ocean retreating from the reef was their signal to run. What does it feel like to literally 'run for your life'? With no clear destination in mind or any thought out plan to follow? What do you see? What do you hear? What will you remember months later?

"Without looking back we ran. I remember it clearly. The adrenalin inside me was huge. I ran. There had been no discussion as to where we would go or what we were running from. We just knew we had to get high. The hill was about two hundred meters behind us and we just ran. Running across the road I saw a man in a lavalava also running and shouting *Tsunami run! Tsunami run!* The adrenaline inside me was so huge that I could not speak. We ran over the road, past a Samoan meeting house, past a shower block, past a house where there was a Samoan lady standing. I couldn't speak. I couldn't even shout to her to run. I don't know what happened to her but I often wonder. I wanted to ask if she minded me running through her garden. I wanted to ask her if I could climb the hill. But all I could do was run and then I was at the hill and I just climbed and pulled myself up. I heard nothing. I didn't hear the water but now know the noise must have been deafening."

Sally and Mike climbed. Go to the mountainside they scaled and you will be amazed how it was possible for anyone to escape up its thicket covered steepness with bare hands and sandals. Stand at the base and let your head drop back as you stare up and up. See the loose earth clinging to roots and rocks. The bushes and trees that balance precariously.

"I stopped climbing when I knew I couldn't go any higher, I was looking for a tree to hang onto and it was at that point that I looked down. The water was there under me, I guess a meter below me. I saw a car floating immediately under me bonnet down. The water was dirty, swirling, muddy, scary, powerful and we knew we had to climb higher. We knew because when I saw this water it reminded me of the news coverage of the 2004 tsunami in Indonesia. I knew that the second wave could be higher than the first and we had to get higher. So we climbed. It wasn't easy as there was no track. We just pulled ourselves up. At one point I lost my footing and started sliding down, but I managed to support myself again. At parts we helped each other. When we felt we were out of danger we stopped. We looked back and saw the destruction."

Mike then took photos of the wave as it pulled out of Lalomanu – the only known pictures of the tsunami that hit Aleipata. The couple continued on up the mountainside, finding a track which took them to a road. They met with several others from their resort. "We came across a distraught Australian girl who had been caught in the wave but had managed to hang on. She was cut and scratched and terrified as she had lost her boyfriend. I comforted her and we decided at this point we should go to the hospital as we weren't sure how injured she was." The young woman had no shoes, so Sally gave her one of her sandals and together they limped along to Lalomanu hospital.

"At the hospital those that were okay sat around and talked in groups. Those that could, went to help. The Australian girl saw her boyfriend walking up the hill which was a huge relief. And then…the trucks started bringing people who hadn't survived."

Those who could move were asked to go further inland to a large, vacant house which belonged to the Taufua family. There the Adyes spent the remainder of that shattered day. Sally helping to feed the crowds of survivors, Mike assisting with gathering names and details for the Civil Defence organizer who arrived from Apia. Sally remembers well, the surrealness of meeting again with those she had just spent the previous evening with – relaxing in the tropical paradise that Samoa can be. "People at the Taufua house had lost children and other family members. We knew some of these people. We had spent the last three evenings talking to some of them, chatting, laughing, enjoying their company. How quickly everything had changed. How dramatically. I think about them often. Gary Cunliffe and his wife. We were talking the night before. They were from England like us and now living in Nelson, NZ. They lost their little boy Alfie…there was also a lady called Isobel from Brazil. She was nowhere to be seen so I assumed she was not a survivor. I had spent an evening at dinner chatting to her. She had been living at the mount on her own and I was concerned about how long it would take to inform her family…"

In the late afternoon, the Adye's found transport back to Apia. "Families with children left first. We were among the last to leave. We will never ever forget the Samoan people that day and how they had lost so much. And yet they made sure we were alright. They gave us shelter, food and drink. They organized everyone and got us safely to Apia. We heard that it was after we all left that they started their own grieving. We will always be hugely thankful for all they did for us. At such a traumatic and sad time. These wonderful people had not only lost loved ones and friends but they had also lost their homes and their livelihood."

Mike and Zana Whittaker-Scott from Auckland, NZ were on holiday at Lalomanu. A shaken Mike was caught on camera a few hours after the tsunami. *There was a big, long quake, the fales were shaking. We thought there might be a wave. We went to the office to check for warnings on the computer. Then we saw the reef…we ran back through the fales trying to get people up. Some people listened, some didn't. We ran back through pig farms and stuff. We just made it. Lots of people didn't. We saw lots of people getting sucked away.* The cameraman asked him if he had any messages for family back in New Zealand. Mike replied, *Hi Mum. I'm safe. Just. See you when I see you. Don't know how long we're going to be here for.*

Further along, the camera caught up with Daniel Rankin from Sydney, Australia. A lean man in a plaid shirt with dreadlocked hair. His family had just checked out of their beach *fale* and were in a car on their way to the airport when the tsunami came. *The first wave took us. The car hit the wall and then came back. We got out of the car with my little baby and my wife. We go up the hill really fast. There was no warning for this village. A lot of kids die. Other villages they ring the bells, but not here so a lot of children die here.* Daniel pointed to the man walking next to him. *He's got nothing. See, he's wearing my shoes. I found them in the car. I've got my things up there on the hill, I've got my passport, but not him. We're looking for his things.*

Rob Bebrouth from Brisbane, Australia was at the end of a week long holiday at Taufua Beach Fales. He was in the shower when the earth shook. Afterwards he returned to his *fale* to resume packing for the next step of his journey through the islands. He heard a child running past, screaming for people to get to higher ground but he assumed he would have enough time to quickly finish throwing the last things into his suitcase and zip it up. Within minutes, he heard "a very loud roar" and then a wall of water smashed into the *fale.* Everything crumpled around him and he was rushed inland in a swirling flood of water and debris. Struggling for air, Rob was taken out to sea and then pushed inland again with the next wave. At some point, when he knew he could not hold his breath any longer, Rob accepted he was going to drown. He relaxed, ceased fighting and breathed in saltwater. "I realized this was it, this was the end."

But it was not the end of Rob Bebrouth's story. His head popped out of the water, "I breathed, it was beautiful." He shouted for his travelling companion and was immensely relieved to hear her answer him from some distance away. He pulled himself up out of the water onto a fridge door and realized then that there was something wrong with his leg. "I looked down and saw a hole the size of a saucer about where my kneecap should have been, I saw what looked like a severed bone." People helped Rob out of the water and up the hill where he was then taken to the Taufua mountain house. He was there until 3pm when he was loaded into an ambulance along with a pregnant Samoan woman who was severely injured. En route to Apia, the ambulance had a flat tyre. Thankfully, a truck pulled over to help – it was a woman from the Australian High Commission office and she quickly organized transport for both Rob and the injured woman.

Rob was at the National Hospital for 18 hours and then airlifted home to Australia. His eldest daughter Melanie Jordan met him at Ipswich Hospital early Friday morning. "He was in an awful state, I was shocked to see him - he was covered in cuts and bruises from his chest to his feet and drifting in and out of consciousness." The nightmare for Rob was only beginning. Septicaemia had set in and he was taken to theatre for what would be the first of many operations. As they wheeled him away, Rob promised his daughter he would *be more with it* when he came out of surgery. It would be several days before she heard him speak again. Surgeons removed part of Rob's hamstring muscle but they were fighting a losing battle as the septicaemia continued marching on, killing live muscle and tissue. Rob came out of the first surgery on life support. "He was on a ventilator, he looked so frail lying there trying to hold on to life. We were told the next few days would be the hardest, that someone else of Dad's age without his good health and fitness would never have made it back to Australia alive. We just had to pray that his spirit would keep him going there and that our prayers would keep him strong." Over the next few days, more surgeries were carried out to clean out more dead tissue. "Every time we visited, I would think…*this is it, this could be the last time we see him alive.*"

Finally, Rob was taken off the ventilator. "Finally after five days I could speak to him again, he was still quite groggy and drugged up, but he smiled at us, said hello and most importantly – *I love you.* Something I was afraid I would never hear again."

For some, the journey through a tsunami lasted longer than others. "On Friday the 6th of November, exactly five weeks from the date of his admission, Dad was released from hospital, weighing a paltry 71 kilos and looking like a stick man, but he was alive and going home. Though only a few months had passed, it was hard to remember how Dad used to be – weighing 81 kilos, being very fit and going about most things in his usual 'ten tenths' fashion. There was a long road of recovery ahead of him that included many hours of physiotherapy to get his leg to the best that it could be. Just three weeks after his release, he was no longer using crutches and was making remarkable progress in regaining the use of his leg. Now he is almost able to bend his leg past 90 degrees and that will be an important milestone in his recovery. His spirit and strength is inspiring to me – I couldn't be more proud of him."

In July 2010, Rob Bebrouth - a self-confessed 'gym junkie' before AND after 29/09 - is training five days a week and doing things the doctors said he never would again. He can bend his knee at a 130 degree angle, squat with 90kg weights and is learning to run again. He writes, "My tsunami experience has not changed how I feel about the sea. I love the water and recently went body surfing at the Gold Coast. It was magic. I envy the Samoans living at

Lalomanu. It is a beautiful location. 29/09 helped me realize that every day, every moment is important. It also showed me that while I love life and living, I am not afraid to die and there are worse ways to die than drowning."

Down the road from the Taufua's is the Litia Sini Beach Fale Resort. Wendy Pearce from Christchurch, NZ was holidaying there with her friend Margaret Coleman. She shared her story with the NZ Women's Weekly magazine. Wendy and Margaret had been jolted awake by the earthquake. The *fales* have communal toilets some distance away. Wendy was in the bathroom, wearing only the *lavalava* she had been sleeping in, when she heard a little girl screaming. *Tsunami! Mum, Dad, quick, tsunami!*

Wendy threw open the bathroom door to see a wall of water coming towards her. "All I could think was, *I'm going to die.* I ran around the side of the toilet block straight in front of the cliff-face and, desperate for higher ground, I started climbing, grabbing rocks and tree roots, anything that would hold me. I can't even climb a ladder but I just kept climbing the cliff face which was at least a few storeys high. I looked behind me to see the wave crashing below. I watched *fales*, cars and people being thrown in the air."

Two of Wendy's travel companions were caught by the wave and sucked out to sea but then washed back in by the next surge. Halfway up the cliff, Wendy was relieved to meet the rest of her friends. They met with the girl who's screams had alerted them to the tsunami – Abby Wutzler. She was huddled at the top of the cliff, crying because she thought she had lost her parents. Abby's father David had gone back into the water to look for his son Max, not knowing that he had already scrambled to safety up the cliff with his sister. Wendy tried to comfort the distraught children. "We kept telling them their parents would be here in a minute…and when they did turn up it was so emotional."

For Wendy, the escape from the tsunami was frighteningly close. "Looking back now it feels like a dream. We were lucky. If I hadn't had that sudden urge for the toilet, I would probably have been on my bed, reading, when the wave hit." And if a young girl hadn't remembered a simple lesson taught by a teacher and acted quickly, then Wendy and her friends might not have made it out.

Once back in New Zealand, Abby's family was quick to call her teacher, thanking her for the lesson that had helped save their lives. Abby was given an award by the Director of the New Zealand Civil Defence Emergency Management who congratulated her, "The way you recognized the signs of the tsunami and acted as quickly as you did helped save your family and others on the beach that morning."

Another family grateful for Abby's warning are the Turton's from Dannevirke, New Zealand – Louise and her two teenagers Shaun and Sarah. It was their first visit to Samoa and they had been staying at Litia Sini's for four days before the tsunami hit. Louise writes –

"We woke about 20 minutes before the earthquake, and were just lying in the *fale* reading. Shaun had got dressed, but Sarah and I were still in PJ's. The first sign that something was about to happen was the birds. They stopped their lovely chirping, and began to screech loudly. Suddenly, they were silent – and then the shaking started. The earthquake was strong, but I have felt far worse in New Zealand. It was not really that frightening, but I was concerned when after two minutes it was not easing. At that stage I told the kids to go outside. I was more concerned about how well-built the *fales* were!

Once outside, everyone else seemed to be doing the same thing – there was absolutely no panic. The shaking stopped, and Shaun then decided to get changed and go for a swim, so he headed back into the *fale*. People were standing around talking in small groups, us included. Sarah and I had our backs to the sea…

Only a minute or so later, a young girls's yells "*tsunami - run*", made us spin around, although I must admit that my first thought was it was just kids being stupid. Thank goodness for that girl (Abby from Wellington NZ). What I saw then, I will never forget. The sea was calm and blue, but the water was receding at such a pace that you would have to have run to keep up with it. The reef was also exposed - a line of shiny black rocks that were normally always under water, with surf breaking on them. Now there was no surf.

Immediately it was obvious what was coming, but how long did we have? We had no idea. I yelled at Shaun to run. He was standing in the doorway of the *fale*, and grabbed my camera bag, which was by his feet. When I asked him later why, he thought that he would have time to video the tsunami!

The three of us, and everyone else on the beach ran for our lives across the road, past houses, to the base of an extremely steep hill which was thickly covered in bush and vines. We only had about 100m to run to get there. About halfway, I looked back, as I had lost sight of Sarah. She was right behind me, but I saw the sea, and that was when I really thought we were going to die. No longer calm and blue, it was now brown and rough looking. The horizon appeared to be rising. It didn't look like a wave, more a rising wall… I wasn't aware of any sound at that stage.

We scrambled for our lives up that hill. We must have been at least 10 metres up when I became aware of the noise. It was an increasingly loud roar

– like an approaching train. As it hit land, the sound changed, and there was crashing and splitting sounds. I was not looking, but it was obvious it was destroying the *fales*, and seconds from getting us.

The sea hit the hill, and stopped about 1 metre below us. It was too close. In the swirling, filthy water were trees, and rental cars being tossed about like bath toys. Chunks of concrete block walls were bouncing about on the surface. And there were people in the water. Most distressing was seeing a car with people inside kicking desperately at the windows.

We stopped climbing, and for a short while thought we were safe. Shaun started videoing the receding first wave. Very soon though, we realised a second wave was on its way. It was extremely frightening, and this time we never stopped climbing, or looked back until we got to the top of the hill. The bush was too thick to see much below us. Gradually, the group of survivors emerged from the bush, tourists along with the local people. Everyone was really silent, apart from a few crying children. The shock on everyone's faces was obvious.

At this stage the Samoan people took charge and led us out of the bush, and to the road. More local people came to meet us, carrying bottled water and juice to share. We were taken to the local medical clinic. Very soon, injured people(and the dead) started arriving. They were rescued by local people and brought up by utes. Some of the injuries were awful. I am a nurse, and offered to help with first aid. Of course there was very little to work with, and the best I could do was disinfect and cover open wounds. The supplies didn't last very long. It as about 3-4 hours before help arrived from Apia. It was then that I realized that Samoa has no helicopters. The only way in was by road, and of course, the medical teams had to follow bulldozers and machinery that had to make the road passable first. Unfortunately, all injured had to be transported by road, on the backs of utes and trucks, no matter how seriously injured they were. That would be unthinkable in NZ.

After the medical teams arrived, we were asked to go further up the road to the partly built Taufua family home. Once there, the local people, despite their own terrible grief, put the tourists first. I was amazed at two young Samoan men who hurriedly constructed a long-drop toilet shed, complete with an All Blacks flag for a door! Within a couple of hours the local ladies had cooked enough food for the 150 or so people at the house. They treated us so well, but at the same time it was heartbreaking to know that they all had dead or missing family members. At best they lost their homes.

The day passed in a bit of a blur. At 6pm that evening we left, and the thought of having to go back down to the coast was frightening for everyone. The damage was unbelievable. Driving through the destruction was so sad.

The worst image I have was a Samoan man approaching the road carrying the body of a young boy. I will never forget that sight. I felt (and still do) very guilty that my kids had to witness death like that. It wasn't the only body they saw that day.

Around 8.30pm we arrived in Apia, and were taken to a motel in a hilly part of town – well away from the sea. There were nine of us staying together. For the next two days we were treated wonderfully by Ricky Craig and his family at Lerichousing Motel, Vaitele. They provided us with meals and drove us into town so we could sort out our travel with the Embassy. Ricky's wife kindly bought us a new change of clothing from the market. We were still in our pyjamas, and now filthy, so it was a very kind thing to do.

We got a flight home three days later. As our plane left, I thought I would never, ever want to return to Samoa. After a few weeks though, we all decided that we want to go back. I need to see how we survived. The day before the tsunami, Shaun and I had been walking at the base of the hill, looking for a path that we thought was there. There wasn't, and I said then that you would have to be mad to try and climb it. The next day, it was the only way out. It's amazing what adrenaline can let you do.

Every day now, I am grateful to be alive, grateful that the three of us escaped unharmed. I hope that Sarah and Shaun feel the same. It could have been so different. Had the tsunami happened at night, few people would have survived."

Becky Glew arrived in Samoa on Saturday the 26th of September, 2009, hoping for a fantastic holiday with a small group of friends. Becky is from the UK but resides in New Zealand. She is a primary school teacher in Manurewa in South Auckland where many of the students are Samoan. "I had heard lots of amazing things about Samoa and was really excited to be going there." In Apia, the group rented a bright yellow car, affectionately christened, 'Buttercup' and on the Sunday, they drove out to Aleipata. "After a night of cocktails on the town..starting as we meant to go on - we headed down to the coast to the paradise that is called Lalomanu. It really was one of the most beautiful places I have ever been. We stayed in *fales* on a white sand beach not more than five meters from the water's edge which lapped against the shore and it was surrounded by a coral reef. Amazing!"

When the earthquake started, Becky's companions were more alert to the possibilities than she was. "Being a little English girl amongst the Kiwis was quite interesting. In New Zealand they have earthquake and tsunami training at school – sadly I had not yet experienced one of the drills yet at our school, if I had, I would have been much better prepared. The Kiwi girls made me get out of bed and out of the *fale* – a good plan as this also meant that I was up and sort of dressed when the next events took place. We went down to the beach and began discussing tsunamis with a family there, the Wilsons, in a very jokey way. After a few minutes, we left their son Max on the beach on 'tsunami watch' and all wandered off laughing."

It was Max's screaming that alerted the women to the ocean. "He and a few others were yelling that the water was disappearing and all panic broke loose." Again it was the New Zealanders in the group who understood the need for speed best. "Sarah threw on her trainers (shoes) and grabbed a five litre bottle of water then began running for the cliff behind our fale. Amy wrapped herself in a sheet as she was just wearing a bra and shorts and also started running…the British girl – me – spent vital minutes finding the car key, grabbing the phone and a pair of flip flops and then a few more moments trying to lock up the *fale* door! (Heaven forbid that someone try to pinch our stuff in the middle of a tsunami!) Thankfully Sara screamed at me to run so I began to follow them and scrambled up the cliff along with many other people who were running for their lives."

Midway up, Becky turned to look back. "The wave was enormous and everything crumpled in its path. It hit the cliff and the water level kept rising and rising. Thankfully for me, it stopped about a meter below my feet. It was full of debris and the thing I remember most is seeing our little car Buttercup floating around in a big spin and bashing into the cliff I had just climbed up!"

Sarah and Amy were yelling for Becky to climb higher because as they well knew, there would be more waves, "I began to scramble up the cliff, holding onto rocks and tree roots and all the time I was thinking how hard it would be to climb back down. As I went up higher, the second wave hit and it washed away where I had been standing before."

The women joined the exodus of people clambering to the top. "We carried children and pulled up people who were stuck and couldn't get any higher…eventually we realized we were at the top and everyone sat around in shock hugging each other and taking care of the kids whose parents had gone back down to look for missing people."

Becky still had her cell phone. "I sent a couple of rather random texts to family and friends to let them know I was safe. I think some of my texts arrived before the news hit the rest of the world though so they were confused a bit at first! Then everyone else started to use my phone to let their families know they were safe too. By the end of the day, I was known as *Becky with the phone."*

Becky's group was directed to the Lalomanu Hospital by Joe, a young Samoan man they had met briefly the day before. "He came into the forest and led us all out to the hospital. That was pretty traumatic as it was a tiny, ill-equipped place and suddenly hundreds of injured people began to arrive. Many of the dead were brought there too." The women moved on to the Taufua house where they spent the remainder of the day. "A wonderful guy called Mika who had UN disaster training soon came and started to organize us…the first food we were given was dry weet-bix. I used to think that it was impossible to eat these without milk and especially hard to eat them when you are parched – but it turns out you can do anything when you are really hungry and you don't know when you will be given food again!" Becky has nothing but praise for the medical team that came to the mountain refuge and for the local people who cared for them. "The team of four doctors and nurses were real miracle workers, they had barely any equipment but they managed to keep some very sick people alive and comfortable. Many Samoans were there bringing us food, climbing palm trees to get us coconuts, looking after the wounded and even building us a toilet outside. And every single one of them was either still waiting for news of family members or already knew they had lost someone. Joe, for instance, had found his dead mother moments before coming into the forest to rescue us and he spent the whole day lifting the sick, running back and forth to the hospital, cooking for us and helping anyone who needed it. They were all amazing."

The priority for all available transport was taking the injured to Apia, so Becky and the others did not head back to town until late that afternoon – after a second tsunami warning had sent many people running back up the hill to the Taufua house. "We got into a complete stranger's car. I had to sit in the hatchback boot which was very scary as it meant I had no way of opening the door and getting out in a hurry. The driver was a lovely, kind man who had spent all day driving the 90 minute drive between the city and the mountains to evacuate all the tourists and the wounded. With great trepidation we drove down the mountain and onto the road next to the sea…mile after mile of flattened houses and villages. We quickly realized how lucky we were to have had a mountain directly behind us as in most areas, it was a good mile or so to higher ground. It was a very sobering drive."

In Apia, the group was taken to Sebrina's Lodge for accommodation where they met up with Buttercup's owners. "They were great. Instead of yelling at us for trashing their one and only car in a tsunami when they had no insurance, they hugged us and bought us beer to help us sleep!" Becky's small offer of comfort was to be able to give them the car keys she had snatched up before her mad dash for safety. The next day, the owners went out to Lalomanu, dug out Buttercup from a tangled mass of debris, put her on a trailer and brought her back home to town. Somewhere in Apia now, there is a little car called Buttercup - somewhat dented, somewhat worn for wear, but still chugging along, cheerily, eagerly - sunshine yellow.

After several days of frustration dealing with bureaucracy, with no money, no belongings and only the clothes she ran away in, Becky finally made it onto a plane home to Auckland with her friends, "relieved but shaken and jittery." For Becky, the tsunami was a pivotal, life-changing experience. "Following some counselling, my boyfriend and I went to the UK to help me put myself back together by seeing family and friends. I also went to put their minds at rest that I was okay. Whilst there, we became pregnant…this is a direct result of the tsunami. It is something we had talked about but I would put it off as 'the time wasn't right'. However, the tsunami made me realize that life is for living and there is not a minute to spare…we are expecting our first child on the 30th of July, 2010. The pregnancy has also made me look to the future instead of dwelling on the horrors of that day and it has truly helped me to move on. I still have fears and nightmares but my recovery from the trauma has been much swifter – it will truly be a magical baby when it's born."

School teachers, Harriet Sherratt and Rebecca Hamilton had just arrived at Litia Sini's at one o'clock that tsunami morning. It was their first trip to Samoa. Exhausted from their late night flight, they had gone straight to sleep in their beach *fale*, barely even glancing at their moonlit surroundings. When the earthquake hit, they were still asleep. "We woke up but were in a bit of a haze and just thought it must have been strong winds making the house move. But when it kept shaking we got up and went ouside. The ground was still shaking while we had our first look at the beach. We both commented how beautiful it was and how we couldn't wait to snorkel the reef. But since we were both still very tired, we thought nothing of tsunami and went back to bed."

The women had only just fallen asleep when loud shouting woke them. *Tsunami! Wake up, run, tsunami!* It was the voice of a young boy. "This woke us up and this is when we realized it was no joke, we had to move. I managed to get my handbag and Hattie grabbed only her glasses, thank God!" Bleary eyed, the women stumbled outside looking around wildly for an escape route. They saw people running towards the cliff across the road behind the resort and so they began following them. "We had no idea of our surroundings since we'd only been there a few hours…it was almost like a dream, one minute you're lying in your *fale* sound asleep and the next you're running for your life. It all happened so fast. Climbing up the cliff it was pure fear and adrenaline running through your body, making you move so quick."

At the top of the cliff, the women were shaken as they realized just how close they had come to being swept away in their *fale*, still asleep. They wanted to know, who was the boy that had called out a warning? Who had saved them?

The Wilson family from Auckland, NZ were staying at Litia Sini Resort. Forty-five year old Christopher and his two children, Max and Poppy. When the earthquake stopped shuddering, Christopher was on guard. He told the children to stay close to him. Poppy wanted to go to the bathroom. "Dad told me to go but to come out if I heard him calling, no matter what. I went to the toilet block and locked the door behind me. Then I heard screams and my dad yelling, *Poppy, Poppy!* I bolted out of the toilet. I was very frightened because of the tone of my Dad's voice and I could also hear other people screaming They were yelling, *Run, tsunami!"*

Christopher had ordered Max to run for the hill and not to look back, as he went to get Poppy. Max rushed to warn people before sprinting to the mountain. Poppy ran from the bathroom block to her father. "Dad said *Run in front of me.* He pointed to the cliff behind the block. He was running right behind me yelling, *GO, GO!* Every time he said go, I ran even faster. Max was already well ahead of us. We ran over a rubbish pile. I dropped my beautiful fan that I had just bought the day before for five *tala*. We started to climb the cliff using the roots and branches of the trees. Maxwell was climbing in front of me, my dad was behind us both, yelling for us not to look back."

As he ran, Christopher was overwhelmingly relieved that both his children had listened to his instructions after the quake, that both were so quick to obey. "I couldn't believe my luck having them directly in front of me and running fast towards the cliff...out of the corner of my eye I caught sight of some pigs scrambling across the road and making for the hill as well. Clever, I thought."

The trio got about fifteen meters up the hill when Poppy struggled. She couldn't reach the next tree root and she screamed for help. "Dad, I can't go any further!"

Christopher called for Maxwell to grab his sister's hand and pull while he pushed her feet from underneath. Maxwell cried out, "If I'm going to die, at least I'm going to die with you guys."

Christopher refused to listen, "Don't be so stupid, just GO!"

The first wave hit the cliff, bringing with it Litia Sini Beach *fales*. And everything else. The sound was immense and it spurred them on. "I heard the roar of the water and the noise of power poles and trees snapping. I climbed even faster. The wave reached up the cliff to where we were."

Poppy looked back. "A yellow car was tossed up against the cliff. At that point I became very frightened. I climbed so fast that I can't remember how I got to the top. At the top we were alone. I was bawling my eyes out. I saw movement in the bushes and I thought it was other people but a stupid pig appeared instead."

Safe at the top, Max looked down. "I saw people grabbing onto anything they could, trying desperately to get out of reach of the wall of water. People were screaming in terror. I saw lots of people getting sucked out."

Christopher and his children were the first ones to reach that point of the cliff but within minutes others joined them. A little girl, and three boys – all frightened and crying. Then two more children, Samoan boys this time. Christopher was now the sole adult in charge of a group of terrified youngsters. "Things were happening fast. I realized I had five children plus my own in this high stress environment and they all wanted their mums and dads. One of the Samoan boys was screaming desperately for his mother. He was beside himself and wanted to go down the cliff to find her. He ran past me and I pulled him to a halt. I told him, no it was too dangerous. Then I turned my attention back to the two little girls who were crying."

While Christopher was diverted, the boy disappeared over the cliff. Max went to chase after him but his dad yelled at him to stop. They never saw the little boy again. Thankfully adults were next up the cliff – the schoolteachers from Gisborne, two young women in their twenties, Harriet Sherratt and Rebecca Hamilton. They wanted to know who had woken them with the tsunami warning? When they met Max, they hugged the slight, dark haired twelve year old. *You saved our lives Max, you did. You got us out of bed, thanks so much. You're a hero.* Christopher asked the women to look after the children while he went back down to search for some of their friends.

"I told Max and Poppy to stay put, don't move, I will be back, I promise. Poppy looked horrified and I could tell she didn't want me going anywhere. I gave her a quick hug and shot down to the edge of the cliff…with any luck, I thought I might be able to catch the Samoan boy before he went too far. But he was gone."

The second wave was pulling out when Christopher began his careful descent, grasping trees and roots for security. There was nothing left recognizable. Lalomanu was now a swirling mass of muddy brown water and assorted wreckage. Christopher saw people clinging to trees at the base of the cliff far below, screaming for help.

"I started calling for our friends Jenny and Sophie. I called four or five times as I made my way across the cliff holding onto trees in case I lost my footing. I was beginning to panic. Then I heard the muffled voice of Sophie answering me back, *We're over here!* I looked over the side of the tree roots where Sophie was and there to my relief was Jenny within arms reach. I grabbed her arm and pulled her up over the tree roots. They both seemed fine so I rushed back to where I had left the kids and they followed me."

Back at the cliff top, several parents had now joined their children. People were hot and bothered, cut and scraped from their mountainside scramble. Jenny carefully measured out capfuls of water for everyone from a bottle she had been carrying during her dash to safety. Unlike everyone else, she also had a handbag. With a cell phone in it. Eagerly everyone clustered round to send text messages to their families, telling them they were alive. '*Samoa hit by tsunami. We are fine. Samoa is not.*' The time was 7:15am.

The crowd grew as more survivors joined them. Some inspiring disbelief from Christopher at their narrow escape. "A naked man appeared over the ridge. He was cut all over his body and I was surprised he could even walk. It was Scott the American artist. I hadn't realized that his wife and children were already with us. Boy, were they pleased to see him." Maxwell adds, "Scott had been in bed when the tsunami hit so he was very lucky to be alive. His family was so happy to see him, hugging and kissing him."

A discussion ensued about what to do next. "Lots of the adults wanted to go back down the cliff. I said that wasn't such a good idea to do that just yet. But some of them were quite insistent in wanting to go down. I shouted, *Don't be fools – we don't know what's happened down there. We don't know if there's going to be more waves. There could be aftershocks. Just don't go down if you value your life."* Everyone stopped when I said that. I thanked what little I knew of God at that point that everyone had the sense to stay put." The time was now 7:40am.

At that point, the group of mainly overseas survivors had no idea how extensive the tsunami had been, whether or not the capital Apia had been hit, or if the airport still existed. For many, the reminder of the Boxing Day tsunami in Southeast Asia was the only example to go by.

Christopher addressed the group, "I told them that this is a real survival situation. What happened on our part of the beach has most likely happened to at least half of Samoa, if not all. We are in the middle of nowhere and can expect very little help for some time. The Samoans would have lost everything. We should stick together as a group and be prepared for at least one night if not two or three in this situation."

With that sobering directive, the group decided to make their way towards an inland church they had seen the previous day on a sightseeing excursion. It was not an easy journey. Most of them were barefoot, some missing items of clothing. The thick bush became taro and banana plantation. A Samoan man came running past. He told them they should go to the district hospital and gave them directions, through the trees, to the road and up past the school.

The hospital seemed deserted with the hum of a generator the only sound in the sundrenched morning. Christopher remembers seeing one nurse on site. Everyone took refuge in the shade to wait. For help. For information. For instructions. Jenny found some stale biscuits in her bag for the children. "It's amazing what you can find in a woman's handbag!" commented Christopher. Some women from the hospital gave them water which they used to clean the worst of the cuts and bruises. The respite was short however. Because within a few minutes, the hospital was transformed into a scene resembling a war zone.

"A ute (truck) turned up at the front of the hospital. They were calling for people to help. I went to offer my help and on the back of the ute were two very badly beaten up bodies, deep gashes that went to the bone. One's leg was pointing the wrong way. Their faces were gouged and cut. They were both making gurgling noises. I thought, *how the hell are we going to help them?* Then another ute turned up with more wounded, then another. The nurse said, *forget about the first two, there's nothing we can do. Help the others.* At that moment, one of them stopped gurgling…five minutes later the other one died as well."

As more dead and injured arrived at the hospital, it ceased to be a haven for the already traumatized children. Poppy was cleaning her scrapes with wet cloths in the shade when a woman got off the third truck carrying a very still shape in her arms. The yard was filled with her wailing. " I saw a woman crying over her dead baby. The hospital got really busy and frightening." Max added, "A couple of minutes after we got there the wounded came in. People died in front of us. I was in shock, feeling queasy in the stomach. The hospital was getting a bit crazy."

Christopher decided to take the family elsewhere. People were going to a half-built house further up a gravel road that belonged to the Taufua family. It sat on the hillside, overlooking the coast and was missing doors, windows and walls. There were piles of timber in a corner and tools – waiting for the carpenters who never came. Max clearly recalls that walk and the remainder of the afternoon. "The road was extremely rocky and it hurt our bare feet. By the time we got to the house, our feet were cut. Our friend Jenny had a piece of stick stuck in her foot. There was no one at the house when we got there but then people started arriving. We made make-shift tables for the wounded out of building supplies. People were crying in pain. One lady had a broken back. I was beginning to feel more stressed. We stayed in that house about four to five hours."

Poppy describes it, "We sat in the house and got some water to drink. After I drank the water, I vomited. Strangely I felt much better after that. We cleared an area for us to lie down as we believed that we would be staying the night. Then some Samoans turned up with woven mats and pillows to make us more comfortable. In the house there were two families who had lost their children. Then a van came with cooking rings and food. They set up a cooking area and prepared us food to eat."

Christopher chose a spot for the children to rest that was as far away as possible from the seriously wounded. His account of the afternoon is vivid - "In the main room we had people with compound fractures. There was a husband and wife – he had a punctured lung and the wife a badly gouged eye. They had lost their only daughter, she slipped out of her mothers arms when the wave caught them. They lay down in the corner…we covered them with tarpaulin… their loss was almost too big for all of us. There was another family who had two boys and they had lost their daughter as well. The father just sat cross-legged on the floor with his head down. The mother would get up occasionally and walk around. I went and held his hand for a couple of minutes when his wife was pacing. There was a Samoan woman with a suspected broken back. She had been travelling on the back of a ute with twelve school kids when the wave hit. Only two of the children came up after the ute stopped flipping. This woman was amazing. When I spoke to her, she would break out in a big smile. Amazing."

People continued to arrive at the makeshift sanctuary. There was painful silence as people gave their details a UN representative. "When the woman with the two boys, who had lost her six year old daughter, got up to register the names of her family, I noticed that her husband remained cross-legged on the floor, still too stunned to move. I jumped up and wrapped my arm around her – I knew this was going to be really hard for her. I managed to contain my emotions as she listed the names of her family that were there with us and then calmly said her daughter's name and then announced her, *missing.* Then she collapsed. I helped her back to where her husband was still sitting and put her beside him. I sat down in front of the husband and held his hands. He looked up. I tried a smile. He tried a smile. I put his hand in his wife's hand and they both broke down, holding each other. I had to leave."

Christopher walked quickly out the back of the house into the hot sun. Sadness choked the clear afternoon. "I walked towards the edge of the cliff. The ground was hard with sharp stones embedded in it which made for hard walking. When I got there, my tears finally stopped and I stood in awe of the view. You could see at least a mile both ways up the coast where there used to be *fale* resorts. They were all gone and from that height it just looked like a deserted beach."

Amy Purcell with one of her grandchildren in their shelter at Malaela village a month after the tsunami.
Photo - Lani Young

A home in Saleaumua completely wiped out.
Photo - Pele Wendt

Lee Palupe stands beside the tree that saved her life, Saleaumua. *Photo - Lani Young*

The Mikaele family – Ave, Mika, Iulai and Phillip.
Photo - Lani Young

Oselani the fisherman pauses in his work weaving coconut frond blinds for his shelter in the bush at Saleaumua village.
Photo - Lani Young

One of the oldest survivors of the tsunami, 92 yr old Suliane with her daughter Faletaulupe and grt grandchild at Saleaumua village. *Photo - Lani Young*

Ian Toomalatai with his daughter Sesa in Vailoa village.
Photo - Lani Young

Jared and Anzonetta Schwalger. *Photo - Lani Young*

Teancum and Abish Schwalger. *Photo - Lucia Hellesoe*

Jerry and Metita Schwalger. *Photo - Lucia Hellesoe*

The twisted remains of the Schwalger home in Malaela village. *Photo - Lani Young*

Sally Taiva'asele of Malaela village. *Photo - Lani Young*

Lesa Foti sits in his temporary shelter at Lalomanu. *Photo - Lani Young*

Rita Romeo of Lalomanu with two of her children. *Photo - Lani Young*

Lalomanu village looking like an abandoned rubbish dump. *Photo - Pele Wendt*

The remains of Lalomanu village five days after 29/09. At the rear, the hills survivors had to scramble up to safety. *Photo - Australian Defence Force*

Sera Adye relaxes at Taufua Beach Fales, Lalomanu a day before the tsunami. *Photo - Mike Adye*

Sina Ale kisses baby Jai Trafford during his stay at Taufua Beach Fales, Lalomanu. *Photo - Sarah Trafford*

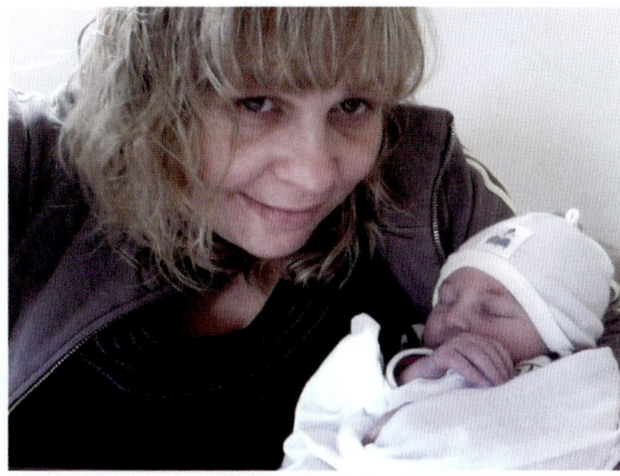
New mum Becky Glew with baby Martha born 4th Aug, 2010. *Photo - Becky Glew*

Mary Ann and Andrew White, Raglan, NZ. *Photo - Andrew White*

Injured take refuge at the Taufua mountain house after the tsunami. *Photo - Mike Adye*

Vivien Hodgins with her daughter in Australia. *Photo - Stephanie Hodgins-May*

The people of Lalomanu prepare food for survivors at the Taufua mountain house after the tsunami. *Photo - Mike Adye*

Rob Bebrouth enjoys a lively dinner at the Taufua Restaurant a few days before the tsunami. *Photo - Rob Bebrouth*

Rob Bebrouth is evacuated by the RAAF. *Photo - Australian Defence Force*

Rob with his children after being discharged from hospital a month later. *Photo - Rob Bebrouth*

Max and Poppy Wilson at the Taufua house after the tsunami. *Photo - Christopher Wilson*

Stunned survivors are driven to Apia in the late afternoon on 29/09. *Photo - Mike Adye*

Otele Taufua meets Olive Loeffen again in Raglan, NZ. Dec 2009. *Photo - Tina DeSuza*

Some of the Taufua family in Raglan, NZ in Dec 2009. *Photo - Tina DeSuza*

A blue van pulled up after one o'clock. "Out piled a bunch of Samoan men and women dressed in hospital smocks. Thank God, doctors at long last." There was bustling efficiency. "They wasted no time at all. They rushed into the house and quickly took control of the wounded. They had boxes of medical supplies and nurses were carrying out the doctors commands, running backwards and forwards with needles and bandages."

A woman started screaming, *Doctor! Doctor, quick!* The man with the punctured lung was convulsing, choking on frothy red bubbles. "A doctor ran over to him and shouted orders to a nurse who quickly went to one of the boxes and got what looked like a packet of injections. The doctor immediately stuck one into the man's chest."

The medical team worked through the rest of the day, checking the injured and sending them to the National Hospital once they were stabilized. With the serious cases out of the way, they then tended to the cuts and scratches, giving tetanus injections. An announcement was made that cars were coming to take all the tourists into Apia. As they prepared for the long ride into town, Christopher gathered his children close. "I told them, *there are families amongst us that have lost children today…I'm so thankful that both of you were so good at acting fast without questions. That is definitely what saved our lives today. We made it. We are alive. Be very proud of yourselves – for listening, for running, for being brave. You both did well, very well, the best there is.*"

The Wilsons flew back to New Zealand the next day. On the 16th of October, Poppy spoke to her Coatesville School, telling them the story of their escape from a tsunami and concluding, "*We left Samoa with just our shorts and t.shirts. We had lost everything else to the tsunami. The Samoans of Litia Sini at Lalomanu lost everything including family and friends, but they still made us their first priority. They are truly kind and loving people.*

A week later, Max shared his tsunami story at a school assembly at Albany Junior High. Both he and Poppy were given special awards for their bravery – the NZ Police 'Area Commander's Commendation for Bravery by the Waitemata District and Rodney Area Police.' The NZ Civil Defence also presented Max with another award, recognizing his bravery and quick actions that helped save lives on a Samoan beach far away.

Rebecca and Harriet are unfortunate enough to have the rare distinction of being people who have run for their lives from a tsunami – twice. The pair was among the last to leave the mountaintop refuge that afternoon. "We got a ride back to town with two men from the Civil Defence and another family of four survivors from Auckland. We had to drive back down where we had come from, past all the destruction. It was like a war zone."

Their truck was sitting in a long line of traffic, not far from where their resort had been when their new nightmare began. A truck was driving towards them with eight men piled on the back. The men stood up, began pointing to the sea and yelling before they all jumped off and started running for the cliff. The driver took off after them. "This was the scariest moment of my life. We couldn't see the sea so we didn't know if there was a tsunami coming but we all hopped out of the truck and started running in fear for the second time towards the cliff. We got to a swamp that the tsunami had made and Harriet and I both stopped. It looked deep and dangerous. It was filled with corrugated iron, wood, dead dogs and rubbish. We knew our only option was to go through it, we were both screaming and we both kept looking behind us to see if another wave was coming. *What should we do?!* I just wanted someone to come and pick me up and take me away from it all. I was so scared. We both started to run through the swamp, it got deeper and deeper. The water was up to our armpits and we had to push logs out of the way. Then we had to climb back up that hill all over again. Exhausted and terrified we didn't stop."

For the women, this second dash for safety was far more horrifying than the first. "Even though no second tsunami came, I think the second time we had to run up the cliff was the worst because we knew what to expect, we knew the sounds and the force of the wave. We knew what it could do. We got to the top of the cliff and back where we had started, at the building where we had stayed for that day. Cold and wet with the sun setting, we hugged each other and cried. With most of the tourists already gone back to Apia, we had to decide if we were going to try again to get to town or stay the night there."

Night was creeping in and bodies were still being brought in to the house. It was with relief that Rebecca and Harriet were able to get a ride into town in another Civil Defence vehicle. "We were taken to a motel in Apia where the most gorgeous Samoan family that owned it looked after us and treated us like family. They fed us and even gave us clothes to wear. I will never forget these people and still keep in touch with them."

Harriet and Rebecca were on a flight home the next night. "It was an emotional moment when the plane touched down. There was relief that we were on our way home and sadness as the doors opened. We saw many Samoans at the airport preparing to get on the return plane, going to help their families in Samoa with the devastation. A lot of them were going to grieve for their lost family and friends that had died. It was very sad."

There are others that Rebecca will never forget, including Max. "He saved our lives. The Wilson family has been to stay with me in Gisborne on our sheep station. It was nice to be able to spoil him and his family on the farm." And Rebecca still hopes that one day, she will be able to finish the holiday that was interrupted. "I want to go back to Samoa and visit that special place, to have that holiday that got washed away."

Ngaire Woodward from Alexandra, NZ had come to Samoa with friends Jill Barron and Matiu Northcroft to celebrate her sixtieth birthday. They had been in the country a week and Monday was their first night at Litia Sini Beach *fales*, Lalomanu. They had a little wooden hut right on the sand, meters away from the water where they could watch the moonlight patterns on the silken sea. There was a palm tree beside their *fale,* close enough that you could reach out and touch the coconuts. "We were so looking forward to our three lazy days on the beach in the sun. We had books to read, some alcohol to drink and plenty of time to swim and just veg out." The dining room was crowded that night, it was school holidays in New Zealand and many families were at Lalomanu with their children. There was no lighting on the beach and Ngaire and her friends walked back to their *fale* in careful darkness.

The group was woken early the next morning by the earthquake. Afterwards, Jill and Matiu were planning on having a swim when Ngaire noticed what she thought was just a big swell in the ocean. She called to the others, *Hey, this wave is going to come up under our fale.*

The next instant, all three of them were slammed through the *fale* by a great wall of water. Ngaire explained, "None of us had seen it coming, it was just a wave that seemed a bit bigger than usual. We were smashed through our *fale* and the next one. It was then that our survival mode kicked in. I don't know how many times I was smashed back down under but believe it would have been 5 or 6 times. The water was really cloudy with the sand and it was as if it was being churned by a large washing machine only there was nothing soft in the water. We were hit by poles, iron, suitcases and a variety of tables, chairs, pool tables, pots, pans coral off the reef, lava rock, cars, trees and concrete. The only way you could tell which way was up was the colour of the water. You swam up to the light and took another breath before being pushed back down again. You had to push away the debris that was hitting you all the time when you had the chance to do so. There was a little green car with people in it, just floating with the tidal movements. They were lucky to get out as they could have been pushed out over the reef. One time when I came up for air my legs were trapped between two large pieces of timber. I knew I had to get my legs out or I would be trapped by the next wave and would drown. I forced my legs apart to get them free. The next time I went under I was under for a long time and had just started to take a breath under water – when it dawned on me that if I continued with this breath that I would drown. I pushed up to the surface and there was a palm tree in front of me. I grabbed this as the wave receded and then was able to look around me. I was about 3 to 4 metres away from a cliff. I had not known it was there previously and there were people up the cliff. Matiu was going slowly up the hill and the people up there were yelling for me to get up the hill in case of another wave hitting. When I looked around, Jill was a short distance behind me and she was also urging me to get up the hill. I was terrified to let go of my coconut palm as that was safety for me. I knew they were right so I tried to walk but this was really painful, so I crawled over all the muck and debris and grabbed anything that I could pull myself up on. Jill was coming up behind me and was pushing me up the hill. Jill's dress was shredded and the force of the water had pulled her knickers off. I still had my knickers on but was near enough to topless. We must have been a bit of a giggle – big boobs and backsides trying to climb a hill!"

Jill added, "When we got to the top of the hill, a lady came up to me with a sheet she had found and gave it to me to cover myself although truly, any modesty had gone out to sea!"

The trio went up the hill as far as the could where they paused. Matiu started vomiting up all the seawater he had ingested. Jill seemed to be alright and their concern was for Ngaire. Her pelvis was broken in four places and it seems inconceivable that she was able to make it up the mountainside. Or down again. People came to help them. New Zealanders Richard and Pip Bullock, Samoan boys Max and Bat. They tried to lift Ngaire, "but this only lasted a short time as I am not a small person. So they helped me walk over debris and muck, concrete, trees, timber and glass." There were many "funny looking fish" scattered all throughout the mess. A young Samoan girl offered Ngaire her jandals but "I explained I would be far better off without them."

At the bottom Jill was thrilled to see her red suitcase, "a sheer stroke of luck!" She quickly grabbed a few essential items from it before continuing on her way to Lalomanu Hospital with the others. There was a hill to climb to get there. Max and Bat ran off and came back with a rickety wheelchair they had found and started pushing Ngaire up the hill. When the front wheels broke off on the debris, the young men tilted the chair and continued wheeling her up the hill. People passed them carrying the body of an elderly man in a woven mat. The boys said it was their family chief, *matai* and they went to help. Richard and Pip continued with the wheelchair pushing.

At the Lalomanu hospital, there was little relief to be found. "It was quite a small outpost, they were not geared up for a disaster like this and the equipment and materials they did have were soon used up. There were not enough nurses to go around and there were a lot people hurting much worse than I was. I lay down on a bed outside in the sun and stayed there." The toilets were not working as there was no running water. The power was out. Everyone and everything was filthy. Jill got a bottle of antiseptic spray and went around spraying peoples cuts." Another Kiwi survivor who had been scouring the wreckage for his belongings came by. "He said, *you know I've found all these bank cards, they were just in a wee circle in the sand and they all belong to the same person, Ngaire Woodward,* and I said, *That's me!*" Someone else found Ngaire's mud stained passport.

Another tsunami warning was announced and everyone was told they needed to move still higher inland. People that could walk started moving. Ngaire was put into a truck and taken to the Taufua mountain house, where she lay face down on mats on the floor, every movement an agony. A medical team arrived from Apia and a local nurse called Elenor looked after Ngaire. There was some panic when everyone was told they were still not high enough and another tsunami was coming. "There were a lot of people who left to go higher but others like myself made the decision, that if the wave got up that high, then that was the end. I couldn't have fought off another wave." Ngaire did not get transported to the National Hospital until late that afternoon. Richard and Pip took her in their brother-in-law's car. Ngaire's 'angel' nurse Elenor went with her.

At the National Hospital, Ngaire met paramedic Steve Williams. He introduced himself, *…and I'm Australian.* Ngaire replied, *Well, I won't hold that against you!* "Steve became one of the most important people in my life after that as he was always there to help if I wanted to talk to anyone. He was able to move around the hospital with ease and could tell us what was happening. He even contacted my family for me." Ngaire eventually made it to a hospital bed in the tsunami ward, sharing a room with Helen, another Kiwi survivor. Elenor continued with her care of Ngaire – bringing her clean clothes to wear and helping her to get clean. "She was wonderful, she washed my hair for me and nothing had ever felt so good." Locals came to the hospital with food and toiletries. "They brought us soap, toothpaste and toothbrushes, there was none of those things in the hospital and they were not prepared for the large number of injured." Carly Flynn from the NZ media interviewed Ngaire and so her children were able to get their first look at their mum, battered and bruised – but alive. They were waiting for her when she was airlifted back to New Zealand three days later. "I think I looked pretty flash when we were airlifted – a lovely lady Marita Wendt had brought me a purple *lavalava* and white cotton top and a beautiful bright bag to carry all my bits and pieces in." Home in New Zealand, Ngaire was in Dunstan Hospital for four weeks and didn't go back to work for four months. "Three months after everything else was healing, I had to have a lump taken out of my back which had grown from whatever was in the water. This wound was really quite large and very deep but it healed very quickly once the mucky stuff was cleaned out."

When Jill arrived at the National Hospital a small, crescent shaped cut on her leg was cleaned and stitched closed. She was then taken to the NZ High Commissioner's residence with other survivors to spend the night. It was lovely to be able to wash off all the tsunami filth and sleep in cool, clean surroundings. But by the next day, Jill was feverish and her leg swollen. She was taken back to the hospital where the stitches were removed but the infection was running rampant by then. Not only that, but she was having trouble breathing thanks to all the septic water she had aspirated. The tsunami ward was full and Jill was put into a room far away from the other New Zealand survivors. While grateful for the care she received at the National Hospital, Jill remembers her sense of isolation and the agitated state she was in the day everyone was getting airlifted out. From her window she could see Ngaire and the others as they were stretchered to the waiting helicopter. "I got really upset because nobody knew where I was, I felt like I could be left behind! Then a RNZAF nurse came in to set me up with oxygen and when he saw how upset I was, he said he would leave his gear with so that I would know that they would be back to get me, they wouldn't forget me. I was so relieved…"

Back in New Zealand, Jill needed three operations on her leg but "time is a wonderful healer" and she has full mobility again. Both women will be back in Samoa for the one year anniversary of 29/09. From Alexandra, Ngaire writes, "We are hoping to catch up with some of the lovely people who helped us during this really stressful time…I have jade necklaces for the young men and women who took care of me, I only hope I will be able to find them. Without their help, we wouldn't be as well as we are today. God bless you all." Jill agrees, "Samoa is such a beautiful island, I want to go back and see all the wonderful people who helped us."

In December 2009, the Cedermans and Andrew White hosted a gathering of tsunami survivors in Raglan. There were survivors from Germany, Denmark, Wales and the UK. They came from Wellington, Gisborne, Christchurch and Auckland. They came from Samoa. Sili Apelu and ten of his Taufua family were there. Tina DeSuza was there. There was good food, laughter, tears and fellowship. Survivors were reunited with rescuers. Olivia Loeffen, her leg in a cast from her sixth round of operations, was able to meet Otele, the young man who had carried her from the waves, up a mountain and to a hospital. People who had last seen each other running across a once-idyllic beach with a tortured sea in pursuit – embraced and shared stories of survival and heartache. Of recovery and hope. Many made plans for their return to Samoa on the anniversary of 29/09.

9
We have to pray.

Past Lalomanu is Saleapaga village. You begin to get a sense of how badly the tsunami affected the village of Saleapaga, when you drive inland to visit them in their new location. The old Saleapaga was at the foot of a steep rocky mountain. Just hugging a narrow shore. There were about five hundred people living in blended extended family fashion on about 800 meters of sandy coastline. Close? Yes. But crowded? That would depend on your perspective. If you wanted to share a bathroom with all your cousins. And a cookhouse with all your in-laws. And a Sunday meal with your grandchildren. A washing line draped between your house and your neighbours. Dogs, pigs and chickens with everybody. Then you would have been happy here. In the old Saleapaga. Where the high tide tickled the stilts of beach *fale*. And nobody minded.

To reach the new Saleapaga you have to drive along the coast until you meet a narrow road that curves up the cliff. There's only room for one vehicle, coming OR going. Once safely at the top, you then have to drive. And drive. And drive. Further up and in. Through ditches. And holes too big to be related to pots. You pass ramshackle huts and taro plantations. And EPC workers sweating to bring power to the masses. And water trucks lugging water where there is none. You see cattle grazing that look too wild to be destined for the grocery store meat aisle. Still the road keeps going. Until you are about to take flight. And when you pause to look back, the whole of the southern coast stretches out before you. Like you could reach out and trace the cloudy swirls on the horizon. And the air has a mountain coolness to it. Here. Where you can no longer hear the ocean. This is where the new Saleapaga is. You stand there and you realize. There is no way a tsunami is EVER going to catch Saleapaga up here.

What happened to the old Saleapaga?

Ask Molesi and Silaumea Pritchard. Molesi is a man of many talents. An electrician who has been busy since the tsunami, bringing light for his neighbours in their flimsy huts and tents. He's a baker who would support his four children with sweet baked treats and hot bread. A musician and composer. Molesi cannot read but Silaumea writes down all his songs. Including the new song he has written about the waves in their villages. Which is playing on the radio the day we come to visit.

After the earthquake ended, Molesi told Silaumea that they should go up the hill because there was going to be a tsunami. "My wife said I just want to be God and know everything all the time. But I replied that God doesn't put anything bad in people's hearts, he only sends good things and we should listen."

Molesi scooped up his youngest daughter and started walking to the mountainside. He was midway up the steep cliff on one of many rough tracks when the waves came. He saw his wife and children washed away. He called the emergency number for help and they told him to get to higher ground, that help was on its way." I said, please my wife has died. I couldn't see her, I couldn't hear her so I thought she had died. Then I called the Red Cross number and asked them please we need help. My wife is dead and a lot of people in our village are dead. All our houses are gone."

Molesi left his daughter at the top and came down to look for his wife and children. Thankfully she and the children were alive. Injured, but alive. He took them up to the hill where the rest of the village had gathered. On the fifth day after the tsunami, Molesi wrote the first song – about his wife. When he thought she was dead and when she was found. About his village of Saleapaga and how the sea washed it away.

Ask Fineaso what happened. A young man with a welcoming grin who looks like he could step into the pages of a tropical paradise calendar peopled with Polynesian runway models. Fineaso has much to smile about. He's a newly wed and the father of a miracle child. His son Narineaso was born two days before the tsunami and the New Zealand media called him – "the youngest baby to survive Samoa's devastating Pacific Tsunami." Yes, Fineaso may be living in a tent with no running water and wearing clothes donated by compassionate givers in faraway lands – but his cheerfulness is catching. Especially when he talks about his son. Who was renamed Tsunami after September 29th. "My uncle saved him. We were so happy and relieved that he was alright. It was a miracle."

That day Fineaso was sleeping in one of the beach *fales* that the family hired out to tourists. When the wave came he ran to the back house where his sister told him to grab their eighty-five year old grandmother.

"As I reached our old lady, the wave hit us and I was trying to keep her safe. It carried us and I was trying to swim until we reached a branch and so we tried holding on to it. I told her that we'll wait here and if the wave comes again and carries us or kills us then so be it. I was just worried that a piece of metal would pierce myself or the old lady, but it was God's will that we were saved."

The next wave was far more cruel. "When it came it was all black and all I saw was death coming. I grabbed the old lady and we went but there was nothing more we could do because it caught us. But I didn't let her go…I took her with me, I held her sleeve tightly in my left hand and started swimming with my right and when we got close to the mountain, I carried her on my back."

Fineaso's grandmother Fuailuma Sagale, is a wizened little old lady whose eyes cloud over when she remembers their bewildering journey in the water."When we got to the mountain he left me there and he went to help the others. I was very worried about everyone. Oh my goodness, this was something so different from anything that has ever happened to our village. My other son helped me up to the top of the mountain and there were so many people up there, small children, elderly people and everyone was congregated there."

Peleiupu Susuga is Fineaso's older sister. She worked as a preschool teacher and has six children of her own. Peleiupu was getting children ready for school when the earthquake began – not only hers, but the children of her siblings."My brother's daughter Fa'amaile who's in Year 3 started crying when the shaking grew very strong and said we should pray. I said – what prayer?! You're supposed to be walking off to school right now! In my mind I thought nothing was going to happen but she understood very well something was going to follow."

It is not difficult to imagine Peleiupu's annoyance that morning. A wailing child wants everyone to pray, while busy adults have a screaming to-do list to get through. Peleiupu tried to shush Fa'amaile." I told her to just sit down and wait. When the earthquake is finished and everyone's ready for school, then we will have time to say a prayer together."

The prayer for Fa'amaile was forgotten. She was sent with the older children off to school leaving only Peleiupu's four year old daughter. Peleiupu was braiding her daughter's hair up into a coiled bun high on her head when she saw the sea."I could see that it looked very strange. It was like it went back and sank under the reef then it started coming very fast and it was spinning. I wasn't too worried because I knew that it would just reach the road and go back again, like other big waves, but then it kept coming over the road."

Peleiupu yelled for her brother Fineaso to get their elderly mother as she turned to grab her little girl."I ran with this little one and as we got to the kitchen the second wave hit us and reached up to our knees. By then I didn't know where anyone else was. We raced toward the pig pen which is quite far from the kitchen and that's where we got hit by the third wave and it was heavy. It hit me on the back of the neck and that's when we got separated. I could hear her crying while I was getting bashed by the wave and I kept falling. I saw my sister's husband Peni in the water and I called to him to get my daughter but then I saw her get washed to the piece of mangrove that was sticking up out of the water, right below the mountain."

Peleiupu's eleven year old son was then able to grab his sister from the wave."The first thing he saw was her hair that I had put up in a bun. It was sticking up out of the water so he grabbed her hair and dragged her up the hill. That's when she *really* started screaming!"

Peleiupu found a strong hold in the mangrove trees before crawling up to dry land and the eventual safety of the mountaintop with the rest of the family. Fa'amaile was there. And so were the other five of Peleiupu's children. But another cousin from their close cluster of extended family houses was dead. And her child Lolani.

"Niusilani was found first and her child Lolani was only found after 3pm that afternoon and she had no head. Because at the back of our house there's a pig pen surrounded by tin and it's very sharp."

After the tsunami, sitting in a rickety shelter in the bush, Peleiupu would be grateful for many things."I am thankful that the wave came while the children were going to school. Because if it came while there was no school and they were all at home, I wouldn't know which child I'd take and which one I'd have to leave behind, especially the small ones. The big kids were already on the road and ran fast to the mountain so they were alright. And this brother here, Fineaso – I am happy he was there. If it wasn't for him my mother probably wouldn't be here because I can't take both my mother and my little girl."

Lomitusi Tuala echoes Peleiupu's gratitude. He and his wife climbed up the steep mountain behind their house with the children as soon as they saw the sea receding,"Because we knew what to do from watching TV, that's how we knew that any earthquake this strong will cause a tsunami. It was very quick. People said it took five minutes but to me it wasn't even two minutes after the earthquake hit that then the wave came."

Lomitusi's family stood and watched the waves hit their house, their shop and destroy their two cars. "The wave was like a savage, a person grabbing everything."

Still, there is gratitude. "I give thanks to God because I know that if the earthquake hit at night, then no one in my village would have survived. I also thank God that the earthquake lasted a long time, to warn us what was coming."

Solonaima Emau is fifty-nine years old. She has been the Principal of Saleapaga Primary School for over twenty years. Her family was sitting in the car about to go to school when the wave came over the seawall. Rather than trying to drive along the coast road in what would have been a vain attempt to escape the wave, the Emau family abandoned the car and ran to the mountain at their backs.

"The second wave was like a barrel rolling along and taking with it sand and rocks – which is what injured a lot of people and houses. It was so loud, it made a noise like a plane makes."

Solonaima believes many children were spared the ocean's wrath because they were at the school buildings already at that time in the morning. Others who were on their way to school were able to dash up the hills beside them.

"A lot of kids were saved because they were at school. They got there early and the wave only reached the school fence so they were all either behind the fence or already on the hill. Also, the tsunami drills that we had last year were helpful, because they knew what to do from our drills."

Even though the school is still standing, Saleapaga is building a new one high inland. One where the sea is but a distant line on the horizon.

In Samoa, the elderly rarely live alone. The family will ensure that someone stays with them to make their food, help with chores, and listen to their stories. To be their child. Someone young. Usually single. Someone strong. A grandchild, a cousin, a nephew. Someone like Tafi Sang Yum. Twenty-three years old. With tattooed bronze arms that paddle in their spare time for the National Outrigger Canoeing team. When they weren't digging taro or making *fa'alifu* for his grandparents, Mata and Su'e Esera.

The Esera's were in their seventies and enjoying the simple rewards of a lifetime devoted to hard work and raising their family. Their eight children were all grown, educated and working – contributing to their parents' retirement. They all worked in Apia, one son was a police officer assigned to the Nafanua Patrol Boat. That September he was overseas on marine police training. Tafi was the chosen grandson who lived with Mata and Su'e in their seaside home. They had three beach *fales* they would rent to visitors. And a taro plantation. A car. When Tafi's paddling team would come for a Saturday swim, the Esera's would make them a proverbial feast. Tender-baked fish that melted in your mouth, steaming hot taro that only the day before had been in the black earth, thick sweet koko from beans that Su'e herself had roasted and ground. Tafi was their child. In that house there was much love and affection.

When the first wave came, Tafi was sweeping leaves in the back yard. His grandparents were in the house. They tried to get to the car but the water was already upon them, swishing around their legs. Again, the size of the initial wave was deceptive. Su'e called out to her husband to come back in the house to shelter from the wave there. Tafi and Mata ran back inside – just before the next wave hit the house.

"The wave took the car. We were bracing ourselves inside the house…I thought the house would stand. I tried to hold on but I felt like my hands were getting electrocuted. I let go and the water washed me out. That's when we were separated"

Tafi is strong – but the wave was stronger. The house was destroyed. "My body felt like I was being beaten by someone because I was battered by trees and rubbish in the water. My body felt numb and I couldn't do anything. The wave took me out with it and I hung on to a coconut tree trunk that was floating. Then when the wave went down, I swam back to the shore."

Tafi stood alone." I stood there and looked to where my house used to be and I couldn't see my grandparents anywhere. I just cried at that time because I had lost hope."

Mata Esera was found several hours later but when nightfall came, Su'e was still missing. "On Wednesday morning we came back early to look for my grandma. We found her just after 8am behind another family's property because most of the stuff from our house was found there and that's where we found my grandma buried under corrugated iron and her skin had turned black. Like she was burned."

Tafi's uncle Ofisa came home in time for his parent's funeral. Crisp, white dress uniforms filled the church as the marine police paid their respects.

Two months on, Tafi helps his cousins build their home in the Saleapaga bush. And clear land for the new taro plantation. Grief is edged with guilt. He is the chosen grandson of Mata and Su'e Esera. Chosen to be their child. Chosen to take care of them.

Leua and her husband Vi'iga are grandparents given a second chance to be parents. Seven years ago they adopted a son. Four years later, another son, baby Kapaneta joined their family. Vi'iga was in Apia the day the tsunami came. Leua was home alone with the children, preparing her eldest son for school. Kapeneta was playing in the front yard.

"I was doing my morning chores and I got a fright when I heard the school children shouting *'the wave'*. I ran outside and held Kapeneta's hand and then the wave came and took hold of me and my child. It took us back to where our house was and then the house collapsed and I was buried under all the wood."

The second wave probably saved Leua's life as it rushed in taking much of the house debris with it. "I was trying to move the house off me...I think it was the second wave that took the house off me and it was so quick and so strong. But then the wave took me to the back and battered me there."

Leua's eldest son grabbed hold of their china cabinet as it swirled past him in the black waters and clambered on top. It would save his life. And keep all the glassware intact. Weeks later, the Vi'iga family cabinet would have pride of place in their bush home. "Take a picture of our cabinet – it should have been destroyed like everything else but here it is all complete and nothing inside it was ruined!" Their deep chest freezer is another example of the Lord's hand at work in amidst their suffering. "Our freezer was taken by the wave and we found it upside down far at our neighbour's yard. When we plugged it in, it still works! It's a miracle."

A head injury would keep Leua hospitalized for several weeks, but in those hours after the tsunami, she ignored the pain as she searched for her youngest son. "I didn't know where my child had gone. I searched for him that whole morning and finally at two o'clock in the afternoon that's when we found him, lying face down in the mud."

Leua's anguish as she carried her dead child to the car that would take them to the morgue was immortalized forever by television cameras."TV3 were here when we found my child...and they put us in the commercial about the tsunami that raises money for aid. I always hear people say *'hey Leua, there's you and your baby coming on TV'."*

Leua is thankful for tender mercies. Unlike many others, her child looked somewhat peaceful in death. "The only wound he had was his eye – like it got scratched by a piece of corrugated iron. And when we found him he didn't have a swollen stomach from taking in seawater, not like the others. His stomach was still soft. He didn't even look like anything bad had happened to him."

This is of particular significance for Leua and Vi'iga because it meant they were able to take a photograph of their son at the morgue. Gently wrapped in a blanket with thick black hair curled over his forehead – he looks like you need to whisper so as not to disturb his slumber. This is the only photo the Vi'iga's have of their son and it hangs on the central house post, draped with shell necklaces. The little grave in the rocky yard has a roof to shelter it from the rain and sun. There are balloons, colourful streamers and a teddy bear keeping it company.

Still, after the death of their child and the loss of their home, the Vi'igas have eyes to see miracles. In a rusting freezer and an old wooden cabinet. And in the generous hearts of many who continue to help them. Three months later, a group of men from their church will help build them a new home in the new Saleapaga. Another miracle.

Continue driving past the old Saleapaga and you will come to the villages of Lepa and Vaigalu. Home of the nation's Prime Minister. A stream separates Lepa from its neighbour Saleapaga, cross over a low ford and rise again to a sandy expanse where villagers have their backs to a rocky hill. Unlike Saleapaga and Lalomanu however, there is a wide tar sealed road that takes you up to the rest of the village, a church , a school, homes and shops. There are also numerous shortcuts up the hill – little paths and tracks well worn and often used by people coming and going. So when the waves came, the people of Lepa were able to get to higher ground quickly. They knew where to go and how to get there. Which was lifesaving because the stream provided an easy avenue for the tsunami to flood in faster. It also helped that at first sight of the ocean pulling back, someone rang the church bell, bringing people out of their homes to ask, *what is it? What's happening?*

At the base of the rocky hill is a clear pool, fed by a tumbling waterfall of crystal water. The village has fenced it and encircled it with stones, making it a favourite spot for bathing. And for the annual ritual of washing and displaying of fine mats, *ie toga.*

Tuluiga Utaulu is a weaver of fine mats. She is the mother of six children. Her husband works in Apia so they only see him on the weekends. Tuluiga's oldest daughter Lele was working at Seabreeze Resort up the road at Aufaga village. Before the tsunami came.

On that morning, Tuluiga was going to the Vaigalu pool with the other women of her village for the ceremony that marked the completion of fine mats. The earth shook, startling them all as several large rocks dislodged from the hillside and rolled to the road in little rushes of dirt. There was a quiet stillness after the earthquake, moments where Tuluiga could feel the panicked pounding of her heart gradually settle again to its regular rhythm. Then the group started an excited hum of conversation as they discussed the shake, still walking to the pool with rolls of soft brown matting in their arms. It wasn't long before the women noticed the difference in the ocean. One of the women exclaimed, *Eh, look at the tide, it's gone right out!*

The women stopped and stared, puzzled but not afraid. Not yet. Then the ocean began to run back towards them, swift and seething. It grew as it ran, as if a creature unfurling from a prone position. The women barely had time to register, to process what they were seeing, to inhale enough air so they could scream, so they could turn and run. They scrambled for the hill, the stronger healthier ones like Tuluiga helping the older women along. Because the road was right there, safety was within close reach. They made it to the church and stood there in a frightened huddle, watching as the sea swept in to their village. But Tuluiga could not stay there. Because down below in their little house she had left her teenage daughter home alone. Quickly she rushed to one of the shortcuts that curved below the church hall and made her way back down to the shore, halting abruptly when her path was blocked by the raging sea. All the houses below were completely covered, only trees jutted from the brown water that was cluttered thick with debris. And clinging to the branches of one of those trees, was Tuluiga's daughter.

"She was caught in the branches of the breadfruit tree, I saw her there, trying to climb up onto the hill and crawl. I was so shocked and afraid. I slipped and fell down half into the water, it was like I had passed out. I think it was the pain in my knees when I fell that shocked me up again. I stood up and crossed the water, swimming through all the things that were in the water, like corrugated iron and wood. I got my daughter and helped her back to the path."

Tuluiga took her daughter back up the hill. They waited there for the water to go down. They were watching when the fire service vehicles came and tried to make their way through their broken road. "The first help we saw after the tsunami was the fire truck…they cleared the road to get through and then an EPC truck came following behind it, those were the first ones on the scene."

The Utaulu family has moved far inland, up in the hills behind the original Lepa village. A road has been roughly cut into the bush so the relocated families can access their new settlement but the houses built with tsunami aid will be slow to reach them. The disadvantage of belonging to the same village as the Prime Minister who is placing the needs of other worst hit areas above those of his own. As Tuluiga explains, "They came and started digging the foundations for our new house but then they stopped work…they explained to us that they were taking the machines to give assistance to Saleapaga village first because the Prime Minister said to give the assistance to our village last." It's the same with the power lines. "They are just starting on our power supply because all the rest of Aleipata is completed and now it is our turn." Those who complain about 'favouritism' for the PM's constituency have obviously not been regular visitors to the disaster zone. For now, Tuluiga cooks over an open fire shielded from the highland winds by pieces of corrugated iron salvaged from the zone.

Her daughters are clearing the tangled grass and weaving coconut leaf blinds for their ramshackle hut. Lele hopes that the Seabreeze Resort will reopen soon and that she can get her old job back. Tuluiga's other daughter, the one she pulled from the tsunami is still not quite right. "We keep taking her for checkups because now her hearing is not good like it used to be. They told us at the hospital to try and keep her out of the sun."

The afternoon sun blazes defiantly on the Utaulu's clearing as we drive away.

Even though a tsunami has destroyed her home and wreaked havoc on her village, Lotolua Niumatapele's faith in God is strong. At seventy-eight years, she is a senior member of the Women's Committee and was preparing to walk down to the pool for the fine mat ceremony when the earthquake hit. Her sons were at the plantation and she was home with three of her daughters and their young children. A housekeeper that came to help with the chores was there too. Lotolua moves slowly, arduously and her daughter had to hurry her out of the house as it swayed in the early morning. Outside, she and another elderly neighbour thanked God for preserving their lives from the *mafuie* before strolling back into their homes both shaking their heads at such unsettling disruptions. In the house Lotolua shook out her fine mat with gentle worn hands, smoothing away the creases, carefully removing several downy feathers that had broken loose from their stitching. It was a stunning fine mat – and Lotolua was surveying her handiwork with pride when the housekeeper called from outside. *Come look at the sea, the tide has gone right out and the sea has turned to mud!*

Lotolua shrugs. "I didn't believe her so I didn't go." Then she heard a neighbour shouting, *Quick, piggyback the old man! Run with him, quick!* There was a rushing commotion outside as people began running past Lotolua's house. Still, she was unfazed. "I wasn't worried, I thought it was just those small waves that come up sometimes. My house is quite high from the sea so I didn't think it would reach me."

Lotolua peered out the window and saw her daughter grabbing her three children, trying to carry them all and run. At that she finally heaved herself from her chair and took her slow steps outside getting her first look at the oncoming wave. "I came out of the house and I looked at the sea. I couldn't believe what I saw…it was like I was frozen." The housekeeper had quicker reflexes. "She came and pulled me to run. She took me to the hillside at the back of our house and dragged me up. We could hear the roar of the wave behind us. I was very tired as I tried to climb. I wanted to rest but the housekeeper told me to keep going, to put my hand around the tree roots and pull myself up. I could see lots of other people climbing up, there were old people at the top who had been carried too. When I turned around to look it was devastating. I didn't want to look at my houses, the waves were huge but the houses were small."

At that most desperate of times, Lotolua's faith gave her strength. "There's nothing that you could do, just cry and ask God to help you with everything but at least you've been saved...most of our village were saved from the wave, even though they were hurt. God helped with saving some of those old ladies that were already at the fine mat ceremony."

When the tsunami was over, Lotolua's family took her to the school on the mountain where many of the village were sheltering. "We stayed there and rested that whole day. We only had a cup of tea because no one was hungry, just thinking about what had happened, but God is everything...the next day I went and stayed at our pastor's house where many people brought us clothing and food."

The Niumatapele children will not build again down by the sea. Lotolua sits in a sturdy plantation shelter and waits while the family uses the tsunami relief building materials to construct a new house for their mother. She has started work on a new mat – nimble, sure fingers weave with patience and faith. "I don't know what God's intentions are for my life or my children... the main thing is my children have seen a tsunami and had it ingrained in them that they shouldn't live along the coastline anymore. They should come inland to where I am so whatever God's will is for me, I will be buried here where we are going to build our new house. We must remember we should always think of God, because He is the one we should turn to in everything we do."

Just past Lepa is the village of Aufaga where the Seabreeze Resort is located. Owners, Australians Chris and Wendy Booth, first came to Samoa on holiday in 1996. They were married at Sinalei Resort in 1999 and they both dreamed of putting down roots in the islands. "We got a list of beaches in the country that might be suitable for a resort, we looked at every single one of them and at the end of the day we knew this one at Aufaga had all the ingredients – it was private, easily accessible and it had a lovely outlook." Initially Wendy wasn't sold on the location that had previously been leased to a school from the USA – "We'd been there swimming several times and Chris would say *oh, I could really do something with this place* and I'd say *right, time to go, quick, get back in the car and let's get out of here before he gets any ideas!* I wasn't keen on it because the buildings there were derelict and I knew it would be major work!"

The decision was made and the couple lived in an open *fale* for a year while Chris started work on the project. He has forty years of building experience and so he led the local team of carpenters as they painstakingly built the Seabreeze Resort. "We built the place from scratch, first we renovated one of the buildings that was there and that took so long that we just demolished the rest." The resort was officially opened on the 17th of November, 2007 and employed twenty seven staff members – all locals from the nearby villages. Seabreeze Resort participated in the two nationwide tsunami drills in 2007 and 2008. "Much to our guests' horror we dragged them all out of their rooms and made them go up the hill. There were many jokes about it because we always took a crate of water with us and they always thought we should have brought a crate of Vailima beer instead because it would take so long after the drill to get the all-clear!"

On 29/09, the Booth's were in their house on the far side of the bay when the earthquake shook. Afterwards, Chris did a quick check of the building and pronounced it "okay". He called out to one of the workers, Tasi, to go and reassure the guests that "it's just a little rumble, it happens all the time." It was supposed to be Wendy's day off from the office, so she was still in her nightie while Chris prepared to walk across to the main block of buildings. "I had finished a coffee and I glanced up and just saw the reef drain, but it happened that quick that the lagoon remained full and outside the reef just dropped and water was just pouring over the dry reef."

Chris said, "Look at that, it's just a brown wall! I've never noticed that before."

Wendy replied, "Oh no, tsunami!"

Chris yelled out to Tasi to run, to get the guests out. "He turned, as soon as he saw the reef, he ran. Looking out to sea, you couldn't see anything coming, it was just dead calm, lower water level than it should have been, but all of a sudden, this thing just hit the shallower water and it leapt out of the ocean just like that. It was going at 30 or 40 miles an hour at least and as it came out of the deep water, all of a sudden it was there and it was chasing Tasi and he was running for his life."

Tasi and fifteen of the guests made it up the hill but two guests - Murray and Maureen Swain -were trapped in their room. The force of the water pushed out all the walls and all of the furniture but the Swains huddled in the cement block shower, fighting the wave as it tried to take them with it. Once the water calmed somewhat they were able to join the others up on the hill.

For Chris and Wendy there was no way to escape from the wave that "sounded like a jumbo jet taking off, like the roar of an engine." They were on the furthest end of the cove, with no ready access up the cliff side. "The first two waves were four and five meters high, quite close behind each other, and they took out the restaurant, three cars that were in the car park and smashed every *fale* suite and everything else there was in the front. As that receded the cars and the restaurant went out to sea and then the other two waves came in about a minute behind those two, but by now the bay was filled with the first two waves going out. So these next two waves were seven and eight metres high, again one behind the other and they had all the water now from the other two and they were coming in with so much more momentum."

As he watched the ocean wipe out four years of hard work, Chris yelled at Wendy, *We're stuffed – the restaurant's gone!*

Wendy's eyes were on the wave building up towards them, *Don't worry about the restaurant, there's a huge wall of water coming straight at us!*

The waves hit the Booth's house. "The water came up through the floor we were standing on, punching holes up like geysers and there was water coming over the deck and over the roof…every piece of glass in the place was blown out, all the furniture was thrown up in the air…we tried to run outside but there was water everywhere and nowhere to go. We were knocked off our feet. Wendy grabbed onto me and I grabbed the side of the building, the corner and water poured over our heads and was pushing us from every direction.

It was a battle to hold on, to even stand. "You were either going to die or hang on, simple as that. If you let go, you were going to die because the water was everywhere…and it wasn't just the water, it was the mud and dirt and sand and you couldn't get any traction because it was so slippery and there was so much debris."

As suddenly as it had come, it seemed to calm. Chris and Wendy stood up and started running for the driveway up the hill. Wendy was barefoot. "I was covered in mud and silt and even had some embedded in my ears…I set off running and there was glass and stuff everywhere and we were jumping over things, nails and glass, and some of the staff came down to help. They were worried about my feet but there wasn't a scratch on them…and it was obvious to me, that we were being helped at that stage."

Gathered at the top of the hill, guests were shaken at their near escape and the loss of their belongings. Wendy and Chris watched as the ocean took out the cars, the *fales* and the restaurant with its stainless steel kitchen. "There was nothing left – but broken timber – we've got the fridge door and a knob off the big commercial kitchen cooker and that's all…we don't know where all the rest of it went." There was no time for falling apart however, the Booths had guests to take care of. As Wendy explained, "I was pretty shaken at that point, but I had to pull myself round pretty quick because the guests were all that mattered and we were determined that we were going to take care of them. We couldn't do anything about a tsunami, but when they left, we wanted them to say good things about the resort and not bad." Chris went to secure the resort site. Many of the village had come running at the sound of the crashing waves. "It kind of became mayhem down below at that point."

The guests were moved to the evacuation point - a large Samoan *fale* further up the road owned by the Tagoai family and Wendy and her staff quickly gathered as many of the guests personal belongings as they could. Wet clothing was rinsed and spread out to dry in the morning sun - on the hot rocks around the *fale*, on the hibiscus bushes. "We carried up everything we could, even cameras, even if it didn't work it didn't matter, at least you could get your things back." The Tagoai's rallied to get food and drink for the stranded guests. "Everyone helped to look after our guests – they cooked for them, washed the dirty clothes. And I was trying to keep everyone positive, I kept saying, *Just think of this as your real Samoan experience! Isn't this fun!?* The family cooked *pisupo* (corned beef) chop suey and they didn't really like that but I reminded them, *This is all part of your TRUE Samoan experience!"* Guests were anxious to leave and get into town as soon as possible, but needed to be patient as the Booths tried to arrange transportation. At that point, the Booths had little idea of the extent of the tsunami. "I just thought it had affected us…I was trying to phone up Tui Annandale at Sinalei for help with our guests so in my brain I didn't have a clue about the devastation, about what was actually happening or how widespread it all was. I just knew that I needed some advice and Tui was my mentor and friend, the most nurturing woman I had ever met who ALWAYS knew the right thing to do."

At about ten am, Wendy was able to flag down a fire service ambulance. "There were two women who had gone into shock and I was worried about them, a pregnant woman Vicky Damon and an elderly woman, Madeleine Lewis. We had them lying there covered with blankets so I was very relieved to wave down the ambulance and ask them to take the women on to the Lalomanu hospital – not realizing that the tsunami had wiped out Saleapaga and Lalomanu! I basically sent them into a warzone, probably just compounded their problems even more…"

Wendy got the DMO contact phone number off the fire service men and was put straight through to Commissioner Tony Hill. Hindsight is an illuminating thing. Wendy was in full throttle wanting immediate care for her guests. "Bossy me – I said to him, *I've just sent two people on to the hospital and I need IMMEDIATE pickup for the remaining sixteen people so they can be evacuated to the NZ High Commission.* Tony's voice was so calm, he asked *Did anyone suffer any injuries?* I told him we were fine and he said, *right, you're not the priority, I'm at Saleapaga now and we've got fatalities here. Keep your guests comfortable and we'll get to them when we can.* At the time, I wasn't happy! But now I know why we weren't a priority and when I think of what Tony must have been going through over there at Saleapaga and then to have us call him like that – well he was fantastic. The calmness in his voice and hearing later about all the work they were doing over there, the stress they were going through - they were brilliant. If that had been me, I might not have been so polite when I got a similar phone call!"

The Booths had to then placate people without panicking them. "Everyone wanted to know what was going on, why couldn't they just get out of there. I told them a few resorts had been hit badly and some roads were closed, nothing too major. Some of the guests had family and friends staying at Saleapaga and we didn't want to worry them unnecessarily either. So, I got the staff to entertain them with singing and dancing, *Right let's have some music shall we? Isn't this lovely?!* When I think of the stupid things I was trying to entertain people with – it was silly!"

Finally at midday, a truck arrived from local company Pacific Aluminium. The owner told Wendy that Tui Annandale had been killed. "It was just so sad and then we had to get all these guests into the back of an open truck and they couldn't take their suitcases with them because they didn't fit so we had to put all their stuff into large rubbish bags. I was saying, *Come on guys, this is your real Samoan experience*... and one man said to me, *I'm sick of your Samoan experience!*"As the truck pulled away, all the staff sang *Tofa my feleni, goodbye my friends* to farewell them and "amazingly the guests waved and shouted, *we'll be back!* So they left on a good note."

Nynette Sasse of the Samoa Hotel Association was the first one to come to Seabreeze from Apia offering help. "She was amazing. She hit the ground running and visited us every day after that until she could see that we were okay."

With all the guests evacuated to Apia, the Booths could allow the devastation to sink in. "That night we were in total shock, you can't even imagine the thought processses – in the morning there was a beautiful place there that we had made our living on with an unbelievable view of the sea – and then by the evening, it was absolutely different, it wasn't there anymore. There was one suite left, the one the Lewis' had stayed in, the one furthest up the hill, that's where we're living now. That night we had a hot shower, we had dry clothes that the village had given us, but it was just so unbelievable to think about it, that while we were having a hot shower here, there's so many people dead, so many cold and homeless, so much destruction along the coastline not far from us…"

It would have been easy to understand if the next part of the Booth's story involved them packing up and moving back to Australia. And yet, there is a team of local carpenters at work building a new Seabreeze Resort. Hammering rings out from the bottom of the cliff side and women sing as they sit cross-legged in the shade of a spreading *tamaligi* and weave baskets and hats. Concerned about the lack of income for her staff during the building phase, Wendy has helped the local women organize a co-operative making handicrafts for sale in stores in Apia. The Booth's connection with their staff, with their village of Aufaga – played a large part in the decision to stay and rebuild. "It's raw emotion that ties you to a place. I mean, how could we just get on a plane and say to our staff, *you're on your own, all twenty seven of you.* We just couldn't do that, they're like family. After the tsunami, the staff helped us clean up. They had a meeting and decided to work for a week for no pay which I can tell you is a very big decision by the staff and it helped us a lot." Most of the male staff were re-hired immediately as construction crew for the rebuilding but there was nothing for the women. "We called everyone in for a meeting, we had lunch together and we had to tell them who we could keep on and who we had to let go and it was hard, harder than the tsunami for us because we love every member of our staff and to say to some that they won't be here anymore…well, it was very emotional for us."

The road to recovery is a long and hard one. For Chris, "the tsunami lasted all of five minutes from the first to the fourth wave…but the living afterwards was the hardest thing. Trying to explain to your family why you're not going back to Australia…they couldn't understand how we could even think about rebuilding and risking it all again, with no guarantee that it wouldn't happen tomorrow, or next week or next month even."

How are they doing it? Wendy explained, "There's a lot of dark days where you really still ask yourself, a*re we stupid?* It's been mentally and physically draining. We decided to just take it a day at a time, sometimes an hour at a time…*let's just get through this day and we can work out the rest of our problems later*. We didn't eat for two days, the village brought us food but we just couldn't eat it. We couldn't sleep either and wasn't because we were afraid of the sea, but because there were so many things to think about, to figure out."

Ask the Booth's about a memory that stands out for them? An experience that stands out for them out of all of this? Chris tells of one that happened a week after the tsunami. A team of scientists were travelling around the island studying the wave impact and they stopped in to interview the Booths. It was the day of the National Memorial service when the country was pausing to mourn as one. All of the staff had gathered outside the one remaining resort *fale* to listen to the Memorial service on the radio. The staff began singing the opening hymn, joining their voices in unison with the radio, with the crowd gathered in the Apia Park stadium. Down in the little cove, the science team stood on the foundations of a shattered resort and listened. "The singing started up there and me and four of the team…we all had tears in our eyes because you could just feel all of the emotion that our staff felt for the families that had lost loved ones."

Chris and Wendy plan to reopen before the close of 2010. "We're not scared, we'll go back and live down in our *fale* in the cove. The tsunami was an act of nature – there's nothing you can do to stop things like that, you just have to be prepared and know what to do. There's no other place that we would rather be or anything else that we would rather be doing. We're happy with what we have here and what we're doing. Some people may think ours is an irresponsible vision, but you've got to make your own vision and we'll take the consequences or we'll reap the rewards."

The Booths are emphatic that their emergency evacuation measures for their staff have totally changed, "knowing what we know now…"

"Everything we taught our staff back then to do – we now tell them NOT to do. We used to say turn off the central gas supply so when the rescuers come they don't have to deal with hazards like gas on. Nope. Not now. Now, you don't turn off your gas supply, you get out of there. Just leave everything and get out, go. We used to say to the staff, lock all the doors, ha. Not now. On 29/09 we wasted time waiting for a warning on our cell phones like during the drills. Now we know, forget the drill, forget the gas and the electricity, forget everything and just get out and we'll worry about the rest later."

In spite of all they have lost, the Booths can speak of the future with optimism. "We've always had one saying *Fortune favours the brave.* Never before in our lifetimes has this phrase been more important to both of us. We are proud to be a part of a community that showed such resilience in the face of adversity. In the months since the tsunami, the generosity of our tourism industry partners and suppliers has been amazing! Now, almost a year on, time has flown, the repair and restoration work is going well and our spirit and determination has never been stronger. Each day we start with reminding ourselves how unique Seabreeze was and with that reaffirmation, we now look forward to our re-opening."

10
Nanny I need you! Where's my nanny?

New Zealander Madeleine Lewis is in the advanced stages of Alzheimers, the disease that slowly but surely tears away pieces of memory and thought. Her daughter Diann organized a trip for the extended family - one last chance for them all to go on a holiday with a beloved mother and grandmother. Samoa was chosen because of Graham's close ties. His father had worked as the Managing Director of the then, Bank of Western Samoa. "We lived at Vaoala and had three really really good years, it was a great time of our lives to be there in such a neat place. All of my family has been back since then." So when Diann was planning a trip for the extended family, Samoa seemed the perfect choice. The group was a big one – fifteen people in total. The Ansell's four children Matt, Tia, Michaela and Conor, Diann's parents, Madeleine and Tony Lewis. Diann's sister Suzanne and her husband Ngatau Hemopo, their two children Teone and Maia. Ngatau's sister Marama and her four year old son Jordan. And the Hemopo's mother, Margaret.

The holiday began with a stay in Savaii at Le Lagoto Resort. From there the group travelled to Aleipata, leaving Madeleine and Tony at SeaBreeze Resort, Vavau while the rest of the party went to Faofao Beach Fales at Saleapaga. Diann explained, "We were quite keen on doing the beach *fale* experience, but we also realized that mum and dad weren't – it wouldn't have suited them at all because it's quite basic at Faofao, but perfect for us, exactly what we were after." Diann recalls that at the time, she was a little disappointed that her parents bungalow was up on the hill, the ones further down were closer to the beach.

The Ansells and Hemopos arrived at Faofao on Monday evening, in time for dinner and then bed. Most of the group however, had a restless night. "We just found it very noisy. It was really windy and the noise of the ocean was so loud so we didn't sleep very well."

The children were up early before the adults, outside on the beach when the earthquake hit, but falling coconuts sent them back into the fales. Ngatau got up to go check on the kids while Suzanne was still in bed. Diann and Graham also piled out to see how the children were taking the early morning shake. Marama and her mother Margaret had gone to the restaurant fale and were sipping tea, looking out over the placid ocean. Everyone was talking about the earthquake. "There was no panic whatsoever. It was like, *wow did you feel that? Wow that was interesting.* But sixteen year old Matt thought about a tsunami and suggested that they act. *Let's go up the hill, if nothing happens then we can come back down, if it does, we'll be up there.*

Months later Diann looks back with hindsight. "We absolutely should have listened to him and done that. But it just didn't feel like a massive earthquake. I guess you just don't think it's going to happen to you at the end of the day. But we did gather the children together, so we weren't all racing off and doing our own thing. We brought all the kids up on to the road and stood there in a group. Except for my sister Suzanne – she was still in her *fale* and Ngatau was saying to her, *it might pay to come out."*

Matt walked up to the restaurant which was slightly elevated from the *fales* and from there he saw the ocean change. "Everything went quiet and the water went really flat..it was strange. Then you started to see the reef and you could see, because the water in the lagoon was trapped, it couldn't go out. You could see more and more of the reef and behind the reef, the water had dropped away."

Matt called out urgently to the group on the beach, *Look! The water's gone flat, we better go.*

It had been twelve minutes since the earthquake. Diann ducked back into their *fale* to get the passports. Graham's cell phone alarm was blaring furiously, set for seven. "I only grabbed his phone because the alarm had gone off and I thought, that's really annoying! So I grabbed it and shoved it in my bag with the passports." Matt's warning gave the group valuable escape time. "Thank goodness for Matt's observations because they gave us a few extra seconds which turned out to be vital. As we turned to run, the waves which normally crash on the reef had come past the reef and were coming towards us." Everyone started running. Matt grabbed the littlest – his four year old cousin Jordan. Diann was running with five year old Maia. Conor picked up his cousin Teone's shoes from outside the *fale* and gave them to him, *quick run!* Graham was making sure all the children were ahead, doing a quick headcount. "I can see all the kids, they're all ahead of us, run!" Ngatau was shouting at his wife, still in the *fale*. Suzanne was in the middle of getting dressed.

"The timing caught me with my pyjama pants on and no top. I took the questionable but modest decision to put a top on. It seemed an absolute age to do this and climb out of the *fale* and start running, all the time Ngatau is yelling at me – *hurry up Suzanne! Hurry the hell up!"*

Meanwhile Margaret and Marama rushed out of the restaurant, Marama screaming for her son. *Jordan, Jordan!* What a relief to see him far ahead of her being carried by Matt. Margaret was running still holding her cup of tea until Marama yelled at her, *Mum, throw that cup away!* Ngatau and Suzanne caught up with Margaret and Marama as they were running along the path towards the hill. Ngatau stayed back with Margaret who couldn't run very fast and tried to help her run faster by grabbing her arm and sort of lifting her a bit. Ngatau sent Marama ahead, Suzanne hung back too but Ngatau yelled at her to go on ahead.

The escape from the beach was one of panic and confusion. Fourteen year old Tia, describes "the fastest run I've ever run in my life…with all these people tripping over, running, screaming. There were all these people in front of me running, falling over and people running over them and trying to climb the hill. We were all running and screaming at other people to run too."

It was about one hundred meters to the hill, but for Diann the run seemed to be interminable. "It seemed like I was running in slow motion and I was surprised that I couldn't run faster. The water was roaring loudly and you could hear the *fales* snapping into thousands of pieces like they were just twigs. It was catching up and it could clearly move much faster than we could…My jandal got stuck between two rocks and I fell. It seemed to take ages to pull my feet out and take my jandal off. I sent Maia ahead of me at that point."

As the children scrambled up the steep hill, the adults were slower to follow. Suzanne never looked back but knew she was in trouble when her "last couple of footsteps were in water. The tsunami struck me. I looked up the hill and only a few meters away I saw Marama, it was a scary feeling to be pulled away from safety that was so close. The water was fierce, there was no control. I was really worried about the debris in the water, a huge chest freezer swirled past me, dangerously close. When the water receded I managed to stand up and quickly get up the hill."

Ngatau and Margaret were holding hands as the wave hit them. Vainly he tried to hang on to his mother, but the force of the water was too strong. He was being scraped and buffeted on all sides so he rolled himself into a tight ball trying to evade the worst of the debris. When the current pushed him above water, he started swimming for the hill. As he dragged himself up out of the water, he was shouting, *I've lost her! I've lost mum!*

Ngatau was covered in cuts, a nail had gone right through the palm of his hand. The force of the water had ripped off his watch and sandals and there were raw welts all down his back, legs and his head. All the children were crying. Tia was holding Jordan who was screaming, *Nanny come here! Nanny I need you, where is my nanny?!* It was an awful ten minutes as everyone scanned the churning water looking for Margaret. "We just all frantically tried to find her and it seemed to take ages. There's wasn't much you could do.

You can't go down there and just hop into the water and go looking for her, it was so filthy and there was so much debris, you just have to stand at the top, looking, hoping."

There was a shout from their right. Two Samoan boys were pulling Margaret from the water. She was clinging to a suitcase, using it as a flotation device. Her leg was broken and she was in a lot of pain, but she was alive. However, the family was still not complete. What was happening to Diann's parents – Tony and Madeleine at the Seabreeze Resort? The day was far from over.

"There was still so much to do. There were still people missing and there were some bad injuries which needed attention as well as upset kids who needed calming. I still had Graham's phone which was unbelievably lucky – he was able to call a work colleague in town, call his family back home and even seek medical advice from his friend Mike in NZ who is a doctor. Some of the injuries were horrific, and many people were in pain. Not just Margaret, but a local lady with a leg injury and another man who had lost his child. I had our passports but I was thinking, *I would give anything to be able to trade these for a first aid kit!*" Diann was sure her parents would have escaped the wave as she remembered with gratitude their hilltop bungalow. But she was anxious about how Madeleine was coping in her fragile condition and how her father was handling the situation. Is there anything worse than being caught with your children in a tsunami in a foreign country? How about being caught with your children AND your aged mother who has Alzheimers? Thanks to Graham's cell phone, the word went out about Diann's mother...*an elderly NZ woman with Alzheimers was staying at Seabreeze Resort, very worried about her, can anyone help, please keep a look out for her, she's wearing an I.D necklace...*

A few miles away in the village of Aufaga, Tony and Madeleine were at the Seabreeze Resort that sits at the bottom of a forty meter steep incline. The resort has two rows of self contained units following the contour of the sheltered cove. A bit further up the incline on a rock outcrop is the restaurant and bar. Other units sit higher on the cliff.

When the earthquake shook, Tony was up and making a cup of tea. "It was about 6:45am...there was lots of noise and quite violent jolts lasting about thirty to forty seconds. I was holding onto Madeleine as she was in bed still."

After the shake, Tony walked down to one of the lower units, in search of milk for the tea. There were three women preparing to leave, loading their belongings into a vehicle. He asked them for milk and one ran back in to get some. Tony stood and chatted to the others about the earthquake. "As we were talking, we noticed the cove water receding, then a large band of grey-white out to sea, then the reef disappeared."

They all knew what was coming. Everyone scattered, shouting *Get out! Everyone get out!* Tony ran back to his room to get Madeleine, helping her to put a sweatshirt on over her pyjamas, sandals on her feet. Together they ran up the incline to where the rest of the Seabreeze occupants had gathered, about ten people in all. The ocean poured over the lower units, rising to cover even the dining room and bar. Two cars floated in the churning white and then disappeared. The Lewis' unit was left intact and the owner of the resort, Wendy Booth, had one of the staff run down to grab their bags – a denim shoulder bag and a suitcase.

The group went to a nearby house. Madeleine was agitated, difficult to keep still. She and another guest began to go into shock. Wendy flagged down the fire service ambulance that was headed towards Saleapaga. Tony put Madeleine's suitcase in the ambulance and followed on foot behind it. Progress was slow with three of the fire service team clearing the road in front of the van. Tony had no idea where the others were staying and clung to the hope that they would all stumble across each other somehow. The long-anticipated family holiday was going horribly wrong.

At Saleapaga, the Ansell family party sat on the hill with other survivors watching as the ocean ran wild through the village below. "It was like a great big washing machine with all this debris. It wasn't like it was wave after wave, it was just like a huge big mulch of water and everything surging and churning. It didn't just come and drain out and disappear immediately, it was there churning for ages."

Tia remembers the locals, "pulling out all this fish. There were just piles and piles of fish and then they gave us a bottle of Powerade and they got coconuts for us to drink while we sat there in the shade. Everyone was really traumatised, shaking and everything. And we didn't even think about all the stuff we had lost, we were sitting there thinking about how lucky we are, we've got our family, everyone's alive."

Eight year old Conor turned to his mother and said, *I know what the most important thing in life is – it's family.*

In less than two hours, the family were startled to see fire emergency vehicles pull up down below. "We thought we were going to be sitting up there for many hours, maybe even overnight, so when we saw the flashing lights of the rescue vehicles, we were really surprised. They were just so quick to get there – this truckload of big, strong Samoan boys from the fire service and they just worked tirelessly taking care of all the people. They were unbelievable."

Everyone was unwilling to come back down the hill. "We were reluctant because there were all these stories going round about another wave coming and it was quite an effort to actually climb down and get to the road because you had to climb all over the debris and you knew that once you got down there, there was no way you could climb back up in a hurry!"

The FESA team began to stretcher the injured down to their fire truck on the road. Ngatau and Marama helped to carry their mother. She was put in the fire truck, ready for transport to the nearest hospital. Ngatau and Marama were standing beside their mother when the ambulance pulled up. With Tony and Madeleine. Tony recognized them with huge relief. He got Madeleine out of the ambulance and she sat on their suitcase under a tree, waiting while they discussed what to do.

From up on the hill, Graham pointed. "*Diann, is that your dad?!* It was hard to tell because they were so far away but I thought it had to be him because the man was wearing sneakers and not many people in Samoa wore sneakers! I got my camera and took a photo on full zoom. It was 9:57am. It *was* mum and dad. We couldn't believe that they were there as it is quite a distance from Seabreeze Resort."

All the family was now accounted for. Just on opposite ends of a disaster zone. But further separations awaited them. A muttering, disoriented Madeleine could not make the climb up the steep cliff and with warnings of more waves – the road was not a safe place to be. Madeleine was put into the ambulance with Margaret and others. There was no room for Tony or anyone else to accompany the women. They stood and watched worriedly as the van drove away. Madeleine kept asking the rescue workers for her bag…*I need the suitcase, I want my suitcase.* The suitcase was left behind on the road.

The fire service team was organizing transport for all the survivors. The family was ferried to Lepa village in a Honda Odyssey. Fifteen people crammed into a stranger's car. "We got dropped off at the bottom of a road that we could walk up to a school on higher ground. It's only then that we realized this kind person was then going back where we had come from to retrieve his own family. He said, *I'll go back for my family now.* We just couldn't believe that he'd taken us first and left his family…such is the nature of the Samoan people."

The college was an evacuation point for the village, crowded with many other survivors. Here, the group met with more kindness. "A Samoan lady took us up to Lepa Lotofaga College and there was a stream there beside the college where we could wash ourselves and have a drink. Then they took us to a *fale* in the college grounds and we spent the afternoon there."

The group met the Principal of the college, the woman they would call "our Mother Theresa" – Tuputala Kaisarina Lene. "Most of her village had been destroyed but she still found time to look after us. She fed, watered, clothed and provided bedding for us as well as seven other tourists and her own extended family. She welcomed us into her house at mealtime, and insisted we use her shower to clean ourselves."

It was Mrs Lene and her husband that drove Diann, Tony and Marama along the coast to the Lalomanu hospital later that afternoon in search of Margaret and Madeleine. "We were incredibly worried about Mum and concerned that she would be very frightened without Dad and wander off. That trip was deeply moving. I sat in the back of the car as we drove through the destroyed villages, not able to speak; what can you say in a situation like this? There was lots of activity along the coast, heavy machinery had been brought in and everyone was looking for the missing as well as belongings which could be salvaged. There was grief and sadness everywhere."

At Lalomanu hospital, the grief seemed intensified. "Mothers were wailing…trucks were bringing in bodies piled up in the back, covered with lavalavas with their feet sticking out. Rigor mortis had set in." Of their mothers though, there was no sign. "the hospital turned out to be a triage centre really – it was tiny and most people were moved onto Apia. There was no record of Mum or Margaret having been there so we had to just assume they had been taken to Apia."

While they were there, an alarm went out – another tsunami. "There was panic. We were really high [inland] but still everyone ran. Everyone was on edge." Afterwards, the trio decided to go back to Lepa, unable to find transport to town and with night fast approaching. The trip back was a nerve-edged one. "It was eerie…it was late afternoon and even though it was still light, the coast was deserted. We didn't once take our eyes off the sea, nervous that another wave could come. There was an empty water container in the back of the truck, Marama had it firmly gripped."

Back at the college, good news awaited them. Graham had received a text from Steve Rogers of the ANZ Bank that Madeleine was safe at the NZ High Commissioner's residence. Steve had seen Margaret in the National Hospital and he promised to be there first thing the next day to drive them all back to town. The family spent the night at Lepa College. Looking back, everyone remembers the generous hospitality, the food in particular.

Conor loved the "huge pot of sticky rice…one of the teachers had cooked it up for his dinner and he just gave it all to us." And the pancakes at breakfast. "So good, crunchy, and hot cocoa too."

Tia remembers the dinner that night. "It's amazing – we were expecting just coconuts off a tree or something but they gave us this wonderful meal – chicken and rice. I used to be a vegetarian but when I was given this amazing food in front of us, I was eating it. Give me the pork, give me anything, I'll eat it! I was just so thankful for all the food that we had."

They slept on the floor in the staffroom. "Mrs Lene brought us thin mattresses, sheets and pillows, even mosquito coils. She thought of everything."

Still, it was an uncomfortable night. "It was impossible to sleep – trucks were going down and back during the night and the sea sounded particularly loud." Both Suzanne and Ngatau had painful injuries that made sleep particularly difficult. "My leg was swollen and infected and my nerves were frazzled…"

Just before dawn, a crowd of villagers gathered in front of the school. A minister led them in prayer and then hymns filled the morning air. Enfolding the empty spaces with love. People wept. On a mountainside rise looking out over the gentle sea as God touched the sky slowly with light. Two children from Lepa village had died in the tsunami – sisters, Lutia and Losivale Tainano. The ringing of the church bell had warned many others of the tsunami. Graham and Diann spoke of listening to the sunrise service. "That was the most moving time I've ever experienced. There was just a huge amount of grief, singing, and my eyes welled up with tears. It was very sad…and heart wrenching."

The next day, the family made their way to Apia where they visited Margaret at the hospital. They were reunited with Madeleine who had spent a restful night at the High Commissioner's residence. And with Madeleine was her suitcase – last seen sitting under a tree on a tsunami-strewn road.

Back in New Zealand, the family put together a tsunami story-book, a journal, giving the children in particular, a chance to record their memories of the holiday of a lifetime.

Teone's Tsunami Story (Age 8)

When the earthquake happened, we were by the girls' *fale*. It was my cousin Matt who told us to run. I looked at the water, it was going fast. It looked like a normal wave when it was on the ocean, but it didn't look normal when it came on the shore. I decided to sprint. I held my shoes and ran in bare feet.

Saleapaga village two hours after the tsunami. *Photo - Diann Ansell*

Peleiupu Susuga with two of her children at the new Saleapaga settlement. *Photo - Lani Young*

All that's left of the front gate to the Saleapaga Primary School the day after the tsunami. Behind the classrooms is the hillside that survivors had to climb up to safety. *Photo - Matthew Leal*

Solonaima Emau, Principal of Saleapaga Primay School. *Photo - Lani Young*

Molesi and Silaumea Pritchard with their family at the new Saleapaga settlement. *Photo - Lani Young*

Fuailuma Sagale grateful to be alive thanks to her grandson Feniaso, Saleapaga. *Photo - Lani Young*

Leua and Vi'iga Faatoia at the new Saleapaga settlement. *Photo - Lani Young*

Tafi Sang Yum of Saleapaga. *Photo - Lani Young*

Lotolua Niumatapele in her shelter at Lepa village. *Photo - Lani Young*

Tuluiga Utaulu rests from her work making a shelter for her family at Lepa village. *Photo - Lani Young*

The Ansell, Hemopo, Lewis family in their van, excited for their Samoan holiday. *Photo - Graham Ansell*

Diann and Graham Ansell in their fale at Saleapaga, ten minutes BEFORE the earthquake. *Photo - Ansell*

Madeleine Lewis rests on her suitcase in Saleapaga while husband Tony and Marama Hemopo discuss what to do. *Photo - Diann Ansell*

The Ansell Hemopo children enjoy the sunrise at Saleapaga the morning of the tsunami. *Photo - Diann Ansell*

A stunning evening at the Seabreeze Resort, Aufaga. *Photo - Wendy Booth*

The remains of the Seabreeze Resort the day after the tsunami. *Photo - Diann Ansell*

The Red Cross drives through Saleapaga the day after the tsunami. *Photo - Matthew Leal*

Some of the victims of the 2009 Pacific Tsunami.
Photos by WT Media

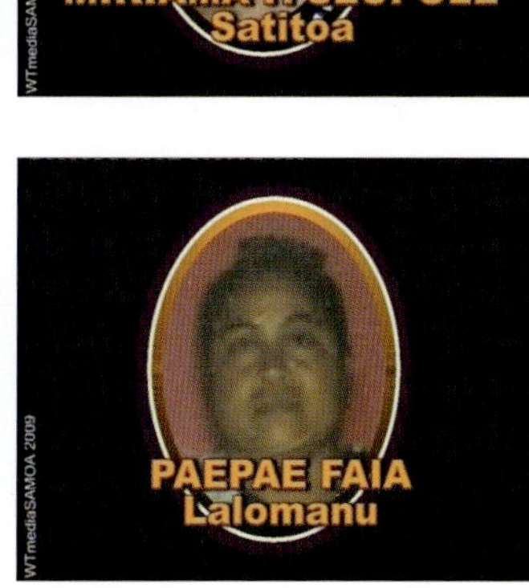

ANITA TAUFUA
Lalomanu
WTmediaSAMOA 2009

MATALA TAUFUA
Lalomanu
WTmediaSAMOA 2009

MOANALEI SARAH LONG
Lalomanu
WTmediaSAMOA 2009

PELESASA TAUFUA
Lalomanu
WTmediaSAMOA 2009

SILI JUNIOR TAUFUA
Lalomanu
WTmediaSAMOA 2009

WENA TAUFUA
Lalomanu
WTmediaSAMOA 2009

SUE TAUFUA
Lalomanu
WTmediaSAMOA 2009

TAUFUA SIO
Lalomanu
WTmediaSAMOA 2009

ALOALO SA'O
Saleapaga
WTmediaSAMOA 2009

ANA IULAI & RACHEL LEUELU
Saleapaga
WTmediaSAMOA 2009

MAKA ESERA TUUMAGA
Saleapaga
WTmediaSAMOA 2009

ANETI ALESANA
Saleapaga
WTmediaSAMOA 2009

SIMAMAO SUANI
Saleapaga
WTmediaSAMOA 2009

IEFATA SA'O
Saleapaga
WTmediaSAMOA 2009

SUI FAIUPU
Saleapaga
WTmediaSAMOA 2009

I was running by myself, my sister and my cousins were all in front of me. I saw Auntie Marama telling Nanny to throw her teacup away.

There were people going up a hill and so I leapt and jumped off a rock and climbed up the hill. I got a cut from the rock.

I saw Mum from the corner of my eye getting caught. I saw Ngatau climbing out of the water. I felt a little bit sad because I had lost some of my best toys. We sat on the hill and then moved into the shade. Someone gave us mosquito fly spray that we put on.

It was scary to climb back down the hill. We drove and walked up a hill to a school. We ate at the school. I had rice and chicken for dinner. I felt a lot safer – the biggest tsunami could only reach up to the bottom of the field so we would be safe there.

We slept there that night. It was a bit noisy. My Uncle Graham knows someone who lives in Samoa, his name is Steve and he got his ANZ cars to drive us to his house. Steve lived in Uncle Graham's old house when he used to live in Samoa.

They took us to the airport. We came back on first-class. The best things about first-class were we could watch movies, listen to music, and your chair turned into a bed.

My scariest memory was when the tsunami nearly got me.

My best memory was getting the flight back home on the second day and being in first class.

Maia's Tsunami Story (Age 5)

During the earthquake I went and got my jandals which were outside of our *fale*. A coconut fell on our *fale* during the earthquake.

When we saw a big wave coming, I ran up the hill. I was running with Auntie Diane first and then Auntie Marama.

I wasn't scared. I saw the wave. It looked like a *koru*. It didn't look big but then it got bigger and bigger.

I sat on the hill and looked at the land when the wave had gone, it looked like a flood. I saw helpers who were finding people.

I climbed back down the hill on my bottom and I dropped one of my jandals, but picked it up again.

The taxi was heavy with all of us in there, then we got out and walked up to the school. It was good the school didn't get flooded and it had some food. We had dinner at the school and we slept in our clothes.

It was good that Suzanne, Ngatau and Nanny survived.

Michaela's Tsunami Story (Age 10)

Run! Run to the hill! I thought, run fast!

A power wave was behind me, making crackling noises as it goes past the *fales*, breaking them to pieces.

I was scared. What if I was caught up in the wave and drowned?

As I got ten meters up the hill, I turned around and saw Suzanne getting caught by the wave, then Ngatau getting pulled to the side. Now I was really scared. I saw water everywhere on the ground with no houses standing.

We had to stay on the hill for awhile looking down on the ground with everything gone, but the terrorizing tsunami.

Conor's Acrostic Poem (Age 8)

T errifying
S amoa
U ltimate
N ature
A ttacking
M onster
I nsane.

When they left Lepa, the Ansells asked Mrs Lene for her bank account number. "She said, *no, I'm not doing this for money!* We apologized and explained, *it's not going to be for you, we'll raise you some money for people in Lepa who need help.* I got back to work and told people the story and so many just came and gave money right away, we ended up raising about sixteen thousand dollars which we sent to her. She divided it up amongst people who she felt were most needy and then had the Prime Minister who comes from Lepa – to sign off and endorse their expenditure. So it was just wonderful for everyone who had given the money, because we could see where the money went straight away, and people who received it were done all right out of it. It was awesome."

Margaret Hemopo had several operations on her leg which was not healing properly. In July 2010 she had her knee cap replaced and has recovered well. Ngatau and Suzanne both needed treatment for their infections and ongoing physiotherapy for tendon damage.

Madeleine Lewis is now in a secure unit in a rest home in New Zealand. Her condition deteriorated quickly after the September tsunami. The family had hoped that her Alzheimer's would ensure she forgot about her traumatic experience in Samoa but she didn't. She continues to mention the tsunami at times, using mainly hand gestures to show how big and strong the waves were.

All the children were unsettled after the tsunami but have improved with time. Tia still has not gone back in the ocean since September and her goal for 2010 is to go swimming. Like many others, she has a new perspective, "[The tsunami] has shown the kids that terrible things can happen, so things that *used* to be terrible for them, aren't. Normally for Tia, having injections used to be the worst thing, now an injection is nothing because she knows a lot worse things can happen."

New bonds have been made. Bonds that span oceans. In January 2010, Graham and Diann returned to Samoa to attend a Lene family wedding. Bonds that bring families closer together. "I think this has been a good wake up call for people in our family, that material things are not important. It's nice to have them, but they're not important when you see people who have just lost everything...while this was a terrible tragedy, it was ultimately a good experience for our family to have shared together, it's made us closer and value what we've got – all our friends and family."

November 2009.

It's been a month now since I started working fulltime on this project. I have met with survivors from Saleaumua. Satitoa. Lalomanu. Saleapaga. Malaela. Lepa. Vavau. And Poutasi. I have seen children who were saved by parents who held them above the water while they were submerged. I have touched trees that people climbed up to evade the waters. I have taken photos of the wooden cabinet a little boy sat on and floated to safety. I have met frail old ladies in their eighties who were carried on the backs of their grandsons to safety. I have listened to mothers weep because they could not save their little ones. I have felt the anger of fathers who could not fight against a tsunami. And I want to listen. And record their stories. And honour their experiences.

But not today. Because today I was afraid. Today we were interviewing survivors in Poutasi village. We had to navigate this narrow strip of hastily thrown together 'road' (A pretentious term for a load of gravel and sand dumped into the sea) to make it through to the actual village. Or what's left of it. The whole village was built on mangrove swamp. Which means it's actually right at sea-level. Which means when it's high tide – as it was today – then the sea is very best friends with the land. And today it was raining. And the sea was high. And lapping at my tires as we carefully drove over the road. And the broken shells of homes looked vaguely threatening. And frightfully depressing. And as I looked out over the ocean, I could totally envision a wave. Coming to engulf us. And I didn't want to be there.

Then we met with a lovely family at the end of the road. The mother was washing laundry at a broken pipe outside. Two daughters and a son welcomed us eagerly, running to bring us a mat to sit on while we interviewed their father. They had to sit beside him and shout out questions because his hearing was damaged in the tsunami.

And while I'm listening to his story, my eyes are drawn to the ceiling. Where the cement beams are cracked. And crumbling. And half the roof is caving in above us. And the reinforcing is gaping through the fragmenting cement. And the ocean is right outside the door. And it looks like just maybe... it could rear up and smash us to smithereens at any moment. And I could totally envision the roof deciding to give up its final gasp. And collapse in a heap of dust. Right on top of all of us. And I didn't want to be there.

I was desperate to leave Poutasi today. I felt panic claw its way up through my chest as I drove back over the narrow bridge. Thinking about waves. Strong enough to roll cars. Like tumbleweed. We only did three interviews today. But it was enough.

Today I was afraid of the sea. And imaginary killer waves out to get me. And imaginary roofs falling down to crush me. And I am sorry that I wasn't up to it today. And I am ashamed too. Because I didn't have to live through a tsunami. And I don't have to still live in a broken house with a fragmenting roof. So what the heck is my problem?

I shall take a breath. And go back next week.

11
The time of Noah is over.

While Tony Hill and his rescue team sped to Aleipata district, Peka Salapo was with the emergency response crew that headed for Falealili district. She threw water, tarpaulin and extra first aid supplies into the back of the pickup before heading out with another fire commander. They paused to let a policeman join them in the vehicle. Driving over the Cross-island road can be breathtaking. The road climbs into wet green rainforest then drops into rolling hills of cattle farm. There are swooping waterfalls to gape at when you have the time. With white terns soaring in the haze where the water meets the earth. The air has a coolness to it. Look back and you can see the Apia coast splayed out on the horizon. Look forward and the southern coast beckons with its promise of a slower, greener escape. Here on the south coast are legendary shores. Like the beach at Lefaga that took a starring role in the olden day glamour film 'Return to Paradise.' The white sand of Salamumu, Tafatafa and Matareva. Beaches that hosted the popular television series, 'Survivor'. Where blonde Americans caught rats for dinner. And squabbled over snails. While meters away, the Samoan forest teemed with orange papaya, juicy mangoes and substantial carbs like taro and yams. Not to mention an ocean thick with rainbow hued protein.

The Cross island road ends at Siumu. Take the right and you come to a line of up-market resorts. Sinalei and Coconuts. The Maninoa Surf Camp hosts the forever young, global wanderers in search of the next best wave. Take the left and the road follows the coast outline through villages and Togitogiga National Reserve. You can't always see the ocean from the road. At Poutasi, one has to turn off the main road to enter the village. There is Poutasi Hospital. A collection of cream-colored buildings. A meeting hall, the nurses' residence sits in fenced, cemented splendour. On a slight rise, you can look out over the ocean from the kitchen. Over the village that sits on the mouth of a quiet river with stately mangroves. Continue down the road and there is Poutasi School. A green field with rugby goal posts. Classroom blocks built on raised cement foundations in case of the seasonal flooding as the mangrove swamp would overflow during the rainy months. Houses line the rocky shore. A Catholic church presides elegantly at the far end of the road, with a statue of the Virgin Mary in the front yard. Her praying hands face the blue sea.

Before the tsunami, Poutasi was an enviably resourced village. A hospital. A beautiful school that called to students from surrounding villages. The new district police station a few meters up the road. They had nearby tourist facilities that offer employment. And an excellent program that gives fruit picking opportunities in New Zealand to the chosen ones who then return home bringing money, containers filled with canned food, televisions sets, stereos, new furniture and even a car or two.

In Poutasi, there is good leadership. From those who work hard to ensure ample opportunities for the village. From people like Joe Annandale and his wife Tui. They were partners in every sense of the word. Seventeen years ago, they stood together at Siumu village a few miles up the road. They looked out over twenty acres of tangled bush and muddy swamp and they dreamed a dream that would be called 'Sinalei Reef Resort'. From an original vision of a boutique resort in a garden setting , a "mama and papa type of place where Tui would cook the food and I would serve in bar and occasionally bring out my ukulele and strum a few tunes…that was the dream." Sinalei grew to become something else – a stunning five star resort with a spa, water sports centre, conference facilities, restaurant and pool. Languidly eased over the ocean were water rooms, honeymoon suites where couples could watch the sunset catch fire on the horizon and see fish dance underneath their glassed feet. Anyone who has been to Sinalei will tell you one of its most memorable features is its gardens. All landscaped by Tui herself. And not just from the air conditioned comfort of an office either. "Tui was a doer, she didn't just sit around and twiddle her thumbs…if she did a landscape, she just didn't do the drawings and say this is how we should do it and then step back, no, no, she did the drawing and then it was shovel and spade and the pick and she was in there planting and at the end of the day she was covered in mud and that's how she approached it."

A year before the tsunami, Joe and Tui took on a new role. Joe accepted a chiefly title from his village of Poutasi – the Tuatagaloa. They built a little house on the Poutasi shore and moved there taking Tui's ninety-five year old mother, Anna Schaafhausen, with them. She would sit on the deck that balanced over the sea and nap in the lilting breeze. "She loved it there because it's right on the water's edge and she thrived there. She's a Catholic and the church is just along the way and the catechist used to come over every second morning and bring communion for her so she loved it." Joe and Tui would commute to work at Sinalei every morning. A fulltime caregiver was employed to keep the bedridden old lady company.

When the wave came, the priority for Joe and Tui, was Anna's safety. "It was really about her, trying to get her out of harms way, why Tui and I got into the car with her because obviously it was the quickest way of trying to get up because Poutasi is quite flat." With strong, sure arms, Joe carried Anna to the front seat of the pickup, Tui jumped in the back with the nurse. They accelerated out the drive and onto the road which ran parallel with the sea.

"It wasn't until then that we saw this huge wall just coming towards us…I mean you couldn't jump out and start running, grandma was right there, you couldn't leave her. So we drove, we ducked in behind Meleisea's house and they were in their pickup truck and they were coming out as we were ducking in, but they were already out in the field actually facing the wave."

Joe pulled in behind the Meleisea's sturdy brick house, hoping it would protect them. "I was thinking that just maybe these houses and the palm trees that were between us and the wave would sort of break it a bit."

It was not to be. The ocean splintered through the houses, tearing through the trees. "It just picked everything up in its path, picked up the Meleisea pickup and the children who were in the front of the car died. It picked up the house we were hoping to hide behind, picked that up, picked us up and there were three other houses in that cluster, picked all of us up and it was just mayhem from that point."

The car was rolled three hundred meters up the road "we were hitting trees and flipping, flip, flip, flip, hitting trees and we were underwater…Tui fell out, the nurse fell out – none of us had seatbelts, at least I had the steering wheel to hang on to and with the other hand I hung on to grandma."

When they came to a halt, Joe and Anna were still in the car. "I managed to crawl out and floated out grandma, she was still alive…ninety-five years old and I don't know how many minutes under water and she survived. I called out to some boys to come down and give me a hand and grandma of course was crying and calling out for Tui."

The boys took the old lady another thirty meters up the road."I told them they had to take her up to safety, my fear was that there would be another wave. They lined a wheelbarrow with some cloth and they wheeled her up the road, the last memory I have of that moment is grandma in a wheelbarrow going up the road. Crying for Tui."

On that morning, standing there reeling, Joe had only one single-minded objective. To find his wife. "It was just confusion all around, total confusion. People screaming and yelling. In that hour everyone just went round in circles, people looking for their family members. I was focused on trying to find Tui and the nurse…I retraced our path. I went back to where we parked our car and tried to retrace the path that our car took and when I traced my way back to where the car was, there were some young fellas standing there beside a line of *fau* trees. One young fella came to me and said, *look I think we've found Tui*. I had this terrible feeling, given the way he looked and what he said…"

He saw her foot first. Dangling from the branches of the *fau* tree. "What must have happened was that she fell out of the car and got swept in the wave and she got hit. She was thrown up against the trees and got trapped there. I saw her foot first and I knew that it was all too late because two of the boys had already climbed up. We had to cut the branches to get her down, but it was all too late…"

They took Tui up the road to where James Belford had set up a makeshift emergency room in the open garage beside a yellow store on the main road. Peka Salapo was there. "Joe was badly injured, he was bruised black all over his face and his back, everywhere on his body. I recognized him but I didn't realize that one of the deceased women was his wife. I kept asking him if he wanted a ride to the National Hospital for treatment because we were transporting the injured – but he said no. Then one of the nurses told me about his wife, then I understood." Peka Salapo left shortly after to go on and join Commissioner Hill in Aleipata, responding to his call for more help.

Joe had a cracked skull, countless cuts, deep bruising, nail holes in his feet, skin with still buried slivers of glass and sediment clogged lungs. But he wouldn't start feeling his journey through the wave for another three days when the pain would begin to slice its way through the numbness. They were days where he was interviewed by international media, met with the New Zealand Prime Minister John Key who was surveying the disaster zone, began coordinating the cleanup of the resort and the village, – and buried first his wife and then her mother.

"Right through from Tuesday to about Friday, it was as if I was in a trance. I remember exactly what happened but in terms of physical pain, I was numb to all of that. Several days later when I went back to Poutasi and I walked around where I had been looking for Tui and I saw the amount of broken glass, corrugated iron and just all sorts of debris that would have caused enormous damage – then I realized why it was that I had terrible cuts on the soles of my feet and the nail holes, but I never felt them. Not until Tui was buried and I started to come to terms with what had happened by about Friday, the adrenaline had started to wear out, then I started feeling this incredible pain in my feet, my knees, my shoulders. I couldn't sleep because the pain was just unbelievable…I couldn't stop coughing…I had to be admitted to hospital for a few days."

Tui was the first tsunami victim from Poutasi to be buried. Early the next morning on Wednesday the 30th of September, 2009. "Tui and I had always discussed what would happen if either of us were to die first. So for me, the arrangements were very straightforward in the sense that I knew what had to be done and the rest of the family had to agree. In our family we have what we call '*tapu'*. No-one goes on ice, no-one is embalmed, we're buried within 24 hours, there's no television or radio notices and we don't eulogise. The funeral is very simple and whoever hears about it is welcome to come and our graves are just a hole in the ground. The person is wrapped in *tapa* and matting and then buried. No fancy graves and that sort of stuff, it's just a lawned area with a rock and some plants. Tui died about 7 o'clock on Tuesday morning which meant she had to be buried by 7 the next morning. At 5:30am we had a beautiful sunrise service at our family home at Tanumapua and by seven she was in the ground and then we had a lovely breakfast. There was no fuss. I was probably severely criticized by members of my village and being the wife of the Tuatagaloa she merits big fanfare with lots of fine mats and all that other protocol – but we had decided and it really lightened everyone's burden, actually there was no burden at all on anybody."

The Head of State said of the simplicity of Tui's funeral – "She wanted to move on with grace. All this is metaphor for moving on lest we dwell too long on death and tragedy, a salutary lesson not only for our family but also for a grieving nation." He went on to conclude, "Tui was a deeply spiritual person. For her, God was not distant and formidable; God was always present and an integral part of loving. He was present when she planted flowers, when they sprouted, budded, blossomed and withered. He was present in her love of animals, especially in her love for her dogs. He was truly present for her when the sun rose and set. He was present when she loved Joe, her family, friends and especially the disadvantaged. He was present when she and Joe prayed in the morning and in the evening. Knowing her, she would have prayed for the last time for the safety of Joe, her mother Anna and Tafa her mother's nurse."

Tui's mother was able to make it to the family service for her daughter but two days later on the Thursday, her condition worsened and on Saturday after the tsunami, she died from pneumonia complications, bringing the death toll from Poutasi to nine. Four adults and five children.

Tui was taken by a tsunami and yet her influence endures. "I like to think that I've never been a quitter and I'm pretty sure if Tui was hanging around, I'm sure she would give me a swift kick up the backside if I were to walk away from here and not continue the legacy she helped me establish." Joe is incomplete but fills the emptiness with a renewed purpose and direction. Where most have moved fearfully inland, Joe repaired his shorefront home decisively within two weeks. "Some people are asking whether I should be doing this and I feel that I want to, it would be a pity for our village to be abandoned, so I'm wanting I suppose to be setting the example."

Joe refuses to be still, to allow the grief to paralyse him. The sections of Sinalei resort shattered by the tsunami are being rebuilt. Bigger and better. And he has a vision for the rebuilding of their village. Bulldozers growl arduously as they clear the way for the new Poutasi. Generous friends of the village from many lands will see their dollars nurture a 'green' model community. Powered by solar energy. A centralized sewage system so the Poutasi springs will be pristine. The coral in the marine protected area will be replanted. The coastal plain will be landscaped so homes will blend in harmony with the sweeping mangrove trees. There will be a new pre-school for the little ones. A memorial for those who died, an information centre with the history of the village, a documentary of the 29/09 tsunami. "What I'm working on right now is a master plan for the rebuilding and reconstruction of Poutasi as a model village. I'd like to start reviving our rich history as well – I know it's a bit in the air but if I don't start thinking about these things and dreaming about it and setting goals then its not going to happen."

There aren't many men who can stand beside a tree that once held the broken body of a beloved wife, look out over the ragged remains of their village - and still, dream a dream. In Poutasi there is courage. Resilience. Strength. They say work is the best antidote for grief. And hard work is something Joe Annandale knows much about. "I suppose in a way, this new Poutasi project will help me stay focused and keep my mind off the tragedy of this whole thing, turning it into these other positive projects that can come out of the disaster."

James Belford is the Nursing Manager at Poutasi Hospital. Because of him, the hospital staff and the handful of patients were all standing on the hill when the wave came. James estimates there was a ten minute gap between the earthquake and the wave. "We stood there and watched it coming. The wave was high and it looked really hungry, angry, like it just wanted to eat everything. And it was a dirty color. It was very loud too. We saw the water cover the village, all the houses."

The injured came to James and his team as soon as the water rushed out again. "Just after the wave, they started to bring in people. The first woman that came up here was badly injured, her leg was cut through, all the muscles were severed from the corrugated iron." They set up a trauma center in a garage across the main road because the hospital was ruined. "Everything was covered with mud, there were dead fish in all the rooms and our equipment was smashed." People came with the dead. "All we could do was register them and then look for their families and take them to a shed on the side to await transportation to town. There was nowhere suitable to wash them or anything. I advised Joe to take his wife to Sinalei, maybe it would still be alright there, better equipped there to prepare his wife."

Tofilau Afatasi drives the Evaeva bus from Apia to Falealili district. He is forty-four years old, married and the father of three children. They lived on the coast with his wife's family, her parents and brothers. Their wives and children. A few weeks before 29/09, the country had made a historic switch – changing their driving from the right side of the road to the left. Tofilau's wooden bus was undergoing alterations at a mechanic shop in Apia to meet the new requirements so he was not on the road at four o'clock that morning doing the slow pickup of waiting passengers. He had however, woken early to pack his bag to go to town to work on the bus, planning to stay there several nights and return in the weekend. But his children had woken early as well, catching him before he could slip away. "I didn't go because usually I have to sneak away. If the kids are awake when I leave its too hard because they cry and ask me not to go. My one year old daughter Maupenei woke up and I didn't want her to cry, so that's why I stayed."

This gentle father who didn't like to see his daughter's tears, was making tea in the kitchen with Maupenei when the earthquake started. His oldest son was showering at the outside tap, preparing for school. The other, three years old, was still asleep. Tofilau's wife had gone to the store to buy bread. Tofilau took his children outside, comforting a crying Maupenei as the house rattled and the earth grumbled.

It didn't take long for the wave to come. Tofilau and the children were back inside the house, getting out school uniforms when it happened. "My neighbor called out, *it's a galu!* It was so quick. My brother-in-law Nonu ran out to start the car and I yelled to the kids to run. I told them to get in the car."

The children were quicker than their father and Tofilau shouted for Nonu to drive without him. "When the car got to the front of the house to try and reach the road, I saw it was too late. I saw the coconut trees lifted above the seawall because the wave had already reached us. The car reversed back and I ran towards them to try and get my kids. I saw the wave take the truck and it hit a tree. I couldn't see anything after that because I got caught in the wave."

Tofilau was taken by the water that churned him and mangled him. His grasping of a floating tree log saved him. But of his family there was no sign. "The only thing I was thinking about was waiting for it to be over so I could look for the kids…I had a feeling that they wouldn't survive."

Tofilau's eldest son was found alive but for the two little ones there was no hope. They died along with Nonu's one year old son. The little girl who would cry whenever her father left for work is buried in the front yard of their new inland home. And the father who could never bear to leave her, is now on extended medical leave. Recovering from his injuries. When we visit, he is making chop-suey with his son. Patiently. Shaking his head with a half-smile when the boy spills the pot of slippery white noodles on the ground. When he eagerly shows off his new Nike sneakers, treasure from a tsunami aid container. Tofilau holds his son close for a photo – "This is my only child left…I just thank God that one of my kids survived."

Moimoi Tumanuvao is a traditional healer. People bring their sick and troubled ones to him for '*fofo*', massage with chosen leaves and coconut oil. As such, he is closely tied to many in his village of Poutasi. He has listened to their worries, felt searchingly for their broken bones, soothed their tensions, gentled children's pains and addressed the spirits on their behalf. He and his family had just returned from early morning Mass when he saw the ocean revealing the reef. "I saw the wave bubbling over the reef and I said *the tsunami is coming.* We ran out of our house and everybody was running. I ran first to the minister and his wife and told them the tsunami was coming and they got up and ran."

Moimoi ran several lengths further, calling out for people to run as the ocean rushed in. Time taken to warn others, was time lost from Moimoi's own race to evade the wave. The first wave pushed him back into the trees behind the village, where he grappled for a handhold. There was another man under the trees struggling to stand, seventy-two year old Ali'i Ula. Moimoi grabbed him and had him climb on his shoulders so he could reach the high branches of the *poumuli* tree. Strength used to lift another out of the ocean's torrent, was strength denied from Moimoi's own efforts to climb to safety.

"The second wave came and the water was starting to get high, but the third one was higher than the trees. As that wave hit me I crashed into another tree, I looked up to find somewhere to breathe and to grab hold of a branch but I missed because the wave was too strong. I fell and was cut by a piece of corrugated iron."

Moimoi's blood mingled with the swirling waters. He was holding on to the tree now with only one good arm, battling the pain that threatened to overwhelm him with dizziness. But it is the screams for help that he remembers most. "At that time I heard women crying from the trees, they were crying for their children…I really wanted to go and help them, but my shoulder had gone, it was cut through, and I couldn't do anything."

Even when the water had stilled and Moimoi was headed for the hospital, the cries of mothers with lost children followed him. "I could still hear them. I stood and looked out and there was no Poutasi. At this time there was iron and wood lying all over the place. I was walking and I saw fish and sharks under the trees. I saw people searching for their families. I reached the hospital and there were mothers and fathers lying there but they were fine. I saw the other minister's wife, she was trying to breathe, but she eventually died. I saw them bring the *faletua*, the wife of Tuatagaloa Joe and she also had died. When I was waiting for a car to take me to the main hospital in Apia, three children arrived and they were dead."

There is great sadness in the healer who knows that there are some things that he cannot fix. Moimoi's children see a change in their father since the sea came into their village. "The reason why he got injured was trying to warn people, trying to save people but he couldn't do anything because not everyone wanted to listen. That's why he was injured. Now he is not the same, his arm is getting better but the tsunami damaged his hearing. Not only that but he seems very old now. He is not the same."

Afia Tauloa is sixty two years old.He has lived in Poutasi all his life, working on the plantation and caring for the family. He has seen many things come and go in his village, and many look to him for guidance. But Afia's forefathers never told him about waves that could cover the village. "I never heard from my parents about any '*galu lolo*', this is something new." So when Afia heard people calling out, *Galu! Galu! Big wave!*, he did not run. Not at first. "I told my wife and children to go inland because a wave was coming. But I stood there and looked out to sea, I didn't think it would cause any big damage."

Disbelief. The wave drew nearer. "The sea looked like a bomb was going off, the sea splashed up and the wave started to come closer but I thought the seawall would stop the wave. I didn't think it would come over the seawall, I wasn't worried."

Disbelief. Not until the wave swept over the seawall, did Afia move. "I got into our jeep, started it up and reversed out so I could get away from the wave…but then the wave caught me. I didn't know where it was going to take me, it was very deep, very high."

Afia was washed out of the jeep and into trees. "I was hanging on there for about half an hour and then the wave started to recede. After the wave, I started searching for a safe place to go."

Afia joined his family inland where they now live. What counsel does he have? "The only thing I can say is to be ready. A natural disaster is not like man-made events...be ready because you never know when a natural disaster is going to hit. Be ready."

Pauta Tuatagaloa ran with her two month baby when the wave came. They were home alone that morning and thankfully, woken by the earthquake, "the strongest I've ever felt in my life." People called out to look at the sea as it began churning but it was a curiosity rather than a fearful thing. "Everyone was just watching it, it wasn't until the sea went right out – that's when people started running. There was nothing else to do but run. The wave came really fast."

Pauta ran up the rise to where the hospital sat beside the main road. "By the time I got up there, I was walking because I couldn't run anymore, I was carrying my son. The wave caught us there, I thought the worst, that we were going to get taken by the wave but then it went back out and I got away."

The waves destroyed the family's home and even wreaked havoc on their ancestors, washing out the family graves in the front yard. "The wave sounded like the roar of a lion and it was so dirty. It dug everything up in its path because the front of our house used to be quite high above the ocean and it's not anymore. All our graves were ruined. We have tidied them up and repaired them but we're not sure if there's anything left in them now..."

Pauta's father was in New Zealand during the tsunami and he has since rebuilt their home in Poutasi. "But me and my mother and brother, for us who were in this disaster and have experienced it, we just don't want to stay near the sea...Our family is building another house inland. People say this will never happen again – but we never know. We never know."

The fourth commandment in the Bible is *Honour thy father and thy mother, that thy days may be long upon the face of the land the Lord thy God giveth thee.* In Samoan culture, respect for parents, for elders is paramount. From them we receive nourishment, instruction, discipline and if we are willing, wisdom. In return, we give respect, security, the first serving of food at *toonai*, humility and love. The instruction for people to leave their parents and cleave to their spouses, placing them above all others, is often a difficult one for Samoans to follow. When you are young or old, when you are single or married, you listen to your parents. Their word is law. Their needs are a priority.

Thirty-six year old Sinapo Faleuli of Satalo found these two givens at war with each other when the tsunami came. And her elderly parents told her to run. To leave them behind. To take her children to safety. To put the children first.

Sinapo has four children. The youngest is a teetering toddler, his curiosity at odds with his desire to stay snug in his mother's arms. The family lived together with parents, siblings and their children. On that day, Sinapo's husband had long left for work in the plantation. Her father was out the back of the house feeding the snuffling pigs while her mother helped prepare the children for school. After the earthquake, Sinapo's brother called for them to look at the sea."Look out to the sea, it's really different, it's gone right out, come see!"The family ignored him, until he shouted that there was wave coming. A big one. "Run, run, the tsunami is coming, run!"

Sinapo and her siblings scrambled to gather everyone, to take them to the bush at their backs. The old man is strong but Sinapo's mother is a frail woman.

"My brother was telling my parents to hurry up as the wave had come. He wanted to carry them. But my parents were saying *go with the children, the main thing is to get the children away.* We were trying to get them to come, but they said no. They said *run with the children!*"

Honour thy father and thy mother that thy days may be long...

It is a hard thing to turn and run, leaving behind the two people who have given you life. But in your arms you hold those who look to you for life. Sinapo ran with her baby in her arms, her sister carried the other little one. Her brother went with the other two. "The wave reached up to my neck... there were two waves and it was the last wave that was so strong and it was that one when I got really tired and I nearly gave up. I heard people screaming and being swept away. I couldn't hear my parents or see them anywhere and I thought they had died because they said *whatever happens to us is the will of God, but the important thing is that the children are saved.*"

Other villagers came and helped pull Sinapo and the children to safety. "The church minister asked us why we were crying and I told them that our parents hadn't come yet. He told us to keep going with the children while he went back to look for our parents, he said he would go with the young men to look for them."

Sinapo's brother went back too, swimming through the water until he saw his father tumbling in the chaos. He pulled the old man to a secure spot in a breadfruit tree and continued searching for his mother. When the water went back to the ocean, they found the old lady clinging to the shattered remains of their cement bathroom block. Alive.

Sinapo is thankful for her parents' willingness to sacrifice their needs for those of her children. And she thanks God for preserving her parents, allowing her children the gift of further years with their grandparents, years to give them honour and respect, to receive of their wisdom.

There are more Samoans living outside of their homeland than there are in it. They live in many nations, working, studying, playing sport, raising families. They go in search of better futures for their children but the ties that bind them to Samoa are strong. Every year, more than 365 million tala (145 millionUSD), flows into the country via remittances from relatives overseas. And every year, Samoans return home, renewing their links with the extended family *aiga*. Ten days before the tsunami, Steph Fa'amatala came from Australia to visit her mother's village of Sa'anapu. Her story, shared with Pacific Eyewitness.Org, is dedicated to Kenape – the boy who held onto her, the boy who saved her.

After the earthquake, people told them to go towards the hills. "The older boys were carrying the elderly behind the houses towards the mountains. The little kids who were all dressed for school, were running up the hill, laughing, happy thinking this is a game. I grabbed my mother's bag and our passports. I looked around and everyone was just standing around talking about the earthquake when the church minister drove past in his car in a hurry and screamed that a tsunami was coming."

It was a frenetic dash for the mountain after that. "Everyone was for their own self. We were running but it was hard…I was wearing jandals and the grass was high, the bushes were thick, the trees and shrubs were sharp but the loud explosion behind us was what kept us going. It sounded like a bomb had gone off. Then there was a rushing sound only much louder. It was the sound of the waves coming towards us."

Steph was pushed and jostled by the crowd. She pushed and shoved small children ahead of her, half lifting them up the incline. She could see her mother ahead of her on the mountain standing with many others, all crying and looking out towards the ocean behind her. Children were screaming and boys were yelling for everyone to run faster, faster. Steph made a grave error. She turned to look at what was approaching. And was caught in dreadful awe.

"I saw the house that we slept in half an hour before look so tiny against the big waves. It was a sight that I'll never forget for the rest of my life. I looked towards the bottom of the hill. There were people still coming…I knew they might not make it. I saw a man hold a woman to a tree and then he threw himself around her and pushed her into the tree. I saw men holding onto each other and holding onto fences that were the entrance to the hill. I knew I had to run *but I couldn't stop watching.*"

Steph stood, still holding a child in her arms and prayed and cried. Her older cousin, ran back, grabbed the child away from her and slapped Steph on the head. *Aikae kamoe! Run!* Shaken from her stupor, Steph stumbled on but she was too slow. A boy called Kenape saw that she would not make it.

"He said *girl, give me your hand, try and hold me tight and I will hold onto a tree.* So I held onto his hand tightly as he pulled me up the hill. Then Kenape pulled me into his chest, slamming my back into a tree. The wave pushed him into me and took all the air out. I tried to breathe but the water washed over us for what felt like ages. I drank seawater, it got into my eyes, my ears, covering us both then it eased and was up to my waist. I looked at Kenape and breathed and I was still crying."

Steph was disoriented but Kenape knew exactly what would happen next. The wave. Pulling back out to sea, sucking, dragging everything with terrifying force. He snapped at the sobbing girl in his arms. *Ku'u ia lou kagivale se, eke mamafa! Stop your crying, you're heavy!* Then he pushed Steph to the opposite side of the tree so they were facing the ocean, her chest against the tree trunk, sheltering her with his body.

"I complained, I didn't know what he was doing, I tried to argue with him but then it happened. The waves started going back to the ocean and then I silently prayed and prayed some more. I lost my footing and so did the boy who was wrapped around me. We were plunged under the water again and I was getting crushed by Kenape. And the pressure from the water going back into the ocean, pressing onto his back and adding pressure on me…"

Steph didn't know it then, but the human shield that was Kenape was being assailed by all manner of debris as he struggled to hold onto a crying girl he barely knew. When the wave eased, he fell to the ground, his back a mass of bloody flesh. "He couldn't move because he had cuts all on his back like someone had given him lashings with a whip."

Kenape and Steph lay tangled in the shallow water, looking up at the impassive sky. "I could hear crying and shouting, splashing as people who were on the hill started coming back into the water to get people out...an arm pulled the boy off and lifted him away. I saw the ocean slowly going back into the sea leaving the houses that did stand. I looked to the sky and it was a beautiful sunny blue – almost like nothing happened. My uncle's face came into view and he held me. I heard my mother's voice and I cried even more."

People from her village were injured, like Kenape. But no one died that day. Steph and her mother stayed on to see the White Sunday celebration in Sa'anapu. A glorious day where the children dressed in white sang to the heavens. "It was a beautiful White Sunday because no one died from our village…including a little girl who was saved by her father. She was caught under the water and when the wave stopped, her father pulled her out. She wasn't breathing. He did CPR and she came back to life in her father's arms. On White Sunday she sang the national anthem at the start of the church service. That day proved that we can still smile even after anything bad that's thrown our way."

Luaoa Ta'ipisa Fonoti is fifty years old. He and his wife have three children, the youngest is a four year old son. They used to live in the village of Vaovai when it was on the shoreline. When it looked out to Nu'usafe'e – Octopus Island. In front of Luaoa's house was a huge *fau* tree "You couldn't even wrap your arms around it, that's how big it was." Luaoa has lived here all his life. He supports his family by working on his plantation.

Luaoa had gone to the plantation early that morning at 4:30am, but the violence of the earthquake sent him running home. His wife greeted him. She was afraid because of the strangeness of the ocean. "My wife was worried because she said there was a noise coming from far out to sea, she said it must be a '*galu lolo*'. I told her *don't worry, there's no 'galu lolo' because Noah's time is over and the Lord promised there was never going to be another 'galu lolo' on this earth."*

Luaoa's assurances died away however as he looked out to the sea – "I looked out towards the coconut trees and there was a wave and it was higher than the trees, so I went and grabbed my sleeping son and we ran while the wave was following behind us. I could see it coming fast and there were other people running too."

Luaoa and his wife took their son to safety far back on the coral dirt path that came from the main road. Then he ran back with other men of the village to help those caught behind in the wave. "We were worried about the elderly of the village that maybe were too slow to run away. As we went, we could see many fish scattered on the road. We met a man who was carrying an old woman that was stuck in a koko tree so we helped him. Then we saw a second wave coming and we ran away fast because we were scared that we might not be strong enough to fight the strength of the wave because it was spinning in. Back inland, Luaoa heard the news that there were children missing. "After the wave left, we all went to look for the missing children. The first one was found right away in the mangroves but the second child was not found for three days. The little boy was swept up with cans of oil and rubbish and the water was all black so it was hard to see him." The two were brothers – two year old Hatonaina and his three year old brother Vaisigano Lauvai. Another two missing from Vaovai were found later. Masela Lafaua had died trying to save her three year old daughter Rosa.

Luaoa knows what a tsunami looks like. He remembers his elders speaking of something they called '*galu afi'*, fire wave. Now that he has seen the tsunami of 29/09, now he knows what they meant. "This wave was spinning as it came and in my mind, I knew this is what you call a '*galu afi'*. You understand how a fire works? When a fire spreads it moves very fast and what I saw was the great speed of the wave coming which is why I quickly grabbed my son. I was shocked though because we had just left the house and the wave had reached it."

Luaoa knows what a tsunami sounds like. "The sound was like trees breaking You couldn't hear waves like when they break on the reef, no. It was like trees breaking the way it came, not like a storm, but a huge bang like breaking. In Samoan we call it '*fafati'* which means the breaking of trees, that was the sound of the wave."

Luaoa knows how powerful a tsunami is. "It was something amazing to my eyes. My house was by the water and we had just finished building the gate, it was brand new. The wave destroyed it. It destroyed the guest house. And all the trees were totally uprooted when the wave came past so that's another meaning of '*galu afi'* because when a fire spreads it just burns everything and nothing is left. The trees that were nowhere near our house – they all ended up on my house. And the giant *fau* tree, it was bigger than the house and the wave ripped it up and smashed it on top of the house."

It is doubtful whether Luaoa will ever read the numerous reports on the 29/09 tsunami compiled by various 'experts'. He has more pressing concerns, like rebuilding a new home for his family, strong enough to withstand the cyclone season that hovers on the horizon. But if you want to know about a tsunami, *really* know – then don't waste your time attending a lecture by a scientist who spends his time studying tsunami ghosts, arriving after the fact to pick through the ocean's leftovers. No, if you want to hear it like it really was, then pay a visit to Luaoa in Vaovai village.

Masela Lafaua was the mother of five children. The youngest was three year old Rosa. Masela was a capable woman, accustomed to caring for the family alone as her husband, a carpenter was often away for weeks at a time working on building jobs throughout the island. Her independence was something that impressed her father, Sia Solialofa who sits in a wheelchair in their plantation house. He was not at their coastside home on tsunami day. "My daughter was a very strong person. She would go out fishing like the men do, she's a very good swimmer. She would even go inland to work on the plantation. She would look after our family and take care of everything."

That is why there is still confusion in Sia's mind as to why his daughter was not able to survive the wave. "I was so shocked, I can't understand why Masela died, why she couldn't escape? She was such a strong woman and a good swimmer, she did all the work for the family. I don't understand..."

After the tsunami, they buried Masela and her little girl Rosa. Then Sia's poor health took a turn for the worse and he was admitted to the National Hospital where he stayed for two months. Today, he watches over the slow rebuilding of their inland home. Masela's four remaining children are under his care now. They serve visiting interviewers cans of corned beef and tuna. Hot sweet tea. The very best for total strangers, Samoan hospitality in its purest form. What message does he have from the tsunami of 29/09?

"The most important lesson is to be prepared and be ready at anytime for such disasters. We must listen to the advice given by the government – like those messages that come on the television by that lady, Mena who is in charge of taking care of our country about natural disasters. The warnings on television and the radio need to be heeded, that's the importance of the message brought by this '*galu afi'*, to be prepared."

Lualei and Fa'aumu Laufatu are sisters. Lualei has just completed her Diploma in Tourism from the Polytech in Apia and the family is looking forward to her first job, to her contribution to the *aiga*, the extended family. Fa'aumu is a fulltime mother to three year old son Kenneth. A rambunctious little boy and the only grandchild, he is lovingly indulged by everyone. The sisters live as one family with their parents and extended family in Vaovai village. Masela Laufaua lived in the house next door with her children. Her daughter Rosa was Kenneth's most frequent playmate. They had an entire beach front to occupy them. And the mangrove thicket behind their houses was an adventure that was always waiting to be explored.

Because the Laufatu home is only a few meters from where the ocean laps the shore, they had a frighteningly perfect view of the wave as it made its way towards them. "The wave came straight on. The sea was dirty and it was moving in a sort of spinning motion. It was only about five or six minutes after the earthquake stopped and then the wave came."

The sisters ran with Kenneth to the next house and climbed on the sloping rooftop. "We stood on top of the roof, it was us and another girl from next door, but then she got really scared so she jumped down and tried to run away. She got hit by the first wave and we didn't see her after that."

The second wave was bigger, more powerful and it destroyed the house they were standing on. Fa'aumu was holding on to Kenneth as they went under. "I was buried under the house with my son. I couldn't see anything, I couldn't breathe. I pushed Kenneth up towards the light and the wave took him inland. Then another wave came and I was freed by the strong surge, it broke the house that I was buried under. All I wanted was to breathe, I was spinning around and around while I was under the water. I was getting cut all over, it was like cement was scraping my body. I clung onto a branch and just held on."

As the water receded, another woman came and helped a badly injured Fa'aumu to go inland, through swamp and bush, where she sat and cried for her son. Where was Kenneth?

When the wave smashed the house, Lualei was taken in the roiling flood of wreckage to the mangrove swamp at the back of the village where she was snagged by wooden tentacles. Sobbing with relief at her escape, she was wading through the moving waters, trying to get as far away from the next wave as possible, when a motionless little body caught her eye. It was her nephew Kenneth, face down in the water, ensnared in tangled roots. He wasn't breathing. Even today, Lualei weeps to recall that moment. "I felt so sad, like grief was cutting through me. I pulled him up and I did CPR on him. Three times I did it and then he came to life and grabbed onto me, coughing and crying. I took him and tried to call for help because more waves were coming. I didn't care what happened to me as long as he lived."

Others heard her and came to help. Lualei and Kenneth were taken inland where Fa'aumu clutched her son gratefully. Her son who was called back with three breaths. How tenuous is our hold on this thing we call life. Both sisters cry silent tears on a rain filled morning. We all sit and watch while Kenneth rides a tricycle over bespattered sand. He laughs at their attempts to call him in out of the rainshower. In this house, there is gratitude for a child who was not abruptly taken, but also sorrow for the family who used to live next door. That now are without their mother, the little girl who will no longer clamber with Kenneth through mangroves in search of elusive silver fish and scuttling crabs. There are so many 'ifs' – they clutter the air and smother us. What if Lualei had not seen Kenneth lying there? What if she had taken her ragged steps just the slightest bit slower and happened upon him a few minutes later? What if she had not known how to do CPR? Kenneth would have been a fourth child from Vaovai to be buried in a hasty grave dug after the tsunami.

In between the two main islands of Upolu and Savaii, are Manono and Apolima. Sixty-nine year old Lesa Tominiko is the mayor of Faleu village on Manono. There are over twenty people in Lesa's family living as one, in a cluster of houses not far from the rocky shore. The earthquake woke Lesa, disrupting his usual morning routine of going to church. He walked outside and stood looking out over the village, checking to see if there had been any damage from such a large shake. His eyes were caught by the ocean. "I looked out to sea and there was this long thing, it came from the side of Falelatai village and came towards us on Manono, but it just seemed like a ripple, a long low thing coming. I stood there and looked at it and then it grew very big and it wasn't long before it was very high." Lesa walked backwards to his other family house, a large European style home that stood several feet high off the ground. He stood in the doorway and continued watching the strange thing that came towards his village on the ocean. "It came to the seawall in front of our village – our seawall is high but this wave was about five feet higher than the seawall, the waves were breaking over the top of it. The wave smashed the seawall and then came and ran into the houses across the road from mine." Houses by the seaside were completely destroyed. "It was so quick, there was no houses left there, even a big *palagi* house was smashed." One person on Manono died. 78 year old Tautua Silupe from the village of Apai. Another little boy from Apolima island, 4 year old Pili Po'o drowned when the boat he was travelling in between the two islands, was capsized by the wave.

Manono is an isle of fishermen that rely on the ocean for life. "The wave took all the Manono boats from the wharf and lifted them and took them to a car park area. They were all packed in there together, all ruined and broken – that's how you know this wave was so powerful and so quick." Several fishing boats that had been out a ways to sea were spared the damage and were in the best position for scavenging as the contents of houses and shops were swept out into the ocean. "That was the horrible thing about this disaster. People with boats went out and waited for stuff to drift to them and they would take it. There was a lot of useful stuff like drawers, chests, furniture – the boats just loaded them up and took them home and those poor people who had lost their possessions really lost out."

Lesa was quick to telephone the DMO for help after the tsunami. "Assistance was very quick. A helicopter came on the first day with water and tents and food for us. You should have seen the children – they were so excited, many had never seen a helicopter before, I had to chase them away! The helicopter couldn't land because the whole field was packed with kids."

Lesa is grateful for everything that Manono has received since the tsunami wiped out half of their homes. "Blessings have come through the many countries that have donated to us. They have given us way more than we ever had before. I'm the one that shares out the assistance and I go to these families and there's no room in their house for all the donated clothing, everything is so full. We joke about it because in those days, you'd have trousers only for the men, but these days, because of the assistance, even the old ladies are wearing pants! And now, families that only had a Samoan *fale* now have a big *palagi* house so there have been many benefits."

Most of the people of Manono ran to the hills when they saw the wave coming. Except for Leiataua Unaite who had been stricken by a stroke some years earlier and could not walk. Dr Joe Williams of the 'Tsunami Samoa 2009' Appeal visited Leiataua while carrying out medical checks for people on the island and was able to hear his family's story, reported in the Event Polynesia website. "It was apparent to the primary care givers for Chief Leiataua Unaite, that they would not be able to shift him out of their house and relocate him to higher ground in time. So Leiataua's wife and his daughter Sema decided to forgo their own safety and stay by his side, whatever happens…within a few minutes the waves broke through the doors, windows and walls of their house. Both Sema and her mother lifted Leiataua up so that he would not drown. When the waves retreated they fought to keep him from being sucked out to sea. All three were able to survive the tsunami."

Dr Williams visited another family – that of Tuilaepa Fua'ava. He was there to check up on the matriarch of the *aiga*, Afuafu who is also elderly and bedridden. The family also had their tsunami story of heroism to share. "Lanu is a woman who has married into Chief Tuilaepa Fua'ava's family and has become quite attached to Afuafu. She calls her 'mother' and Afuafu calls her 'daughter'. Lanu is Afuafu's primary caregiver. When the tsunami hit Manono island, Lanu was already on higher ground with other family members when she realized that Afuafu was still back at the *maota* (chief's house) on the beachfront. Lanu ran down towards the rising sea, single-handedly carried Afuafu to her wheelchair and then started their run for life back to higher ground. The waves were around their knees now but Lanu did not give up or turn her back on Afuafu." Both women survived. Dr Williams remains moved by their story, "Afuafu and Lanu remind me of two women from another story from the ancient world, about a mother and a daughter-in-law and their commitment to each other. Their names were Naomi and Ruth. *I will not leave you, nor turn back from you…"*

November 2009.

Today we went to the village of Vaovai. We drove inland on a dirt track to find families who have relocated and are slowly rebuilding. We met a man heading for the plantation with a basket of coconuts on his shoulder. He was happy to speak with us but the rain sent us scooting for shelter into the nearest house. Another tsunami survivor family. They graciously welcomed us in. The head of the home is an older gentleman in a wheelchair. Silver-haired and handsome with a roguish smile. During the conversation, he sends his grandson to the store to buy two cans of pisupo (corned beef). Sitting in a ramshackle shelter, we are served a sumptuous lunch - rice, pisupo and tuna thick with oil, sweet hot tea. The family sits there and watches us eat. I feel terrible. We have come to gather stories from people who lost everything and they are feeding us the very best. We apologetically explain that because of our religious beliefs, we can't drink the tea. The man laughs. He jokes "Don't worry, I won't tell on you...see, look at our roof? You are sitting under it and God can't tell if you have some tea today!" Then he sends his grandson back to the store. To buy large bottles of Coke for us instead. I feel more terrible. We eat. I haven't had pisupo for about three years now - trying to be low-fat and healthy you know. But it would be even more terrible if I didn't eat. They bring bowls of water for us to wash our hands. We depart well-fed. With many good wishes and blessings, hopes for our work to go well this day. Another dirt road. Another interview. This time with several men resting in the shade beside cut trees. Yes they were in the tsunami. Yes their houses were ruined. But life is good. Nobody in their family died. The main speaker – a princely man in his sixties – has a tiny house in the bush that sways in the wind. But he has 40 head of cattle which he points out proudly. He winks, "See, I'm a rich man, any of you ladies single?!" They get us fresh niu to drink. And give us another basket of coconuts to take with us. "Better than spring water! You take these to keep you refreshed during your work." We drive back to town. Humbled. Grateful. For Samoan hospitality. For generosity amidst devastation. For the kindness and courtesy of strangers. We ask a lot of people as we visit - please welcome us into your home, please share your tsunami story, please allow us to see your loss, please let us witness your attempts to rebuild. And time and again, they do. With grace. With warmth. With patience. And sometimes, with pisupo.

Thank you.

12
We had nothing except our lives.

In the village of Siumu, there are three different tourist accomodations lining the shore. Sinalei Reef Resort and Coconuts Beach Resort. Sandwiched between them is the Maninoa Surf Camp.

Sose Annandale is the general manager of Sinalei which has been in operation since 1996. In 2008, Sose attended several tourism workshops where the DMO spoke about Samoa's tsunami risk. "It caught my attention when they talked about the fault line, the tectonic plates, the possibility of a tsunami and all that sort of stuff – it was sufficient to get my attention. I invited them to come out here and talk to our staff, give them some training."

The DMO visited Sinalei and carried out tsunami awareness training with the resort staff. Sose and her team did regular drills, practicing an evacuation procedure for their resort. But still she did not feel like they had done enough. "I then went and looked for an alarm system, some sort of siren. It was very expensive but my feeling was that if you have the safety of your guests in your care then you need to go the extra mile, you need to make sure that you take these things seriously. I hoped of course that a tsunami would never happen in our lifetime, but unfortunately it did, so I'm very glad that we made that decision to install a siren in 2008."

When the earthquake ended, Sose mobilized the staff. "We assembled everyone here at the main office. I told my son to go and get all the cars ready out front, we got the rooming list out and gave a list to each of our security guys so that they knew which rooms to go and knock on the doors. I got the office staff ready to phone all the rooms but I was waiting for that tsunami alert that we were told during the national drills would come over the cell phone. I didn't want to alarm our guests too soon in case there was nothing to be worried about and that's why I was waiting, I had my phone in my hand, waiting for the warning to come…we must have lost at least two minutes waiting. "

With everyone poised and ready to move, Sose's teenage son shook his head worriedly, *Mum I don't think we're doing enough, we need to do something more*... Sose made the decision to act. "We pulled the alarm and started getting everybody out. Right at that moment, I looked out and the sea was beginning to go out…"

There was controlled panic as staff moved quickly to escort guests to the waiting vehicles. "There was no difficulty getting anyone out, we all knew what to do. The office girls called every room to make there was no one left, we sent security around to check every room. We checked everyone off using the room lists. We got everybody out."

All thirty-eight guests were evacuated safely. Including a couple staying in the honeymoon suite over the water. "Somehow they didn't react, they just didn't react at all. After two phone calls and one of our security guys going down and knocking on their door, they didn't react, they just stayed in their room. I think they may have been asleep. One of our security boys saw the wave coming and he just belted down there because he knew that one couple was still in their room. He broke the door down and pulled the two of them out of their bed. He got them out just in time before the house was hit and it just disintegrated."

The guests were taken inland to the mountains. Later they would be transferred to Aggie Grey's Hotel in Apia. Shaken by their near escape but unhurt. "They were panic-stricken, very grateful that they had gotten away safely and of course very emotional as well." Seven resort units were wiped out, along with the restaurant, bar, water sports centre and spa. No staff member or guest from Sinalei Resort was caught in the wave. No one was injured.

Coconuts Beach Resort

Coconuts Beach Resort has been in operation since 1990. Lina Bell is the general manager. She has been with Coconuts for over fifteen years so she has watched it grow from a restaurant and 3-stool bar to a spectacular boutique style resort that embraced its breathtaking natural surroundings.

Lina is one of eight staff who lives on the premises. That morning, she was in her house when the earthquake hit. She immediately went out and walked around the resort. "I checked the property to see if there was any damage from the earthquake, but nothing was destroyed and all the guests were okay. I spoke to some of them while I was going around the property, I told them just to keep an eye on the ocean because at that time we hadn't had any tsunami warnings yet. Then I went back to the office because I wanted to check the computer to see if there were any warnings and also to call the meteorology office in town."

Before she got to the office, Lina heard the siren. "I looked over to the restaurant and saw the staff there were laughing and going about their normal routine, so I called out to them to get out of there but then we figured the alarm was just a warning."

Thinking there was time to organize an evacuation, Lina called to one of her staff to start telling the guests to move. She asked another to bring the resort cars from the back so they could evacuate the guests. She then continued inside the office to look for keys to other vehicles. "As I came out of the office I saw everybody was running around and there was only one car that was parked right next to our office and one of our construction managers was inside the car already and staff and guests were loading into it."

The first wave had arrived. "I saw this big wave breaking through the windows of the *fale* beside the lounge room…I started running."

Lina's boyfriend was screaming at her to get in the car. She jumped in and the vehicle roared away, just as the ocean began to bulldoze its way through the resort. People screamed as the waves caught them. After the waves, Lina came back down to the resort. "We started looking for people, making sure the guests were safe. Some of them we found here in the property and we took them right away to the nearest hospital, MedCen. The others who were not injured we relocated them to another hotel in town. When I was up at the hospital, talking to all the guests, that's when I found out that there was one lady missing, her husband was there injured. So I rushed back down to Coconuts and my staff had already found the dead body. She had been washed to the back of the resort.

There were forty guests staying at Coconuts Resort that day, including an Australian couple, John and Maree Blacker who had come to celebrate Maree's fiftieth birthday. After hearing the siren, the Blackers had run from their suite and were in the car park when the ocean caught them. Mr Blacker told his story to the Australian media.

"All of a sudden we were hit by this bloody massive wave…I couldn't hold on to Maree. I was getting rolled like a bowling pin, I thought I was a goner. I couldn't grab on to anything, until the last second. It was a palm tree. I was just able to wrap my arms around it. The debris was still bashing into me. I took in a lot of water, but there was no way I was going through all that again, I wasn't letting go, I simply wasn't letting go…I must have been unconscious for awhile."

John Blacker suffered a broken jaw, cracked ribs and cuts to legs and feet. He was airlifted out of Samoa a few days later. Two weeks later in Tasmania, more than five hundred people attended Maree's funeral where Bette Midler's song *Wind Beneath my Wings* played as the mourners entered the Longford Baptist Church. A eulogy from Maree's fifteen year old daughter Laura was read, *My mum was always there for me and I saw over the years how much she cared for others. Mum, we didn't have long enough together, but I'll treasure our memories forever. I love you.* The Reverend said of Maree, *She wasn't afraid to call a spade a shovel and to hit someone over the head with it if need be...John we know you're hurting and we are hurting with you.*

Rachel and Jeff Lynn worked at Coconuts Beach Resort. Rachel is a runner and was training for the 2010 Boston Marathon. That morning, Rachel was doing her tempo running workout when the earthquake struck. "My vision seemed blurry, and I thought I was faint, but oddly I felt fine. I noticed the ground ripple in front of me. I stopped running and put my hand to my head, when I looked up at the power lines I saw them swaying…we were having an earthquake. I started running for home."

Jeff met Rachel on the way, coming to check on her in the car. Back at the resort, Rachel hopped in the shower while Jeff went to check the office computer for a tsunami warning. As she stepped out of the shower, the sirens started wailing. "My heart sank. Jeff ran to the front door and yelled *we have to go, a tsunami is coming! I'm going down to the resort to warn the guests."*

Rachel ran from the house half dressed. "I ran as fast as I could up the hill. The kitchen staff was moving up the hill as well. They were beginning to tire and slow down. I kept screaming at them to run. I knew we had to get as far and high as we could…we still weren't on very high ground yet. I ran for my life. Yes, I ran for my life and everyone else around me did too. Sometimes you have dreams where you are running and not really moving. I had always wondered if given the situation where my life depended on my ability to move, if fear would paralyse me. It didn't, I was able to get away. I was full of fear and adrenaline."

Down by the beach, Jeff was running yelling for the guests to run. He saw the wave coming and ran back to the entrance of the resort. The water caught him so he climbed a coconut tree. He went up about twelve feet and still the water washed higher and over him. "He realized that if he continued to hold on to the tree he would drown. He made the decision to let go and take his chances floating…he washed down almost a hundred yards and then swam to the two storey block of the resort, he tried to go through the lower floor trying to get to the land side of the building. He found himself floating against the ceiling on a mattress. He heard voices on the second floor and punched a hole through the wall and called for help."

There were some guests sheltering on the second floor of the building. A wave had taken Jeff's shorts so he asked them for a replacement. Then they tied some sheets together to try and pull him up, yelling that they could see another wave coming. As Jeff grabbed the sheets, the hastily tied knots came undone and he fell back into the water. He screamed, *please don't let me die!* As the water level rose, Jeff somehow pulled himself up to the second storey where he collapsed exhausted. When the third wave had gone, the group came down and started running up the hill, Jeff hobbling behind them.

After the first wave, Rachel ran back down to the resort, worried for her husband. "I could only make it to the entrance, everything was destroyed. I was screaming Jeff's name, praying for a response. There was none. Some of the villagers were down there with me. They mentioned broken electrical lines and I waded out of the knee deep water. Then they pointed to the ocean and we all saw the next wave coming. A wall of black water. I ran for my life again, up the hill, filled with fear and despair."

Rachel joined others after the second wave receded, cautiously walking back down to the resort. She had lost her shoes and slowly made her way through the wreckage, calling despondently for Jeff. The sirens began their wailing again. A third wave was coming in. Again, Rachel ran up the hill. This time another staff member called to her from his SUV, *get in!* "As we were making our way up the hill I saw a wet man with a familiar Kid Rock t-shirt on. I didn't recognize his shorts but I was pretty sure that it was Jeff. As soon as we stopped the car I ran back down the hill. It was him! He had a bed sheet wrapped around his head and an unknown pair of swim trunks on but he was alive. He had a huge gash on his knee, he was coughing up water, he could barely walk. I asked him to stop and lay down but he was deathly afraid of the next wave. He wanted to get far away from the ocean. He was starting to go into shock."

At the top of the hill, Rachel helped other staff organize the injured. She and Jeff got into a taxi and raced up the mountain to MedCen hospital with horns blaring. Jeff's cuts were stitched and he was hooked up to an IV. Rachel went around the hospital checking the rest of the guests from the resort. "The feeling at the hospital was an overwhelming sense of gratitude to be alive. Everyone there was beaten and battered, but alive. One guest, John was still missing his wife. They had come out of their room and started to run when they heard the siren. Then she fell. He picked her up and started to run because he thought she had broken her leg. As he carried her the wave came. He can't swim. He held on to her as long as he could but she slipped away. That was the last time he saw her alive. Later that afternoon he was taken to the morgue to identify his wife's body. "

Another man wanted to thank Jeff for helping to save his family. "They had been staying in one of the beach *fales*. They had been standing on the beach, dazed after the earthquake, they heard the siren but didn't know what it meant. They thought it was just another earthquake warning. When they heard Jeff yelling they started to run into the woods. As the wave came he got his family up in a tree and he started to climb up as well. He saw a car come floating straight towards him. The car hit him and pinned him against a tree. He grabbed for his son as the water got higher. He held on tight and his family survived."

The Lynns went back to the resort later that afternoon. "At this point we had nothing…nothing except our lives. No wallet, no money, no underwear. Nothing. We drove over the hill in silence." The sight of the resort was a blow to the solar plexus. Like it hurt to breathe. "The utter destruction was unbelievable. The over-the-water-*fales* were completely gone. They were found the next day in a village a mile down the beach. The villagers there were moving into their beachfront homes. The restaurant was tilted, the furniture was gone. Nothing was left of the bar but the floor. There was a twenty foot fishing boat on top of the whole mess. The office was gone, the walls and windows of the beach *fales* were blown out."

Rachel and Jeff's beach house had been wracked by the ocean, lifted off its foundations and floated some distance with the wave. Looters had already taken what the ocean didn't want. "All the clothes in the closet were gone, yet the hangers were still hanging. I commented to Jeff that we were going to start seeing people in the village wearing his shirts..."

Lina Bell is still troubled by what she saw that day after the tsunami. As people rushed to the resort. To steal anything and everything useful. There were even police officers looting the weights room, putting gear on their truck. "I came back from the hospital where I was trying to take care of the guests and the whole place was filled with different people. I just don't know where they all came from! So many of them. It was so difficult to stop them from taking things. We tried to keep them out of the property…it was so embarrassing for me, being a Samoan and it's our island, our people. I couldn't believe it."

In the late afternoon there was another tsunami warning. "We ran for our lives again, only this time it was harder. Jeff was slower, we had to dodge debris and dead fish. There was an eel dying slowly in his new hole on a sand walkway, thirty feet from the ocean. We ran across barbed wire and wood, torn plant life, and soiled machines. But we ran."

Rachel writes, "Words cannot describe the utter terror of the situation. Driven by a primal fear for survival I ran faster than I knew possible in a pair of flip flops. Knowing that my husband had gotten washed in a tsunami I desperately called for him. Seeing buildings flattened, torn from their foundations, flipped over, I knew his chances of survival were slim to none. His own primal will for survival led to the ultimate decision. Fortunately the risk was worth the reward.

We felt despair at the realization that everything we owned was gone – if not washed away by a tsunami, then stolen by the villagers."

Several days later, Rachel and Jeff drove over the hill to Coconuts. "We each began to cry as we saw the ocean. It is so beautiful, that drive over the hill. The ocean is such a rich aqua colour. It looks so inviting but it is so deadly. If it wants to take you, it will. I am only grateful that it wanted us to live."

Maninoa Surf Camp was a collection of casual beach huts scattered over the sand, a restaurant and office building on a little rise as you drove in the gate. It's where those in search of good surf would come to stay. People like UK surfer Tom Gogola and his girlfriend Rachel Cant. New Zealanders, Charlotte and Beau.

In September 2009 Tom and Rachel had come for a four week surfing holiday. Tom is a heavy sleeper and Rachel had to wake him when the surf camp guide came at around 6:30am to announce the surf boat would be leaving in five minutes. She stood on the shore and watched Tom wade into the ocean to climb on the boat. Sunrise embraced the surf party as the boat churned out towards the break in the reef. Rachel waved them out and then went back to bed.

"I awoke to the most terrible feeling. At first I thought it was a dream, the *fale* was shaking, dust was falling from the ceiling and objects were dropping off the banisters. I got out of bed fast and went outside…Charlotte was staying in the next house. She came over to me, looking just as confused and nervous as I was…the telephone wires up high were swinging gently side to side and the parked cars in front of us were shaking. Then it all stopped. It was silent again – just the sound of the waves on the beach."

The guests clustered outdoors, talking agitatedly about what had just happened. A few of them tried to laugh away their nerves but one of the Samoan staff members looked at them and shook his head, *it's not funny, that earthquake could lead to a tsunami!* Hearing that, Rachel and Charlotte decided to walk over to the hotel next door, hoping they would have more information about any possible waves. They had just grabbed their bags when the sirens went.

"The alarm sounded – a piercing sound, like a world war two alarm. I ran out of my *fale* and looked out to the ocean. The sea had completely sucked out, the reef was on show, it looked like the sea had gone out for miles...all the Samoan workers started running so fast away from the shore, the man who had warned us was screaming, *Run, Run! Tsunami coming!"*

Rachel was going to sprint after the men, but Charlotte screamed at her to get in her rental car. It was a little jeep, parked awkwardly on the sand, facing the sea. There was jostling confusion as the women tried to get in. Charlotte couldn't get the key to open the passenger door so she ran to the driver's side and got in that way, with Rachel banging on the still locked passenger doors. Charlotte unlocked the door from the inside and Rachel clambered in, banging her knee on the dashboard. Then Saga, a young Samoan girl who lived at the camp, ran over and tried to open the boot door so she could climb in. Charlotte hopped out and yelled at her to get in the driver's side. Saga crammed in beside Rachel while Charlotte again got in the car and fumbled with the key in the ignition. By now, all three women were sobbing, heaving harried gulps of air, *please, please, God please!* The nightmare took on horror movie status – Charlotte turned the key, and the jeep wouldn't start.

"I was crying *Go, go, go! Please!* The jeep was facing the ocean, I looked out to the sea and there it was – a massive roaring wall of white water, far out to sea but travelling fast. I had never been so scared…I wanted to get out of the car and run, but I froze."

In the next moment, the car started and Charlotte struggled to put the stick shift in reverse. Success came with a lurching jolt as the jeep revved backwards, hitting a coconut tree. More screams. "Charlotte turned the car around and drove fast up the slope, we were driving past so many people that were running. As we approached the gate, we saw it was shut, most people were just jumping over it, but then one guy managed to open one side, not big enough for the jeep to fit through. Charlotte floored it, slamming through the gate and we were out and going up the dirt road…it was so rough that I banged my head on the ceiling several times. We were getting thrown around in the car."

At the top of the dirt track, Charlotte turned right on the tar seal road that ran parallel with the not so distant ocean. "I looked to my right into the jungle where, even though you couldn't see it, the coast was not very far away. I was just waiting for that huge wave to take us from the road, there and then."

The women sped through villages that seemed oblivious to the drama that was unfolding. "We drove past unsuspecting villagers walking along the road, we waved and beeped the horn, screaming, *Run! Run! Tsunami!*" Some people misunderstood their calls and waved back with huge smiles of greeting. Others got the message, grins turning to alarm. The jeep slowed at a pedestrian crossing where teachers were walking about fifty plus children down to their coastside school. "We screamed to who we thought was the teacher. Immediately he turned to the kids and pointed to the hill. They all started running. We kept driving. I looked back to see large numbers of people running now, maybe not even knowing what they were running from, maybe they had heard the alarm?"

All three women were praying aloud. Charlotte spoke in Maori, with her eyes trained on the road. Saga implored the heavens in Samoan. Rachel's prayer tumbled thick and fast in English.

Would God make sense of the muddle? Would He hear them? Onward they drove, unsure – how high was high enough? How far was far enough? Finally, they pulled in to a driveway, turning the car to face the faraway horizon. "We all looked at each other, all scared and knowing by the look in each other's eyes, what we were each thinking of. Saga's brother was down there, asleep in his *fale*. Me and Charlotte both had our boyfriends out there surfing. We all cried and held hands…*God please bring them back safe."*

Where was Tom?

There were eight other men on the surf boat that morning with Tom. The boat was an eighteen feet fibreglass vessel powered with a 30-hp Yamaha engine, more than enough to take them the fifteen minute ride to the shallow coral reef that edged this part of the island. There was a break in the reef where the river emptied into the sea. "By the time we got to the channel and put the anchor in, the sun was just starting to rise above the beautiful landscape to the east of us. The surf was glassy, clean and breaking at the four foot mark – it was looking good and the rides were up to one hundred meters on the good sets. This wave, as with most on the island can only really be surfed from mid to high tide as any lower, it will wrap around the reef too much and close out on the end section, not only that but its way too shallow to the shelf, meaning that if you fall hard, you'll be licking your wounds for days to come!"

Tom was first out of the boat, paddling on his board about a hundred meters out. The day seemed perfect. "I caught the first few waves of the day before the rest of the guys had paddled out and was stoked about surfing better than I had the day before."

Another boat arrived – Neil Lumsden with two young boys, Manoa and Kealoha. Neil anchored the boat, leaving the boys to play while he went to catch a few waves. Friend, Darren Greenwell soon joined him. There were now ten surfers out in the water. None of them felt the earthquake. "We had no warning, being out in the water outside the reef pass we didn't feel the earthquake happen and everyone was caught completely off guard."

An unusual wave was Tom's first sign that something was wrong. "Everything was good and calm until out of nowhere this small wave came in. When it hit the reef it jacked up in height but it had a real square and unsurfable look to it. I was paddling back out at the time and duck dived through it. As it broke over my head I looked at the reef…the sea was bubbling as if a ten foot plus wave had just detonated on the reef. We were all puzzled… the sea kept fizzing and surging violently, then a guy from the other boat said casually, *I think the waves are done for the day, we best get off the reef, I think there's a tsunami coming…*"

As soon as Neil saw the reef shelf sticking up out of the water, he "knew instantly what the situation was, this was extremely serious…" He and Darren looked at each other as the water dragged out around them, their thoughts in agreement. Tsunami. Neil's driving concern was the two boys in the boat alone. He needed to get to them, fast. Panic was rising in a swelling surge but it was not a fear that he wanted the boys to see. Hence his 'casual' warning to the other surfers before a speedy paddle back to the boat. Darren explained, "It's normally like a five minute paddle from where we surf to the boat, but it took us about three minutes! Within those minutes there would have been four and a half meters of reef sticking out of the water. It was like the reef was rising up."

The two men made their way back to the boat that was now sitting in a fast-shrinking pool of water. The boys were calling out fearfully, unsure what was happening. "After the ocean receded, the surge began to fill back in within about a minute. The water hit the boat and began pushing it toward the shore, but the mooring was still hooked and the force of the surge pulled the line taut until it began pushing the nose of the boat down into and under the water. There was only so much length on the line and as the water rose above normal sea level and continued pushing the boat toward shore, the nose was submerging…the groms were thinking the boat was going to sink and were ready to jump out and swim, so I yelled at them both to stay in the boat."

At the boat, Neil and Darren struggled to climb in, fighting against the current. "The force of the water was so powerful that even though I had gotten to the boat…I couldn't pull myself out of the water because the current was so strong sweeping past us and pulling on my legs." Darren could only get in the boat once he took off his surfboard leash. Instantly his board was gone, swept away by the current. Manoa and Kealoha helped to drag the men onto the boat that was now dangerously inverted. "The water levels were starting to rise again, we had trouble getting the boat off the mooring, there was no slack in the line and the two of us physically couldn't pull the boat forward or back to get some slack so we could unhook the rope off the bollard. We didn't have a knife either so we couldn't cut ourselves free." The boat was sticking out so high now that the engine prop wasn't even in contact with the water.

Darren laid himself along the rear of the boat, weighting it enough so they could get the engine to power them, giving enough slack on the line so they could release the boat from its mooring. Once free, the men roared the boat out to sea and safety.

Behind them, the other surfers continued their struggle with the ocean current. Tom and three others had the same idea – get out to deeper water. He was still stung with disbelief, "denying the fact that a tsunami was on its way" but still he paddled furiously away from the reef, away from the distant shore. Not everyone agreed with this safety strategy. "I looked for the other surfers and to my surprise, saw them paddling back in, towards the boat, towards the shore, swimming like they just saw a shark…we shouted for them to turn around and come back, but they didn't."

Tom and the others would remain there for another hour occasionally paddling against the current, "waiting for one of two things to happen, either for it to finish and level itself out or for a big tsunami wave to kill us all like the one in Indonesia…but there was no monster wave like everyone imagined there would be where we were, but the tide height dropped like someone pulled the plug out of the ocean for kilometers around so that the reefs that were once one meter deep, were in the space of half a minute meters above sea level and because they had millions of gallons of water on top of them, they created violent waterfalls."

Some distance out to sea, Neil and Darren felt the wave roll under the boat. "It was more like a big rolling hill, or crossing over a huge hump. It was about thirty meters in width of water coming towards us and we went over that, we just rolled over the hill." The men could see the waves hitting the shore. "We watched as the surges powered toward shore and absolutely levelled everything. Complete devastation the likes of which I have never witnessed before, everything was washed away into everything else – boats into huts, cars into trees, with water surging a mile inland in some areas. Then the muddy water pulsed in and out about four times before finally receding back into the ocean."

The pull back took all the debris with it. "There were pieces of decks, three seater couches, a lot of stuff from the resorts, even full walls with plumbing still attached, flooring with toilets still standing – all swirling out by the boat." The men were dodging debris and still trying to calm the two boys. They unobtrusively started scanning the water for any people who may have been swept out. "The real tough part was worrying about the kids – this was the biggest thing I'd seen in my life but you're trying not to distress the boys, trying to make sure they're okay…we were looking for bodies in the water, looking out for people but there was no one."

The same concern had occurred to Tom and the other surfers with him. Once the surges seemed to have calmed, they paddled back to the surf boat that was "thankfully still good" and made their way back east and up to the surf camp, meeting pieces of wreckage. "One of the first things we noticed was a couple of kayaks that had been clean snapped in half and were bobbing upright way out to sea, then we saw the luxury restaurant roof float past us…and most unbelievable, the two butane gas tanks for one of the restaurants that would have been two meters in height, spinning, bubbling and hissing as the gas escaped… that's when we realized that if anyone had been snorkelling or kayaking on the inner reefs they may have been swept out to sea or already dead, so we decided to check for people before we headed to shore. We scanned everything as best we could, but after an hour of searching, we couldn't see anyone."

The ocean continued to alarm them though as it dropped again several more times and surged in again, "We played the waiting game for a while, staying in the safety zone until the surges seemed fewer and further apart."

Tom came ashore at Maninoa. Or what was left of it. "The only things left standing were the coconut trees, the reception building that was on high ground and part of the eating area which was built from thick timber poles. Every *fale* was spread from the sea to a hundred meters back up into the bush, nobody could recognize whose roof was whose."

There were assorted police and locals there, but of the families and friends they had farewelled a few hours before, there was no sign. Nobody seemed to know what had happened to them. Tom and the others started looking through the destruction. "It wasn't our belongings that was on my mind, it was the fact that nobody knew where Rachel or the other girlfriends were. The other surfer with me was frantic about his wife and two young boys. I must have searched under roofs, bits of wooden walls and fallen coconut trees for a couple of hours in the blistering heat, not knowing if I would find Rachel or anyone else under them. There were rumours they had gotten out safe, but not enough clarification to be satisfied. I found various items of ours, nearly none of them which was of any use…ripped, dirty clothes to old pictures of home. The only thing I found that was of any good was my wallet – emptied of all the money, but I still had my cards. Then miraculously, I found our folder of important documents about a hundred meters up from where our *fale* once stood."

Tom was relieved to see the surfers who had opted to swim for shore. "They told stories of how they rode the second surge and breaking waves, all the way up into the thick jungle. They wrapped their arms around coconut trees and climbed them to stop from being sucked back out to sea. They had all made it except for one guy who was still missing, a Brazilian surfer." The police had been told about him and far down the coast, the Nafanua Patrol boat was scouring the sea.

Finally, the group decided to head inland in search of their friends. The ocean had ruined all their rental cars though, so they were happy when an English couple drove down the Maninoa road. "They had heard the news about the tsunami on the radio and wanted to help. Once we explained our situation they offered us a lift up the mountain to a school where people had evacuated to."

The surfers piled in and drove inland. Tom's consuming thought... where's Rachel? At the busy school they didn't find the girls, but they did see the missing surfer. "He was sitting in the corner of the school field, I told him to jump in the car and come back to the camp with us, *everyone thinks you're dead!*" Tom and Beau took their lost surfing companion back to Maninoa, hoping the girls had shown up.

Rachel and the two women had driven further up the mountain, following the exodus of vehicles and bewildered people. "We got out of the car and searched frantically through the crowd. Everybody seemed calm and relaxed, there were tourists in their pyjamas, everyone was looking out to the ocean, it was so far away, it looked so normal, like nothing had happened. Locals were gathered around a car listening to the radio, trying to find out what was happening down there. Saga translated for us...*the wave was big, it has taken everything, it's still too soon to tell of any deaths or what damage it has done."*

More people arrived, lumbering trucks and buses emptying and driving back downhill to get more people. "I kept a close eye on every vehicle that came along, hoping and praying that Tom was in it. Charlotte looked at me – *they're going to be fine, you'll see, they'll be there waiting for us when we go back down.* I looked in her eyes and I could tell that she was finding it hard to believe herself. I felt sick and weak. *Tom's dead, Tom's dead*...kept going through my mind over and over...*will I ever see him again?*"

The radio began telling tsunami tales. "They were talking about more bodies being found, there were more and more emergency service vehicles speeding down past us, heading to the shore. I broke down many times in tears, not believing that Tom had made it out of that. The devastation reports on the radio just sounded too awful."

When the tsunami warning was finally cancelled, the women drove the long drive back to the surf camp. "As we arrived, it was unbearable to look at, destroyed, everything was gone. The beach was covered in things washed up on shore. I saw the boys surf boat still in one piece and my heart lifted. Then I saw Tom. I can't describe how I felt when I saw him safe and alive. I cried as he held me close. Charlotte found Beau and Saga met up with her brother. "

By now it was mid afternoon and Maninoa Surf Camp was a busy bustling place as looters crawled over everything, "looking for washed up fish, wallets, alcohol, anything." Tom and Rachel loaded what they had recovered of their belongings into Beau and Charlotte's jeep. "We put five surfboards on the roof, four or so bags in the car as well as six people into the tiny Suzuki and headed for Apia, feeling exhausted and drained, all we wanted was to get out of there." In town, they went straight to the New Zealand embassy office. It was crowded with others in similar situations as them. Washed out, no clothing, no passports, no money and no place to stay. It was a frustrating few hours. "We were called through to a conference room where we filled in forms and pondered on where we would go from here. The embassy told us we could use the phone to make calls but just within Samoa and we couldn't use the internet to tell loved ones in other countries that we were alive!"

There was wild panic at about 4:30pm when the siren went off again. "The radio announced another tsunami warning. Staff at the embassy ran out of their offices and shouted to everyone to get to higher ground. We ran out of the office and piled into a pickup truck that belonged to one of the staff. Everyone was evacuating to the safety of a school further up which is where we sat until sundown with no idea if anything had actually hit or not."

That night, new found evacuation friends offered them a place to stay. Two days later, Rachel and Tom finally arrived in New Zealand. "We were still wearing the clothes we wore on that devastating day. On our plane were a lot of other people who had the same experience as us." Rachel's story concludes – "For me and Tom to still be alive is an incredible thing. When I thought Tom was dead was the longest, most horrible three hours of my life. I will be forever thankful for our survival, but my heart goes out to the people who have lost family and homes. I hope they can get their lives back together soon."

Neil Lumsden and Darren Greenwell cautiously brought their boat in at about ten that morning. Darren found his surfboard floating in the channel. Neil's car was a quarter of a mile inland, stuck in some trees "crushed like it had been in a trash compactor, flipped upside down and sitting on top of another car's roof." Darren's van was " in a tree, a bit shaken up but still salvageable." They didn't like to hang around though as the beach was filling up with looters. "We walked past Coconut's Resort and it was a mess, unfortunately people weren't doing the right thing, there was a lot of looting and stealing going on. At first I thought they were trying to clean up, like they worked for the resort, but then I realized they weren't. I didn't want to hang around there – people were getting drunk on stuff from the bars and getting kind of violent. So we got out of there and walked up to the main road hoping to get a taxi home."

Darren hadn't told his wife *where* he was going surfing that day. It had now been several hours since the tsunami had hit and she was frantic with worry about him. She was driving along the Siumu road when she passed the bedraggled group. They had just stepped out of the bush, Darren clutching his surfboard under one arm. There was tired relief. Exhaustion and gratitude. Neil summed it all up, "It was an intensely depressing day, with everyone in shock at how to begin the recovery process from this horrific event. Without a doubt this was the gnarliest experience I have ever gone through in my life."

November 2009.

Today for the first time on this project, I felt a gut wrench of personal loss. Fifteen years ago we stayed at 'Coconuts Beach Resort'. Three days. I remember the lobster, stuffed with wild rice and seasoned with garlic, baked to perfection by Mika - the best chef on ***any*** *island. I can still taste the crab cakes we had for breakfast. We ordered every single cocktail on the menu at the 3-stool bar. Midnight blue fire over crushed ice and hot Kahlua sweetness underneath an endless night sky. We got hot and tired paddling a two seater paddle boat into the mangrove swamp and I was afraid of crocodiles. And swamp monsters. (My new husband's first glimpse into the paranoiac muddle that hides in my overfed imagination!) We've been back several times since and in 2008, we took the whole family there for the weekend. The baby bathed in the spa tub fashioned out of black lava rock where a waterfall gushed out of the wall dotted with green ferns.*

And now there is nothing left. Well, that's not true. There's a huge mess left. The gecko-shaped pool we swam in is now filled with boulders, coral, trees, and pieces of the beach villa that we stayed in. The 3-stool bar is gone and the restaurant is a heap of rubbish. The staff cry as they tell me about their tsunami day. I cry with them. They were going through wet linen, muddy towels, crockery picked from the chaos – trying to find anything salvageable, locking it away from the scavengers.

I am stunned. And I do not know how anyone could possibly recover from this. Where do you start? If you had built this with your own hands from the ground up, pouring your soul and your bank account (as well as the banks interest-laden accounts) into it – why would you want to start again? It's like standing on the threshold of a truly horrific bedroom that has been completely trashed from top to bottom and not knowing how to begin to fix it. So instead you just step back and shut the door. Firmly. With finality. And walk away.

They tell me the owners will rebuild. And I am stunned some more. Take the insurance money and run. Far away. Would be my impassioned retort. Shut the door. Firmly, With finality. And walk away. I cannot even begin to imagine the commitment, the strength, the money, the resolve required to even contemplate rebuilding. It moves me.

Today I realised, I had been underestimating the courage required to rebuild. To carry on. Everywhere in the disaster zone, survivors are making the decision to move forward. To keep living. Everyday a decision is made to keep cleaning up. Keep rebuilding. One day, one step, one breath at a time.

13
The First Response

It was nearing 9:00am on tsunami day, when a white pickup carrying Filomena Nelson and a policeman, slowed to a stop under a breadfruit tree in Saleaumua village. She had taken the long route through Fagaloa Bay, checking on the status of coastal villages like Ma'asina and Uafato on their way. A dog had rewarded her with a vicious bite for her trouble and now she limped around the village, her leg wrapped in a makeshift bandage that seeped red.

No amount of training or hours of preparation – not even the 'knowing what to expect' could have prepared Filomena for the first sight of the disaster zone. And it was surely no comfort at all to see that which she had been warning us about had finally, truly come to pass. Three months after the tragedy the 'Samoa Observer' newspaper would name her one of the nation's people of the year, calling her 'Samoa's Noah'. The magnitude of the task that now lay before her must have made her blanche, and the keen observer would have seen her dark eyes glisten for a few moments. But she wasn't appointed to this position for nothing. Yes, the situation looked overwhelming. But how does one do the impossible? One step at a time. And now, all her multi-tasking talent came into play.

Everyone was calling Filomena and the DMO staff that morning. Filomena had three cell phones and a wireless radio ringing incessantly. The police wanted to know what was needed first. The overseas media wanted to know what was happening. The Red Cross wanted to know what was needed most urgently. The Health service was anxious to get an estimate of the fatalities so they could prepare for the influx. A callout had gone over the radio for road construction crews to head for Aleipata to help clear the roads and construction companies wanted further confirmation of the needs. The Prime Minister Tuilaepa Lupesoliai Sailele Malielegaoi in New Zealand was trying to find out what the heck was going on in his country? And crying, desperate survivors were still somehow getting put through to Filomena's phone as they called emergency numbers for help, please help us, is anyone out there listening? Is anyone coming? And who better to give them all a straight answer than the Principal Disaster Management Officer?

Fielding several anxious calls at once, Filomena started directing the initial relief efforts. She had already called the Electric Power Corporation to shut down the power to the whole district before she left town. With the flick of a switch, countless numbers were spared electrocution as they slogged their way through muddy water and tangled wires. All Water Authority vehicles were asked to head for the zone to assist with transporting the injured. And the dead.

"It was lucky that I took a policeman with me in the car because they need to identify the dead bodies, officially. He identified the first victim at Saleaumua, a woman. It was sad to see the husband and her son there, her little son. They were just standing there looking at their mother..." Filomena shakes her head at the memory, "She was safe you know, she had gotten away, she was supposed to live and you know what happened? The family told me, she forgot her phone and she went back to get her phone and that's where the wave pulled her down."

Filomena would hear many such stories that day. Women who stopped to lock all the doors and windows of their houses before evacuating, only to get caught by the wave that smashed their house to smithereens. People who turned back to get an MP3 player, a cell phone, a wallet.

Continuing on their way was a slow and sad process that morning. Seeing the DMO blazoned on their truck, people came running. To bring their injured and dead. *Please can you take us to the hospital?* To seek reassurance. *Please, will there another wave?* That was a question Filomena couldn't answer and they continued to warn people that further waves could hit and it was not safe to be on the coast yet. Tsunami waves can continue to recur up to twelve hours after an earthquake. The whole while Filomena was getting updates from her contacts all over the island. A co-worker in the Land Management Division who lived at Lefaga called in to report they had safely evacuated the area after the earthquake, all the way down to Lapata'a.

"The poor guy, he works for our ministry, he was busy evacuating people and then he looked back and his house was gone!...People we had worked with in the past, from the women's committees and the *pulenuu's* (village mayors) – they were on the phone with me the whole morning saying, 'we evacuated this area and that area."

In those first hours after the waves, the strength of the traditional village organization was evident. Yes there was initial confusion, but leadership came from where it always had. From the village council, from the *pulenu'u,* from the women's committees. At Saleaumua, Filomena gave the essential instructions to the *pulenu'u* – search for the injured and get them to the roadside for transport to the hospitals. Prioritize the injured. Get the numbers of the dead. Watch the sea for more waves. Keep the children and elderly inland.

Over at the village of Matautu in Falealili, there was collective action. After the earthquake, a church minister Amosa Vaetoe rang the church bell to warn the village. The *taulealea* (untitled men) carried the elderly on their backs. Mothers ran with their little ones. Tilasi Popese estimates the wave hit ten minutes after the earthquake. "Absolutely no one in our village was caught in the tsunami." Their neighbours were not so fortunate. "We could hear the screams of the other villages that they wanted to live and they needed help because next to us is the village of Vaovai where four people died." The young men were sent to help – rescuing several elderly and a child caught in the mangroves. That night, the village camped together in the bush. They made one huge *umu* (cooking pit) and everyone ate as one. The unity continued in the following days. "If anyone wants ice water then Sene's family get it because they have a lot of fridges in their shop that wasn't destroyed. It's all transported back to our camp by the people who own cars…there's none of this separating of families. Everyone works together as one, everyone is all family."

In most areas, the extended family, the village family – were the first responders. The first helpers. Church ministers organized search parties to return and look for the missing. The *matai* (chiefs) mobilized the young men to erect rough shelters in the sweltering forest so the elderly could get out of the hot sun. No one had to face their crisis alone.

While Filomena led the Disaster Management team starting from Saleaumua village, Tony Hill was fighting his way through from the opposite end of the district. He had started meeting up with his team as he made his way over the cross island road. He was in a pickup truck. Behind him was a fire truck. Following was the fire service ambulance. The team had no idea that Falealili district had also been hit by the tsunami. So they were taken aback when the coast road was busy with people. Some running, others carrying children and elderly.

"People were bringing their injured out to the main road to get help. That's when we first knew that Poutasi village had been hit as well. We met our first victim there on the road. She was a girl about sixteen. We pulled over and managed to resuscitate her before continuing on. Then they brought a baby. The baby only had a very slight pulse so I told them to put the baby in the ambulance and take him to Poutasi hospital."

Tony made the decision to keep going – knowing there were more teams following close behind. He continued to Aleipata district, coming to the village of Lepa first." I looked and I just couldn't believe the damage that I saw. We couldn't see any people because most had already gone to higher ground."

Progress on the tsunami-ravaged road was arduous. "We had to start clearing the road of rocks and debris to get through. We cut power lines and I had to use my truck to pull fallen power poles out of the way. There were only ten of us and it was tough going."

The rescue cars stopped at the next village - Saleapaga. It was unrecognizable. One member of the FESA remembers it vividly. "We found destruction, like it was a total cleanout. I had just been there two weeks before and it was nice. The homes – they were beautiful big houses, but when we got to Lepa and then Saleapaga, it had all been wiped out. I couldn't believe that in only a few minutes, everything was gone."

Tony's story is heavy with the sense of isolation and helplessness they felt as ten people tried to do the work of several hundred. "We were the first ones out there and for about three hours – we were the *only* ones out there. There was no other help. We were the only ones and everyone was looking to us to do something. I was really shocked. Survivors were already pulling out dead people and bringing them to the road. There were a lot of wounded people, calling out for help. And a lot of *palagi* people there too."

It is difficult to comprehend the helplessness the helper feels when confronted with an impossible number of people all desperately in need – of help. There is an image in Tony's mind that he will never forget, that of the children of Saleapaga."We were standing there surveying the damage and I looked up to the mountainside and there were all these children, just standing there and looking down at us. And I couldn't believe it, you know? I couldn't believe these children had survived. They had to run over rocks and stuff to escape and to climb that cliffside – I can still see them standing there looking at us and looking at their village that was gone."

Assistant Commissioner Tipaula Laupu'e remembers,"There were people walking around, sort of in a daze. They had no shoes – some of them had no clothes." Tipaula remembers one woman in particular. An elderly tourist with grey hair, tall and thin. Wearing an orange sweatshirt and pyjama pants. When the rescue team arrived she was on the side of road – sitting on her suitcase. "How she managed to escape the tsunami with her suitcase, I'll never know. But she was determined to take her bag with her. She was very agitated. I thought at the time she was just suffering from shock, but later I found out that she has Alzheimers. When it was time to evacuate, she wanted to take her suitcase in the ambulance with her but there was no room. She had to leave it behind and she was really upset."

The team of ten worked at Saleapaga digging people out from the wreckage, administering first aid with only the bare essentials, and helping to carry the dead to lay them beside the road for later transportation. The injured were put on top of the fire truck and taken back to Lepa, back to the hospital in town. At about 10am, Tony got a phone call." It was my wife and she said,

It's just come over the air, theres a good chance youre going to get hit again, maybe about 1 oclock. At that time, we were right into it, digging out people. At that time we had no transportation. I was pretty upset."

Tony had to decide. Send the survivors up the steep mountainside for safety and get his team out of there? Or wait and try to evacuate them all via the coastal road? One of his team pointed out that the villagers were afraid of the mountainside collapsing. Huge chunks of the hill were already eaten away by the ocean and didn't inspire much reassurance. They decided to evacuate all the survivors immediately. The word went out. *Leave the dead. Bring the injured. There's another wave coming soon.*

The news invoked terror. Understandably. People flocked to the two rescue vehicles, clambering onto the fire truck and cramming into Tony's pickup. And one Good Samaritan's taxi. "We were really lucky because there was this one taxi driver who kept coming back – I think he was trying to get through to go find someone he knew. And I pulled him over and told him I really needed his help with transporting all these people. And he looked at me and he said, *okay Commissioner, I'll do it."*

If it hadn't been so fraught with fear – it would have been a comical sight. "This poor guy and his taxi..." Tony shakes his head smiling at the thought, "He had all these people in his car. Some *palagis* were hanging off the back of his car, he had the back door open. And I was trying to tell everyone that if they didn't fit, not to worry because I would make sure they did get out later, but they insisted on climbing on his car. The bumper fell off, there were people sitting on the hood of the taxi. When he drove off you could barely see his car at all. But you know, after he took them all uphill to Lepa, he came back. And did it again...he was a big help that day."

An ambulance had arrived from National Hospital. But it was so full of doctors, nurses and medical gear that they couldn't transport any survivors out. They were treating people right there on the ground. Right on 'Ground Zero' as Tony called it. Another group of emergency service personnel arrived – Peka Salapo was one of them.

In the midst of trying to get the mass of people out of Saleapaga before a possible next wave came, there was another welcome gift. Or three. "We were very lucky, these three vans arrived from the Mormons. They said to me, Tony we're here to help, how can we help. There was a lucky break again. So they started taking people uphill. We used these vans all that day."

It wasn't easy, but after another hour, most survivors had been evacuated inland. It was then that Tony received further bad news. The Number Two Fire Truck had crashed while on its alert mission on the opposite coast, killing a firefighter. Four others were in hospital. Ata Suluape was one of them. His wife Natasha was a fire-fighter working alongside Tony out at Saleapaga. But like so many other rescue workers, she could not leave her duties to see to her own personal disaster.

The rescue team was shaken as they heard about their colleague's death. The critical state of the other fire personnel. But the focus had to be the task at hand. Tony decided to move on to the next village Lalomanu. "We were about to move on when I saw this young fella walking round and round in circles. He said he was looking for his family. I asked him where he last saw them and he pointed in the direction of the mountain. There were all these people we didn't even know about up in the mountains, some of them injured. I sent some of the team to go get them but when we brought some of them back, this fella had gone into shock. He didn't even know his family. He pointed at them but he didn't recognize them."

There were a lot of people in the mountains, many who refused to come down. Several of the fire team escorted them further inland to a distant road and walked them back down. To the Lalomanu Hospital. To the school. To emergency shelter. Peka Salapo was in the team that went inland to find survivors. "I thought we were lost because we kept going so far up the hills and we didn't see anyone. Then finally we found everyone so far back in the mountain. It was tourists and villagers. They'd been standing up there all day with nowhere to go and no shelter. They were hungry and thirsty. Some of them had injuries. Some of the old people were sick before the tsunami and now they weren't doing too good. We had to call for a van to come and get the tourists and the sick ones."

Tony continued on to Lalomanu. As the team loaded up on the fire truck, one of the fire team picked up the elderly woman's suitcase and appealed to the Asst. Commissioner. *Boss, can we take the old lady's bag? Maybe we can find her somewhere later?* The team stowed it safely away. Tipaula smiled,"We all agreed that we could make space on board for the suitcase. She had been so upset…it just didn't seem right to leave it sitting there on the road."

At Lalomanu, further carnage awaited them. The bulk of the dead in the entire Pacific tsunami, were from this one area. Fifty nine people. Here Tony and his team would stay for the rest of Day One, never making it on to the rest of Aleipata.

One section of Lalomanu curves back into the cliff-side and when hit by the tsunami, it was almost like it created a bottleneck, a big chute that channelled all the debris into a pile at the end.

"The sea had come right in and taken everything back into the mountainside. All the vehicles that must have been on the beach were right at the back there and the damage done to them was incredible. These were brand new vehicles, I just couldn't believe it. They were mangled beyond belief. There must have been more than twenty vehicles back there. The first victim we came across was a young girl – Sili Apelu's granddaughter Moanalei. They had just pulled her out of a car. Then we just kept looking for bodies and more bodies just kept coming out."

By now it was after noon. The one o'clock deadline for another tsunami had come and gone. But still the rescue team worked with a heightened awareness of the ocean at their backs." You're working trying to dig for bodies and look for survivors, but all the time you're thinking about the ocean behind you, and what it's capable of. It was a pretty tense time."

Worrying about what the ocean could still do, hampered the recovery efforts of the Nafanua Patrol Boat. Once her Commanding Officer Petone Mauga had arrived at the wharf, they had headed out to safety in the open waters. From there they had been called first to go to Siumu. Several surfers were missing, feared swept out into the open sea. They spent several hours vainly searching for the men – who unbeknownst to the Nafanua – had already come to shore several miles down the coast. The searching was not without its risks as floating debris met them about two nautical miles from shore - trees, the roof of a *fale* Samoa, a chest freezer. It was slow, cautious navigating. At 3:30pm they were ordered to head for Aleipata to assist with the rescue work there. They had just arrived at the Saleapaga coast when another tsunami warning sent them back out to open waters. By the time the warning was cancelled and they had come back in, it would be nearing dark. They launched their small boat anyway and began searching a crowded ocean. On that first day they recovered the body of one elderly woman caught on the reef at Saleapaga.

By the afternoon, cars had started arriving from Apia. And not all of them were welcome ones. Yes some came to bring assistance from town – like EPC trucks there to fix power poles and others coming to help their families. But others were spectators, come to gawk at the chaos. When you've been digging through muck, pulling out one crumpled body after another, hearing a mother sob as you hand her a muddy, lifeless child and a grown man weep as he covers his mothers body with his *lavalava* – you have little patience with people who are out for a scenic drive through a disaster zone. Tony made a harried phone call to Filomena.

"*You have to get on the radio and stop everyone from coming across here, people are just driving around taking photos, they aren't here to help. You have to cordon off the road.*"

And so listeners glued to their radio's that afternoon were given a special announcement. *'The DMO is asking everyone to please – don't come to Aleipata district. It's dangerous. The police and rescue workers are trying very hard to do their job. Don't come out to look. Only go there if you are going to help. Go in your work clothes and wear boots and go help."*

For a rescue team that had been out in Ground Zero for over six hours without respite, without food or water - there was frustration. And exhaustion fuelled anger which was intensified by the next lot of visitors – scavengers. It seems inconceivable that there could be those who's first instinct after the tsunami, was to come running to look for… stuff. Food, clothing, household goods, money and more. It was all lying there. For some then, as Tony said resignedly, "I suppose it was like Christmas with all this stuff lying around. There was a lot of people coming and just helping themselves. There were guys who were drunk from the alcohol they had taken from the resort bars. We were helping injured people and there they were sitting on a van piled high with gear they had collected, choo-hooing and shouting."

Nynette Sasse is head of the Samoa Hotel Association. A powerhouse of efficient action, she was at Lalomanu trying to find scattered tourists and ferry people to safety and was disgusted by the sight of young men eagerly ripping into resort alcohol. "I was so angry, I couldn't believe that they could do that when people needed help. If I'd had a machine gun, I would have turned it on them!"

People without missing and injured family. People from outlying areas unaffected by the waves. They came in droves to pick through the mess. Everyone disappeared though when another tsunami warning came over the cell phone network at about 4:30pm."It was like a cartoon – there were all these cars parked alongside the road and within seconds they were all gone. And the people, everyone was just sprinting for the mountain."

Peka can't forget those panicked moments. "This voice kept yelling, *run, run there's another wave coming!* So we all left the beach and started running for the hills and it was like nothing mattered. Even the iron roofing in the way, I didn't care about it. We got uphill and some of the boys helped us up. My husband was there with me because he was with the police that had come out to help. And we were both sitting in this tree and I was thinking about our kids. Our six kids. And what would happen to them if we both got killed in a wave?"

Pepe Christian Fruean, CEO for Digicel Samoa remembers that run very well. He and others had gone out to Aleipata with food and water supplies. "There was a little bit of a shake, an aftershock and the sea went flat, the waves went out again and it was very scary. We couldn't get out because the cars were all backed up so we abandoned our car and we ran up the hill and there was a little bit of an embankment and basically there were at least two hundred people who couldn't get out, they all went up those hills and were hanging off trees…it made me realize that while our intentions were to help, we never recognized what actual danger we had put ourselves in, because there's always the possibility of secondary waves and stuff."

A rather shell-shocked, Tony remembers, "I looked up the hill and there was this line of pigs just looking down at us…what must they have been thinking of us, everyone scrambling, well at that point in the day, I just didn't know what to think."

With the warning cancelled, Tony's team went back to the seemingly endless task of recovering bodies. Would the day never end? "I'll never forget that. Pulling out all those dead people. It was unreal. And people standing around just crying, and they're looking at you – you know, to do something. Finding those babies. And little children. We put people on a kayak to float them out to the road. And most people were naked and I had to tell people to give me their *lavalavas*, to give the dead a bit of dignity, you know?"

Peka Salapo is the mother of six children that live on the island of Manono. She had not been able to reach any of her family on the phone and only knew that waves had destroyed many houses in their village. She tells of her feelings when assisting with the recovery of the body of young child floating in the foul waters at the back of a broken house. "His body was bloated, it had expanded because he had been in the water for about two or three hours…for me, I couldn't take it. I had the picture of my children in my mind, because we live on the coastline and at that time I hadn't heard from them. So I had it in my mind that my children could be in that situation. Every time someone called out that they had found another person, I just couldn't actually bring myself to go to them."

Some lay like they had drifted off to sleep. Like a little shake was all they needed to wake them. Leua Vi'iga's baby looked like that. With only a scratch on his forehead and a bruised lip. One young girl was found wrapped in a sheet of corrugated iron. Still wearing her pyjamas. Others had far more obvious cruelty done to them. Staff at the National Hospital morgue would be grateful for the Disaster Victim Identification team who came from Australia to assist with the grim task of identifying the incomplete dead. The Head of the National Hospital Laboratory Vaomalo Kini explained, "We had difficulty identifying some of the victims because there was a lot of sharp debris out there in the water with them. There was one young boy who was claimed by two different families because part of him was missing from the torso up. The team used DNA samples to identify which family he belonged to and so that family was able to take their son home for burial." The DVI team was also able to identify several of the lonely dead – those travelling alone. A German woman, Lieselore

Rock. And another visitor from Brazil, Ana Isabel Pinheiro da Silva would not be formally identified for another ten days, using dental records.

Even for a seasoned veteran like Tony, the time spent recovering victims was very difficult. "I saw some awful things – and I thought I'd seen it all. But even for me it was really hard. I was more worried about my team though. Many of them were very young and new recruits. Witnessing death is part of our job, but on that scale, its very traumatic."

On that Tuesday, Tony's team struggled, doing the best they could, feeling increasingly isolated as the day wore on. They were not to know that other sections of Aleipata had suffered similar fates to Lalomanu. That Poutasi and Vailoa villages were struggling to find their dead. That Malaela village had no houses left. That Satitoa village was crying for its missing. That many others – volunteers and government workers had started their efforts from the opposite end of Aleipata at Saleaumua. Or that the repeated 'new' tsunami warnings throughout the afternoon meant many cars kept turning back on coastal roads, fearful they too would be caught by the ocean. The sun was setting when Filomena arrived at Lalomanu with her DMO team. In time to bear the brunt of Tony's tired frustration. Filomena had come with a truckload of tarpaulin and other relief supplies. She asked Tony for his teams help to distribute it.

"I was furious. Where are all the DMO people? And the government people? I said we're not here to distribute stuff. My boys and I we've been working here all day. We've had no food or water. We're tired. We're going home. I was really upset. I felt like we had no support all day. I told her we've been working out here alone all day. Where is everyone? We could have used so much more help. Where were all the government vehicles that could have helped with transporting people? I was really disappointed that we hadn't gotten more help. Where was everybody?"

A lesser woman would have reacted. Badly. Especially if she were a woman who had spent the day up to her eyeballs in relief work and body transportation. A lesser man who had endured a day like Tony's, might have followed through with his initial response, packed up his troops and gone home. For a well-earned rest. And both could have felt justified in their choices. But its moments like these that one can see heroism defined. Not those dramatic pulling people from a burning building moments. But those far more seemingly mundane situations. Where one is required to put aside personal agendas. Leave egos at the door. Forgo a sense of personal rightness.

Tony's team piled onto their trucks again and started pulling out of the zone. As he carefully navigated the remains of the tsunami road, Tony's attention was again drawn to the top of the hill. Where the children of the tsunami still sat. Watching. Waiting. Tony's phone rang. It was Filomena. A quiet plea... *please,can you come back, we need your help.* Tony and his team turned back and the search and rescue team became the first relief distributors.

"I said to Filomena, I'm very concerned about these children up there. That's all I ask, that we take care of these kids first."

Working together with Filomena and her limited staff, Tony and the team distributed clothes, tents and food supplies. They gave rides to the hordes of children sitting at the top of the cliff, taking them inland to find their families. They passed others doing similar work. Police working with Red Cross volunteers to survey the numbers of displaced villagers up in the bush, assessing the immediate needs. A truck giving out hot food ordered from Pinati Restaurant in Apia. It was a dispiriting task. Seeing families huddled under a square of canvas. Or a sheet of mangled roofing iron scavenged from the beach wreckage. Many of them still wet without a change of clothing.

Finally, at ten o'clock that night, fifteen hours after they had set out to search and rescue – the first response team began the long drive home to town. About the same time, CO Petone Mauga and his team arrived back at Apia harbor, carefully delivering their cargo to a waiting ambulance.

They would all be back again on Wednesday morning. To do it all again.

There are those, who at the first hint of tragedy, will rush with cameras and recorders – to get the story. It's their job to take the photos. Of headless houses. Crumpled trucks. Dead people floating in a sobered ocean. The photos that will move us. Inspire us. Horrify us. It's their job to get the story. They film the mother weeping as she clutches a baby covered in bruises. They ask her *what happened?* They zero in on a man who grips the fence posts for support as he tries to process seeing his wife crushed underneath an angry barrage of debris. They ask him *what happened? Tell us, tell the world...* They watch the man who finds the body of his mother in a cluster of fallen trees, watch as he weeps, hugs her, takes off his lavalava and covers her from the sun, shooing away the flies. They watch and through their lens – we watch. Curious, concerned and captivated. On 29/09, the world wanted to know what was happening on this cluster of islands in the middle of the Pacific Ocean. And local media were rushing to respond. TVOne and TV3 Samoa both had teams out in the disaster zone on the first day. International media were mobilizing and Australia had the first overseas news team arrive in Apia on the eve of 29/09, quickly followed by many from other countries.

Yet, the first images and news of the Pacific tsunami did not come from the organized media. Thanks to the wonders of modern technology, the first eye-witness photographs were taken by mobile phones and uploaded on the internet. The first news of the devastation was tweeted, blogged, and blackberried. Nola Gidlow is a doctor who drove to Aleipata with Lucia Hellesoe and her husband – to find Lucia's parents and to see what she could do to help the injured. Images from her Blackberry and others, ripped through Facebook and the worldwide network of Samoan family '*aiga*' got their first look at what was left of their coastal homes. With the grainy pictures was Nola's plea… *The people of Aleipata need help. They need food, clothing, water. The destruction is unbelievable.*

Social networking sites were buzzing as locals posted updates of the situation and as the world tried to find news of their families and friends. Bebo, MySpace and Facebook cried out, *Can anyone tell me what's happened to the Taufua family at Lalomanu?I can't get through on the phone...Has anyone heard if the village of Vaiusu was hit? I'm trying to contact my grandmother...* From Australia, Melissa Jenna went on YouTube asking the world for news of her friends in Samoa, sharing photographs of smile-filled nights on the town, *These are my friends Lynelle and Tim Betham and this is their sister Roda...can anyone tell me if they are okay? I'm also worried about my friend Jonah who's a DJ on the local radio station. Please, if you know anything about them, contact me.*

It didn't help the worries when the initial reports of the disaster announced dramatically that the capital Apia had been 'wiped out'. That the entire island had been wracked by the ocean. Tweets twittered with facts and rumours.

@LATimescitydesk: Tsunami advisory for California coast, with higher than usual tides possible around 9pm.

@tunasupport:In Samoa my guy Ziggy just said that officials are checking the airport now to see if the runway is operational. A lot of chaos still.

@Mami206:I'm devastated. Hopefully it's not true that 40 are missing from Lalomanu. Please God keep them safe.

@tunasupport: Ziggy says a big mess. No internet or phones. Many houses gone. Town of Pago Pago hardest hit. Several died, many missing.

@LeicesterCook: Saddened to hear a family friend died trying to escape the tsunami which struck Samoa. Thoughts are with the Annandale family.

@Taihape72:My brother borrowed a phone. They had to run for their lives from their resort.

@yo_c: Just had a msnchat with my mother in law on Samoa, upolu. She saw sea level resorts and fales on the south coast of Upolu.

@ciiusa:Our prayers go out to all the victims of this tragedy.

Almost immediately, hotels in Samoa had widespread booking cancellations as holiday planners changed their minds about a trip to what they thought would be a country brought to a complete standstill. Nynette Sasse of the Samoa Hotels Association explained, "We've got an online booking system and there was this mass cancellation, just like that. Hotels were going so well and then by the next evening everybody had cancelled. So the disaster was really bad for those who were affected physically but it also impacted on those who weren't physically affected because one minute you had these full hotels and what have you, next thing, everything was empty. It kept coming out in the media overseas that Samoa was pretty much wiped out and we ourselves were slow to get the right message out - I know I was interviewed right, left and centre during the disaster and that was always the message that I always put out *this was only one small part of the country that was affected and we hope that people will not panic*."

Samoa had made world headlines only a few short weeks before with their roadswitch, and on 29/09, the world was watching again.

14
The Day After

What greeted people on that Wednesday, twenty four hours after the tsunami as they drove from Apia and made the familiar drive over Le Mafa Pass, descending to Aleipata?

It was eerie. Up until right before Aleipata – Samoa looked totally 'normal'. People were playing volleyball, gardening, still selling their flowers, fruits and fish by the roadside. There was no preparation for what was approaching. There was no indication that just over the next rise, entire villages lay reeling from an ocean that had gone insane just the day before. Then you arrived at the entrance to Saleaumua. One minute you were in a tropical paradise. The next it was hell, the kind of horror that pictures cannot capture. The first thing that hit you was the lack of colour. Gone was the endless array of greens, blankets of grass shaded by the swaying trees, the fiery reds of hibiscus and ginger. No more yellow alamander, wine purple *ti* leaves, the delicate pink tinged frangipani. Everything was brown. Grey. Washed out. Leached colourless. There was a distinctive smell. Not immediately one of decay, but more of a dank, stale seaweed smell.

The sea had claimed the land right up to the road. Houses and shops were a mass of rubble. The one constant was the road - threading all the way round the island – thanks to the big machinery that was working overtime to clear a path through the wreckage. In some sections they were trucking in loads of sand and stone to reinforce the road so cars could fit through safely. Everywhere there was a multitude of blue overalls and hard hats – the EPC crew righting fallen power lines. They worked in amidst those few survivors who picked through the mess, searching for something, anything left from their lives. Most villagers were inland. Here on the coast, there were only bedraggled pockets of people moving through the wreckage. Boys carried pieces of scavenged timber and the least dented pieces of corrugated iron roofing – building blocks for the rough shelters going up everywhere in the bush. The mind thought back to the villages on the other side of the mountain where cheerful groups played volleyball, *they should be here, cleaning up, looking for the missing, nobody should be happy anywhere. Nobody should be playing anywhere. How can anyone be happy when this is so awful?*

The tsunami had blurred all lines of ownership. One woman's kitchen was now scattered through another woman's vegetable garden. One man's store was now littered over several hundred meters. Your furniture was now in my breadfruit tree. My refrigerator now sat on your shorefront. It was difficult to assert claim to much of anything. Not even to the battered land where your house once stood. Just past Saleaumua, two men were arguing about the timber strewn over one's yard.

'This is my wood, from my house that was broken and washed over here...' asserted one.

'No, this is my land, where MY house was so that wood belongs to me. He shouldn't be coming here on my land to look for stuff!' argued the other as he appealed to passersby to be jury and judge.

'But look at all these things here on this land, the wave brought them all here, they don't belong to him. We all need useful things, we are all looking for things for our families. It's not my fault we have to come here on his land, it's the waves fault."

Further along, a mangy brown dog stood and stared at the mutilated muddle that was Malaela. It's head cocked to one side, confused? A month before, relay teams had run past here and children had lined the road outside a basketball court, cheering the weary runners on. '*Go, go! Fa'amalosi, be strong!'* Now there was nothing where the court was. The brain screamed, '*where are the children?'*

There was an old woman standing on the waters edge at Satitoa. With wild grey hair blowing in the wind. Barefoot, a shapeless sack of a dress, she was shaking her fist at the ocean. "Where are my children? Where are they?" she cried. And again, "Why? Why have you done this?!"

There was a truck chugging slowly along the dusty road bringing food to the zone. Everywhere they saw people, they stopped. As soon as they heard there was hot food available, the hungry came from all directions. Many had been given a sack of rice, cans of tinned fish, boxes of speedy noodles from the DMO and the Red Cross. But where were the pots to put them in? Where was the fuel to cook them with? So they came to the soup truck. Police searching for bodies took a break. Truck drivers ferrying water pulled to a stop. Bulldozers and diggers rumbled to a halt. And incomplete families seeking the lost paused.

At Lalomanu a man wolfed down two hot dogs before describing their escape in the hilltop behind the mangrove swamp, climbing trees, hanging on. And now he's there to find his wife's cousin and her children. There was a woman sitting under a tree on an upturned crate – waiting. *"Im waiting for them to find my baby. And my sister and her baby. All dead."* She politely accepted soup but didn't eat it. Her mouth smiled but her eyes were empty.

What do you say to people who have lost everything? What do you give someone who has nothing? Who has not even the bodies of her children to cry over? One soup truck worker spoke of her apprehension when trying to give food to scattered survivors. "I was so humbled by our people's attitude. I think I would have been cursing the heavens if it had been me digging through my house to find my children. I was almost afraid to offer people food in case they lashed out at me. I imagined them screaming at me '*Are you crazy! Of course I don't want a #@& cup of soup – I just lost my entire family here and you want me to have some chicken soup?!'* Instead, we got nothing but appreciation, gratitude. Blessings. '*Faafetai mo lou agagalelei'. May the Lord bless you. Thank you so much. You're so good. Thank you, thank you.'* Everything taken away in one swooping wave and there they were giving thanks. And asking God to bless ME."

The road was very busy. Trucks go past with cloth covered shapes in the back and people wearing masks. There were trucks piled high with sacks of rice and bags of clothing, flapping loosely in the breeze. White vans filled with health workers. Everyone made way for a cavalcade of government vehicles inspecting the zone. Already there were television crews galore, congregated at the worst hit site, Lalomanu. Each country had their little territory marked out. The Japanese were there, standing on the newly tiled floor of what used to be a kitchen – the cement block all that's left of the beach resort that once catered for over a hundred guests. John Cameron was there from TV3 New Zealand. White shirt and slacks. Reporting from beside a pile of rubbish where a recovery team poked and prodded with long sticks, slowly, unwillingly searching for bodies. Australia was there interviewing a tired looking man. In a *lavalava* (sarong) and singlet. Another man in uniform was saying that it is difficult to know just how many tourists are missing because most of the beach accommodation either didn't keep records, or because the paperwork is a hopeless lost cause. As Rita Romeo explained with a shrug, " We don't know how many people were staying here or what their names were. They come and pay for the night or the weekend, they don't always tell us how many will stay with them."

Further along at Saleapaga village, people were desperate for water. To drink. To bathe. To cook. Trucks from private construction companies carried giant water tanks on their backs. There were two vans – seats piled high with tarpaulin and canned food sitting patiently outside the shell of a church. Looking for a congregation to receive it. Trucks loaded with water bottles and crackers, sheets and towels. The tsunami zone was flooded with helpers.

30th September, 2009.

We arrived back in Apia at eight o'clock. Drained, exhausted. Grateful for the embrace of children and refreshing cold showers – but with our thoughts back at the disaster zone.

The nights here have been very cold lately. I'm shivering under a blanket and I hope the people of Aleipata have something to keep them warm. The food is finished. We left the last of the bread and curry at a makeshift hospital at Poutasi. And the rest of the lollipops. They looked rather dismal in their surroundings. My daughter said – it feels good to help. Did you see how happy they were to get some food? I'm glad she came with us. But I'm not sure about the happy part. I keep thinking about the woman sitting on a broken crate under a tree. Waiting. I gave her a cup of soup. Then went back and hugged her. I cried. But her eyes were dry.

Do you think she's still waiting there?

15
The Searching

Wednesday 30th September, 2009. For the FESA and from a recovery perspective, the second day was better. As Tony Hill explained, "Wednesday we organized ourselves a lot better. We had surgical gloves and breathing apparatus. I knew by the second day people were going to start to smell. We took a water pump so we could clear out the flooded points. We set up a tent. We did our marking and gridding of the areas and everything started to go smoothly."

Day Two saw the recovery effort benefit from much needed assistance. Several local companies came to help, bringing their workers and some diggers – saying '*we're here to help, what can we do?*' Like Vernon Wetzell from Apia Concrete Products, Laurie Burich, Kyle Keil and the Arco Chemical team, a band of men bearing chainsaws from the hardware store Gold Star.

Vaughan Simpson from Ca'bella Construction was one of those who came on Wednesday and every day afterward until the searching was called off. Relying on only fragmented reports of the damage at Aleipata, he had mobilized his team on tsunami afternoon and they were on the road before daybreak that Wednesday morning. Thirty men loaded on four heavy duty trucks with ropes, bush knives, generators and other supplies. "We wanted to see what we could do to help. We didn't really know the full scale of the disaster, we thought we would just drive until we saw someone who seriously needed our help. We were looking for the end of the line basically and we found it with Tony at Lalomanu. We arrived at half past eight and the fire brigade had just got there. We looked at Lalomanu and we just knew – this was the place to stop."

It was a devastation that film cannot truly capture. You had to be there. Not on the road driving past, but really there. "From the road you couldn't really see the thick of it. The houses had been pushed through the trees, leaving a shield of debris, like a shelter or blanket along the tree line. The natural lay of the land also meant the ground fell away from behind the trees…we called this area, *Ground Zero.* It was so thick. There was roofing iron, toilet outhouses, wire fences, furniture, fabric, dead animals, locals and tourists belongings, vehicles, kids toys – all pushed through the trees back and forth with each wave like clothes in a washing machine. Most of what was left in piles wasn't much bigger than a meter long and so densely compacted that in some areas it would take three men, several hours to clear even one square meter."

And of course, no media reports could capture the smell. "Anyone that's been to Samoa will understand the heat. You can nearly chew the air as you come out of an airplane, and when you add water like rain or in this case, the saturated ground due to the waves – as the water evaporated up from the ground it was intense. In that heat, things go off really fast. Pigs, fish that had washed in with the waves and all the food that was in the many fridges and freezers, many which were full…frankly, the smell of some of these things made you want to vomit. But you kept searching. Some of the guys wanted to burn the bigger animals to stop the decay but we wouldn't let them because the smoke would sit in the search area and hamper our efforts. Then, as the days went by, smell was the best way to locate the problem areas that needed searching. "

The searching began with a prayer before the group spanned out. "We would walk in line to a grid pattern , left to right, back and forth, power pole to power pole, coconut tree to coconut tree. We would paint marks on trees to give us reference points or stand temporary markers so we could know exactly what areas had already been searched and continuously radio in where we were and which areas needed to be marked off."

It was never as simple as just 'walking in a line looking for a body' though. Not with so much debris to get through. "Apart from the search line we also had five or six gangs with about six people in each that would work with chainsaws, bush knives and a small one ton excavator. These gangs were used to pull apart the very dense concentrated areas that were as big as half an acre. It would take them all day to get through."

The searching was tinged slightly with apprehension. Major earthquakes are always followed in resultant days and weeks by aftershocks, some of them quite substantial ones. For a people already staggering from the previous day's disaster, the earth shaking was a cause for alarm. For those on the tsunami sites as well as for those loved ones worrying and waiting for you at home. Throughout the day there were several aftershocks. Immediately cell phones would go off all down the search line, shrill and piercing, twenty or thirty of them as wives and family would hit speed dial. *Where are you? Are you anywhere near the water? Did you feel that? Are you alright? You need to get to a safe place…Be careful…* Heads would whip back for an assessment of the ocean, *is that a wave? Is it pulling back? Were those rocks showing before?* Then everyone would turn to look at the hill, gauging the distance that would need covering, scrutinizing for the best routes up. Then huge breaths would be taken to steady tensely strung nerves. And the work would continue.

What drove the rescue teams? On that Wednesday, there was still the hope that they would find survivors. People alive underneath the saturated wreckage. An elderly man trapped under a pile of housing scraps, dazed but breathing. Hoping for the hands that would pull him free. Maybe a young woman who had been on the beach holiday of fantasy when the tsunami came but was now caught in a mesh of corrugated iron… cut beyond belief but waiting for trained hands to lift her out. Or a child that you could pull from a tangle of trees and hold gently in the shelter of your arms, the look of tired fear on his face replaced with a fragile smile. So you could take those few steps to hand him to a waiting mother. Sitting on an upturned crate waiting. Waiting for her child to be found.

Yes, on that Wednesday, there was hopeful anticipation. It was an unspoken almost furtive kind of feeling. As if by voicing it aloud in the horror of the surroundings, you would be forced to confront the absurdity of such a hope. Because really, who could have survived such a dreadful thing? So they did not speak of it. But it was a quiet hope that made them work with urgency. With care. With thoroughness. As the day trudged on though, and as more and more dead were found, hope faded. Unlike an earthquake that would trap you under rubble, where you could then theoretically, wait it out in sprinkled pockets of refuge, this had been tonnes of ocean. People were crushed, cut and slashed. And then, they drowned. Tony Hill and his team found no one living. They slowly, unwillingly, let go of those trembling hushed hopes. Let them slip loose in the stench-ridden air to flutter skyward in the sunlight. Butterflies that live for a day. Midway through Thursday, the reality of their task was unavoidable. The search became a body recovery mission.

Vaughan's only uplifting memory is of the Friday when they found a live turtle about 300 meters inland. The men paused in the sweltering day and walked to the water's edge to return the turtle to the sea. "It took off like it had a missile tied to its back, all four propellers going…it was the one moment we had something positive from the situation we were in."

Gradually throughout those few days, others came to volunteer. To walk the line. To dig, forage and hack their way through Lalomanu. Matthew Leal came as a Red Cross volunteer. "We were assigned to go to Lalomanu and help with 'clean up efforts'. We later found this was a euphemism for picking through rubble searching for the deceased." He quickly realized this was not work for the faint of heart –or of body. "Searching for the dead is difficult in many different ways. Most rubble is heavy and haphazard and jagged. Sifting through, lifting and dropping and finding footholds is strenuous and Samoan humidity doesn't help. The scene is surreal. Dead fish, left behind by the ocean litter the affected area. And then there's the emotional leap required to look for dead bodies."

Matthew had been one of a relay team that had run through Aleipata only a few weeks earlier on a 103 km race from Siumu to Apia. The difference between then and now, was mind-numbing. And digging through the personal leftovers of people's lives, felt intrusive. Almost criminal.

"More than anything, I felt like an intruder. I'd driven by the area we searched many times before the earthquake, but never did I stop to search a family's possessions or prod through their kitchen. And yet there I was today, finding family photos and Quiksilver baseball caps and notebooks and novels and gin. Two days ago, it would have been completely unacceptable for me to trudge into these people's lives. It's as though with the loss brought on by the tsunami comes the loss of one's dignity. I felt like I was snooping."

There were of course those who came with every intention of 'snooping', who's purpose was not to locate the missing – but to see what they could steal. Whether it be from the dead, the missing or from the absent, those too terrified to return and scavenge for their own possessions. There were those with backpacks, picking their way slowly and methodically over the beaches and through the shattered terrain. Eyes to the ground, shifting debris here and there. Aha! A wallet is eagerly rifled. A purse turned inside out. An intact bottle of whiskey is stowed safely into the backpack. Who were these profiteers of misery? Many came from villages and areas unaffected by the tsunami. Some even drove all the way from Apia in search of the mother lode. Those who had seen the wave, watched it rip loved ones away in a deadly swirl of grimy water had other concerns to busy them. One survivor whose one year old son was never found shakes his head resignedly. "Those people looking for things to steal were not people from my village. Us, we were too busy looking for our missing children to worry about looking for valuable stuff. We didn't care about that stuff. We were busy burying our families, busy trying to find our baby." There were even a few thieves among those that were sworn to protect and uphold the law. On that Wednesday, a little cluster of policemen were standing in the shade of a *tamaligi* (tamarind) tree, comparing their booty. Two were adding up a handful of sodden bank notes. Another had scored himself a pair of sturdy black work boots. They hung over his shoulder, loudly proclaiming their innocence.

Sili Apelu from Taufua Beach Fales was lying in his hospital bed a day later, watching television footage of the aftermath when he saw someone walking past the camera with a familiar red suitcase on his shoulder. "I saw somebody carrying my suitcase and when I saw that I said, *hey, there goes my $1,700 New Zealand dollars!* I was referring to a DVD that was in the suitcase because I had just returned from Auckland the weekend before the tsunami. I had gone to have a brain scan and it cost that much and the DVD was meant to be given to my doctor here. People just came and grabbed whatever they could take."

Treasure seekers were mingled then with searchers of the missing. Vaughan recalls the emotional impact of finding passports, photos, driver's licenses, intimate personal belongings that all had their own story – all reinforcing the sense of loss. They unearthed a family's bag that contained various photo identification, clothing and also, children's school books. You could just imagine that conscientious mum making sure her children took their homework with them on their holiday, no matter how unwilling they may have been! "A short while later, we found the mother's body. It was awful, just awful."

The heat was grueling. The stench nauseating. The work backbreaking. The finding of bodies, shattering. "We treated every casualty with the utmost respect. They would be identified as male or female, estimate an approximate age and possible nationality. The location would be plotted onto the search grid map." The line would pause. The machinery would still. The teams would go quiet as a prayer was said and the victim transported out of the area on a stretcher. First to Lalomanu Hospital and then later to the National Hospital in town.

The work was not without its conflicts and frustrations. While Tony and his assorted teams were focused on the north-west side of Lalomanu, there was another group working separately on the eastern end. This group was using heavy machinery to charge through the wreckage. "They were just using the diggers to load stuff onto trucks without even looking at what they were digging up. As soon as they left, we moved straight into that area to give it a thorough check and came across a European female corpse embedded into the base of the hill. We had to dig her carefully out of the bank. I was angry, so angry that these idiots who were simply too lazy to lift stuff by hand were going to squash corpses into the mud and we wouldn't have a chance to find them and deliver them to some form of resting place…we asked the group to leave and they didn't return. We found another body in that same area the next day."

On another day, Tony's team of volunteers encountered a group of search workers who had just arrived from overseas - cadaver dogs with their handlers. Vaughan had hoped they would bring their expertise to assist in the area they had been scouring for the last many days. "We were trying to explain to them that the majority of the body count had been found in the area we had been searching…but we were told that they were under their own authority and that the dogs were hot and they were going to take them back to Aggie Grey's Hotel to cool down. Cool down!? It had been a long, hard and trying week and I wasn't sure if they were joking with us or what…maybe the dogs needed to cool down because they had only arrived the day before and hadn't acclimatized yet. The handler proceeded to try to stress this by getting his dog to drink from his gloved hand. I thought that was an impressive waste of taxpayer's money. I had this image in my mind of the handlers sitting at the tables around Aggie Grey's pool eating steak and eggs while the dogs swam laps in the pool to cool down. Tony and I looked at each other in disbelief and walked away."

Another challenge for the searchers was the media. It is difficult to report on tragedy when it is so raw. One treads a fine line between compassionate concern and invasive scrutiny. To reconcile the 'public's need to know' with an individual's need for sensitivity and privacy. There are those who felt that line was crossed. Vaughan Simpson had his limits tested as journalists flocked to Lalomanu, eager for stories, for pictures of the body recovery. "Some of the media were a little frustrating. Every time we found a body, we would radio them in to Tony. The media would come charging into the bush and they're trying to take photos or they're trying to film the dead people and I would tell them to *f... off!* I would chase them out. The dead, they're people, like your brother or your sister, they could be your kids and these pricks want to come in and film it. People aren't supposed to do that. Some of the film crews were shocking, they were relentless. I chased them all out one day cursing at them and they reluctantly left. They probably went to the next village. I know they have to report on the world news, I suppose, but when you're out there, working with the dead – it's not nice. The media have got a job to do I guess – they've got to make money somehow, don't they? I found it disrespectful, but I suppose it records history."

The media weren't the only ones wanting pictures. "Every time we found a body, there would be some prick in amongst the people that was looking on, that would race up and take photos from their phone. One guy in particular was the operator of one of the diggers and he was just like a kid with a toy. He'd just go up there and take a photo of a body. What for? I have no idea. In the end we kicked him out of the site. But that was just amazing to see that. Why on earth would anyone do that? Crazy."

Probably the greatest challenge though, was not knowing how many people you were searching for. "The hardest thing was the not knowing if you were doing everything physically possible...was there something we could be doing better, was there somewhere else we could be looking, or doing differently? When were we going to stop looking? No one knew the answers. No one seemed to be able to tell us if there were more people missing."

The big question after a disaster is – how many dead? How many missing? And every foreign country with nationals in the disaster area is driven to get answers. In those first few days, rescue personnel had no idea exactly how many people were unaccounted for. Locals with missing family came to sit and wait for the search teams to find their loved ones, but the consular offices could not confirm how many of their people *might* have been in the tsunami areas. Samoa is not a troubled hot-spot. People don't feel the need to register their presence there, or even to advise their families where they will stay and how long they will be. The NZ High Commission office in Apia was receiving hundreds of phone calls from worried relatives. *My uncle Johnny is on holiday in the South Pacific, I'm not sure where but he did mention he would be stopping in Samoa...My friend and I were back packing in Samoa last month, I left early to come back but she's still on holiday there, we both loved Lalomanu, there's a chance she could have been there...My family was staying at Saleapaga beach fales, I'm not sure which one, I haven't heard from them...Our mother is missing, they last saw her in an ambulance headed for the Lalomanu hospital, she has Alzheimers, can you please find her?* At one point, the High Commission had over seven hundred requests logged in of 'missing' nationals and staff were working overtime trying to track them all down.

David Dolphin is the Deputy High Commissioner who was managing the huge task of finding New Zealanders "Even on Day Four after the tsunami, we still had about 150 people who we just couldn't account for. We were working round the clock – calling every hotel in the country trying to find out whether these people were staying there. We contacted airlines for passenger manifests, in case these people had left the country by their own means and we just didn't know about it. People at our office in Wellington were calling back those who had lodged the enquiries, two, three, four times to find out if they had received any more information...because people make the inquiry and then when their family shows up safe they don't always bother to call back and tell us! We just had to work as best we could to get that number down as quickly as we could and we also had to apply a little bit of commonsense – if 140 Samoans in total had been killed, there was no way that another 150 New Zealanders had been killed as well! So it was a matter of using commonsense.

Arne Schreiber is the German Honorary Consul in Samoa. Without an embassy to back him up, he alone was tracking down 175 German visitors that were logged in through Samoa immigration prior to September 29th. Four Germans were in hospital with tsunami injuries and needed to be airlifted out. They required new travel documents and visas to transit through various countries. Arne was working closely with local tourism authorities to cross every name off his list. On his list was a Ms Lieselore Rock, a single woman in her sixties travelling alone. A week after 29/09, he made the 45 minute drive to the airport to wait for Lieselore, who was scheduled to fly out after a three week holiday in Samoa. Arne had hoped to see her catch her plane. But Ms Rock never showed up. She never called to cancel her flight. Arne's search carried on.

Back at Aleipata, the searching continued. For every searcher working the lines, there were countless others supporting them in a myriad of ways. Vaughan remembers them with gratitude. "There were so many volunteers – it's hard to say how many as we didn't necessarily see all of them. There were those who brought the endless supply of water and food that fed our team and the nurses that would come down from the local hospital every couple of hours to treat anyone that had been cut by the debris. There were so many people that would give several hours of their time to walk with the line, like this group of Japanese tourists who were visiting Samoa at the time, they hadn't been affected but because they were in the country, they came out to help. And there were US Peace Corp volunteers who came out as well. Many people were coming out and just dropping off food at the base for us. There was a guy from the Church of Scientology in Australia – when he heard about the tsunami he caught a plane immediately, he was cool, he turned up there every day and helped us. Then of course there was the support we all got when we went home to our families every night."

There was an aerial search happening too. An Air Force Orion plane had come from New Zealand. One of the first actions necessary after a disaster, is an impact assessment. Aerial surveillance is essential for identifying the extent of the damage and those areas worst affected. The Orion arrived in Samoa at about 5pm that tsunami Tuesday and took the first sweeping photos of the affected villages. The following day, it worked closely with the Nafanua patrol boat to search the ocean for possible casualties. If you had been at Aleipata that day, you would have seen the plane dropping intermittent smoke flares out over the sea. And survivors looking for their missing on the shore would be watching and waiting hopefully. Petone Mauga reflected, "It was so helpful to have them here. Most of the bodies were mingled with the debris in the sea and it was very difficult for us to find them. They were spread out over a huge area too. The Orion would drop flares and show us where to go to find the people."

The Nafanua team had recovered the dead before. Drowning victims. A swimmer. A snorkeler. But never so many. And never like this. Thirty-seven year old Petone is somewhat new to his command and father of a handsome little boy. "The first victims we found that day in the ocean, were two young kids and they were in school uniform. It was very sad. I was devastated by the fact that they were going to school happily, not knowing what was going to happen. I was very emotional…but I had to be strong for my team. We've done lots of drills but dealing with reality is very different and I could see my boys, they were nervous and afraid when we first came to deal with the bodies, but we got through that. It was not easy."

Several of Petone's team had grown up in the disaster area. The Nafanua team included Ofisa Esera who was overseas on training when the tsunami occurred. Both Ofisa's parents were killed at Saleapaga village. It was impossible for this recovery mission *not* to be personal. In a community where *'aiga'*, extended family means everything, searchers were not searching for strangers. They were searching for family.

One of the missing was Mika and Ave's son, Aleki. Mika asks, "Do you know what that's like? To still be waiting all those days for your son to be found? I'll tell you, everyday we go to the hospital first and we wait there, to see if any new bodies have been found in the night. And everywhere there is the stench of the deceased. And then after the hospital, we go back to our land, back to where our village was and we try to look. But it's so hard. I was disappointed that all the searchers – the fire and rescue, the police, the army people – they were all focusing where the tourists stayed. They only started searching our area on Friday, three days after the tsunami. And when our daughter Malo Maryanne was found, she was in bad condition, face down and buried under trees."

Mika haunted the hospital at Lalomanu. In the month after the tsunami, he was a regular visitor to the morgue in Apia. Every time someone was found, Mika would go to the morgue. To see if that was his son. Every time a body part was found, Mika would go to the morgue. To see if it was a remnant of his son. Mika is a man who has looked at many dead people with the careful wish that they will be his beloved. More than any one man should have to in their lifetime. Mika asks, "Do you know what that's like? Do you know that when they find a body that late, they put it in not one, not two – but three zipper bags? And when they bring that bag into the hospital, it's like a falling back, everyone moves away, everyone covers their mouth and nose, everyone moves away. And when they open the bag, the smell is so terrible, and the sight is so terrible. Their skin is peeling, the eyeballs have come out, I'm sorry to tell this to you…and to this day, my son is still not found."

On September 29th, the ocean took people emphatically. The furthest body was found five nautical miles out. Others floated in the deep channel between the off shore Aleipata islands. Where sometimes, whales come. There were two that week afterwards. Keeping watch. Keeping the Nafanua patrol team company in sentient silence. Images.

A man with skin powder white from his time in the ocean. His tattoo (pe'a) a stark contrast of black patterns.

A young boy lying on his back in the silken blue. Peeling, patched skin. Arms outstretched in the sunlight. Gently lapped by the water as a whale slides heavily past.

The Nafanua Patrol boat continued searching for another week after 29/09. Bodies were recovered on Wednesday and Thursday. And then no more. They persisted but still nothing. CO Petone sensed his team's discouragement at the futility of their mission. "I saw the boys were starting to give up. It was hard to keep looking and not find anyone. But I told them we need to think about those people who lost their loved ones. They can see us still trying, not giving up and it gives them comfort. We need to make sure that we are doing everything possible to find their families."

Look closely into the darkness of a natural disaster and you will find the gentle gleam of kindness. Compassion. In a rescue team that remembers the agitation of an elderly woman and safeguards a suitcase through the chaos of a death-filled day. On a patrol boat that carries on, ploughing through the white capped waters off the Aleipata coast. Looking for those that the sea will not return, ensuring families can have some small measure of hope. Searchers who dig through filth with gentle hands so that strangers beloved by those unknown,will not suffer further desecration from the claws of a digger. Who will then pause in the tropical heat, to pray – for peace, for comfort. For the strength needed to carry on. Kindness. Compassion.

2nd October, 2009

Today in the zone, it was blazing hot, dusty with sea silt. And it smelled. Bad. Very bad. Not just wet, dead fish smell. The air was thick with what I can only describe as lots of dead things smell. Dead dogs rotting by the side of the road smell. It seeped through the windows. Into your clothes. Your hair. Making you want to cover your nose. Or hold your breath for the three hours it took to drive the zone and give out food. But you didn't. Because it felt rude. Disrespectful. To all those picking through the pieces of their lives. Those pushing a wheelbarrow down the road with a rice sack and a water canteen. Those sweeping sand out of a spindly hut thats surrounded with hills of tsunami trash. Those that ask politely - is there a spoon to eat this soup with? While they're wearing one jandal and sitting on the cement foundation of their swept away home. Those that announce with a sigh of relief - 'No, theres nobody missing from our houses here. Everybody is found. 8 bodies all found yesterday. Nobody missing anymore. '

The zone is not an easy place to live. Or to work in. But extreme circumstances require exceptional effort from ordinary people. People like EPC linesmen. The day after the tsunami they were putting up new power poles. Clearing away fallen wires. Today, they were still there. The same teams because lots of them recognized the soup truck and greeted us like long lost friends. I will never flip a light switch again without a greater appreciation for the work that goes into shining light in darkness. Amazingly, power poles are up. Lining the road clear through to Lalomanu.And as we drove away at 5pm, still the linesmen labour on. I doubt they're calculating their overtime pay. And certainly none seemed to begrudge their hard work. As the lights start to go on at Aleipata - we say thank you.

People like - the road crew. Immediately after the waves hit, a call went out on the radio for all and any heavy duty equipment companies to please, please come help clear the roads so rescue teams could make it through to the disaster zone. They came. Bulldozers. Diggers. Dump trucks. Excavators. And with them came the machine operators. The truckloads of road workers. They cleared the road. They started trucking rocks and sand to fill in the holes so cars could make it through. They're reinforcing the seawalls. Moving rubble. Shifting concrete blocks and fallen bricks. Helping to find the lost. You can drive smoothly and pothole-free right from disaster zone start to finish. I got home at 6pm, eager for a shower to erase the grime of the day. I wondered

- what time will the road crew get to sleep tonight? As the relief trucks roll in on beautiful strips of road - we say thank you.

People like - the Samoa Police force. The Fire and Emergency Service team. On the eve of tsunami day, the radio called for 'all police to report at 4am in work boots and heavy duty gear. Bring a change of clothes because you will be swimming and diving to recover bodies. Come ready to work.' And they did. On Tuesday, police and emergency service workers were everywhere. Doing everything and anything. Today, they're everywhere. Doing everything and anything. At Saleapaga, over 60 of them are scouring the ground for the 7 people still missing from that area alone. They found one crumpled 4 year old today. They'll be there till late tonight. And they'll be back again tomorrow. Until they've found them all. We got to them at 4pm. They were tired. But cheerful. As the relief effort continues, organized and orderly - we say thank you.

Extreme circumstances require exceptional effort from ordinary people. And the results? Extraordinary.

16
The First Help. Samoa's Response.

While the body recovery efforts trudged on, others made their way to the tsunami zone. They came to bring water, food, rough shelter materials and medical supplies. They came to clear the road, straighten power poles and reconnect water lines. Aleipata and Falealili were barely visible under a wave of workers – both government and private. Samoa's response to the disaster has since been praised on many fronts. The United Nations Development Program Apia-based coordinator, Neelima Noble said afterwards, "I have rarely seen – and I have served in many countries – such a complete sense of the community coming together, the sense of self-reliance. Within hours, within days of the tsunami, you saw the community in action, clearing up the debris, tending the injured…in very few countries have I seen a government respond so quickly and decisively following a natural disaster."

At eight o'clock on tsunami morning, the National Disaster Council had their first quick meeting at the Faleata Fire Station, *what's happened, what are we doing about it.* Lavasa Malua arrived at the station, to find it crowded and busy. "Mena's people from DMO were already there, the emergency people were there – things were moving." With both the Prime Minister and the Deputy PM out of the country, the Minister of Revenue, Tu'u'u Anasi'i Leota was in charge. A few minutes after Lavasa, Tu'u'u walked into the station, out of breath and red-faced. "He doesn't live that far from Faleata and I think he had been out for his morning walk when it happened. He popped into the station in his shorts and running shoes and he said, *I felt it.* So he had hurried straight to the station to check what was going on…we needed to start making calls because we had to declare a state of emergency."

Vaosa Epa was there. She had called the Prime Minister to update him on the situation. He had just landed in Auckland, New Zealand returning from a trip to New York city. Faumuina Liuga, the Minister of Natural Resources and Environment was there. Tuisugaletaua Sofara Aveau, the Minister of Works was there. At that first meeting, details of the disaster were still sketchy but the council knew enough to start organizing. As Lavasa described it, "Immediately the thinking was – let's not waste time talking, let's move."

Vaosa added "The overriding priority then was life. We were getting reports of a few deaths and injuries and so the important thing was that we had to get people who were affected, to the hospital as fast as possible. We didn't know how high up the waves had gone or how badly damaged the roads were, but we were just making sure that teams from the police, from the health department, the rescue teams, that they were all already on their way to get the injured out of there." The meeting was a short one so people could get to work as soon as possible. Walking out of the meeting, Vaosa saw Nynette Sasse standing in a fire house empty of fire-trucks. "She was crying and that's when I first heard that Tui Annandale was one of those killed in the tsunami and then more reports were coming of more casualties and that's when the enormity of the whole thing started sinking in for us."

The day continued with a further build-up of clues to "the enormity of the whole thing." Vaosa received a phone call from her daughter Gardenia who works at the Samoa Tourism Authority. There were van loads of tourists arriving at Aggie's Hotel in town, shaken and stirred. No clothes, no luggage, no money, no passports - but brimming with their stories of near escape from a tsunami. Resort operators were calling town for assistance with transporting their guests.

Reports from Filomena in the disaster zone were not promising. "She said, *people here are suffering, they need food, they need medication...* So we knew what we needed to do. The problem was, all the wholesalers were closed, there was no place to get bulk food. We also didn't have fast cash at the time and all the government ministries were closed so there was nowhere to process payment." Lavasa's team found a wholesaler willing to supply the necessary goods.

Another challenge was transportation. Relief supplies needed to be taken the hour and a half drive to the zone. Along with heavy machinery to clear the roads. Wounded and dead needed to be transported back to town. "We called on all the members of the Disaster Council to avail their vehicles and their staff to help with the transporting. We also had a lot of volunteers coming in to the station that were helping. We had to bring bodies on the same trucks that were taking out food and I was telling our people, *hey, make sure you clean those trucks properly.*" The team got a call from a Greenpeace ship that had left the Apia harbor several days before. "They offered us assistance and said they were going to turn around and come back. They have a little helicopter and when they were in flying distance they sent the helicopter ahead to us. They were the first helicopter to get here. It stayed a whole week and helped take supplies out to the area which was a great help because we needed something more than the trucks that we had. Then when the Greenpeace ship arrived, they gave us big water containers and also fuel which we used for the trucks."

While Lavasa and his team were coordinating the logistics of the relief operation from the nerve center at Faleata, the Ministry of Works was attacking the damaged roads with the help of private construction and road companies. The EPC were headed to the zone to deal with the power situation. The Water Authority was taking trucks of water. At the Red Cross Headquarters at Motootua, Tausala was organizing her volunteers to respond to the crisis. "Our role is an auxiliary one because we support government in whatever they are doing in a disaster and our priority is the most vulnerable people. On that first day, our people went out to help build temporary shelters to house the people with families lost. We were giving out water supplies, blankets and tarpaulins. We were doing our assessment to find out how many people had been affected, how many needed help."

At 11am on 29/09, the tsunami all-clear was issued and at midday, the Acting Prime Minister, went to the disaster zone with the Head of State. A second Disaster Council meeting was held that afternoon at 4pm at the Police Station, this time with Samoa's partners, NZ and Australia and others. It was a meeting where "everybody was there ready to assist…NZ was calling out *we've got the Orion on it's* way and Australia telling us *we've dispatched a medical team, they'll be here early tomorrow morning* and everyone there was saying, *this is what we've got, we are on standby, just give us the nod and we will come…*" The meeting came to an abrupt halt however when sirens screamed out another tsunami warning. In the ensuing rush to evacuate inland, Vaosa received a text from her daughter Gardenia. *I'm here in Lalomanu. We are being told to evacuate.* "My heart sank and the selfish me came through, I was so upset, so worried." A frantic mother called her daughter. *Who told you to go out there?! And why didn't you let me know that's where you were going? You shouldn't be out there!* Gardenia's reply was calm but unrepentant. *Mum this is part of my work, we had to come out here to try to locate the other tourists and help get them out.* "When she said that, I had to check myself, I knew she was right. We all had our work to do. I asked her to please go up to the hills and then I called her sisters and brother." Gardenia's sister Aurora is studying in Australia and her brother was holidaying in the US with their father who is Samoa's Ambassador and representative to the United Nations in New York. Vaosa sent them all a quiet plea. *Please pray for Gardenia's safety,* she's *out there at Lalomanu and I can't reach her physically. Please pray for her and everyone else out there so that there won't be a repeat of the morning. Please pray.*

Vaosa then drove out to the airport to meet the Prime Minister, following a winding coastside road through empty villages while people stood high inland and studied the evening ocean with worried eyes. "It's one of those times, you're doing your job and you're feeling also for so many other people and especially for your own child. When you're that far apart and you know what took place just that morning and you're so afraid that it could happen again."

Vaosa briefed the PM and at 6am the next morning, the National Disaster Committee assembled in front of the government building. All Cabinet Minsters, CEO's of various ministries and representatives from NGOs drove out to view the disaster zone. As the procession of vehicles began pulling out of town, Vaosa received a text from her daughter Aurora. *Pe solo mai mauga, luluina mea uma, tumau lou faatuatua i le Atua...Even when the earth shakes and the mountains fall, we will put our trust in God.*

There are formidable women like Vaosa at the helm of many of Samoa's governing bodies. Women who are powerhouses of strength, resolve and efficiency. Many of them were in that procession inspecting the disaster zone that morning – Tausala of the Red Cross was there, Brenda Heather-Latu, former Attorney General and current Chairwoman of the National Health Board, Gatoloai Amataga Gidlow the Minister for Health, Palantina Toelupe the CEO of Health, Fiame Naomi the Minister for Women's Affairs. As they stood in amidst a broken Poutasi and gazed upon the desecration of muddied churches, the ruins of a school and a hospital, hearts wept. "It hit us, and we just cried. We cried because we knew there so many people involved and even if you don't know them personally but you feel their loss. To stand there and to see the devastation at the hospital and the school there, one minute it had all been there and the next it was gone...well, we cried and we just clung to each other because it was an overpowering thing to be there and to see it, to feel the reality of such a complete loss."

A tremor rocked the earth that morning at Poutasi, and Palantina's son Vaitoa called his mother with an urgent reminder. *Mum, there are still tremors, you've got to be careful out there, you've got to move inland.*

The procession continued on through Lepa and Saleapaga, encountering the emergency response teams at work. "We stopped in Saleapaga and they were finding bodies. A woman had just been found and brought to the road. It was just devastating, the family was crying and trying to claw at the body and we tried to offer words of comfort and prayer, but it felt very helpless. And the whole drive through there was very emotional but you realized that you must stop crying, you must hold your emotions in check and do what you have to do and deal with the people who are very much affected."

At Lalomanu Hospital, a row of bodies lay on tarpaulin at the back of the building. The Minister of Agriculture and Fisheries, Taua Kitiona was there working alongside his wife Dr Esmay, working alongside the people of his village. Taua had been bestowed with a second chiefly title on the 15th of September and many of his extended family had come from overseas to be there for the occasion. Visiting family members were still at Lalomanu when the tsunami came. Seven people from his *aiga* were killed, four adults and three children. If ever a family needed their *matai*, it was now. If ever a community needed their leader, it was now. Taua and Esmay camped at the hospital that week. He led the young men of Lalomanu as they handled the dead – many times without gloves, masks or protective clothing. "Yes, I knew most of the dead. They were people from my area – my village, my constituency. Many people worked together because of the numbers of people dead. I didn't really want to see their faces…there were big cuts from roofing iron, nails, wood. Trucks came with the dead people and we had to make space for the injured people, we had to put them outside because there was no hope for them. The most stressful time was the three days after the tsunami, because we were still looking for missing people and there are some still missing to this day."

Palantina Toelupe remembers some of that 'stressful time'. "When we got to Lalomanu hospital that day, there were some bodies that had just been found – about seven adults – and they were starting to smell, even their relatives, their mothers and fathers couldn't get close to them that's how bad it was. The pressure on people working there had got to the stage where the nurses and the police were starting to have a good fight, because the police were saying, *we're not the ones that have to deliver these bodies in cars to town* and the nurses were getting frustrated with the police. So we got there then and we were also quite affected emotionally too and then I realized that by law, I have a responsibility to do something about burials and things like that. So I asked the nurses to get the village leaders, while Brenda Heather-Latu was trying to talk to the police about the law and what they needed to do so they would provide some assistance with transport of the *palagi* bodies. I said, *Look, there's a way about it, we'll get the village leaders to facilitate family endorsements and then bury their loved ones immediately – there's no place in Apia for more bodies*. It was important we do something while there's still integrity left about the dead. The other thing that bothered me so much mentally was that while everyone else was getting tired and frustrated – there was so much love being shown by the young men of Lalomanu led by the Minister of Agriculture. They were the ones running around on the trucks getting the bodies and they were the ones that were lifting them, even without gloves, even without gowns. And I said to myself, *this is the reality.* Even with the good intentions of officials, they also get tired and protective of their own safety and security, but the real people that were making it possible, were the Lalomanu men themselves, the young generation. Later on in that week, somebody was saying,*it's turning out really bad out there because the young men are all getting drunk and how could they do such a thing at a time like this.* I told them, no – I don't think so. Those were the very boys that were handling the dead and the smelling people, the remains that were there and therefore, that's their way out, let them be, they will come round – and they did. It was their way of coping – horrible as it was, but you can just imagine having to persevere despite the smell, despite the sight, despite everything else, and they did it."

When the Disaster Council inspection was completed, Palantina organized her staff to go to Poutasi Hospital to start the cleanup. "We went to do what we could to support our colleagues that were directly involved in the clinical care…we took everyone who works in policy, in administration. That whole week, we went in trucks and buses early in the morning to clean so that by the time Dr Dean and his staff are ready to go back, it would be ready for them. We were in those coastal areas by 3:30 in the morning, it was still dark and you imagine what it felt like when there were all those aftershocks…it was scary, quite scary." For Palantina, credit for Samoa's national response to the disaster must go to the planners. "I have to say that our health response goes a long way back to other sectors. The MNRE had been quite assertive about disaster planning…government had the Disaster Council and they were assessing the laws and the processes involved and setting up all the plans and we were sitting back and going, *okay, plan plan plan!* But we complied and did our bit and were able to pull together all the different sectors that we are responsible for – and it wasn't easy because this is where you have to have clearly defined roles and make sure that everyone knows what they are doing so that we don't intrude into the professional boundaries of other colleagues,,,but we know where we should operate from and how best to support."

Lavasa oversaw much of the national response from the Faleata Fire Station. Ask him how it felt to be MNRE CEO for only a day and then have to work on Samoa's worst natural disaster in living memory. "It was definitely a once in a lifetime experience but I reminded myself, this is Samoa, there are always people you can turn to for help, there are always those you can depend on. I was able to get through because I could see all these skilled people, and when you see someone standing around and you know that person has the necessary skills, you say, *why are you standing around, go make use of yourself!* Also, this disaster was a new experience for everyone – it was not just me – so I wasn't really concerned about not knowing what to do because I know that everyone else doesn't know what to do too!"

It was the National Disaster Management Plan which guided them. "The emergency process was very clear, everyone had to operate according to the response plan…I think co-ordination was good because people knew what to do and it was a matter of making sure they get what they want to be able to do their work. By the end of the first day, it was amazing because a lot of cleanup had already been done and the roads were good enough that you were able to get through to the whole island." On 29/09 a decision had to be made whether to close off access to the disaster area. "In other countries when a disaster is declared or a state of emergency, then security tightens as well, so not every Tom, Dick and Harry can go into the disaster area. We didn't do that – because we had families that wanted to help each other, there were families in Apia, they had family out there that were stuck and you're not going to prevent them from going over to assist. So I said, why should we stop people who want to help?" Lavasa admits there some problems with that openness, " There was a mix of people going out there – some going to help and some people going to look for things they could get for themselves…and police were more focused on assisting the coroner's job of getting the dead people out of there so security was a issue at times."

The Samoa Red Cross worked closely with the government every step of the way during the national response. The day after the tsunami, the Red Cross was out in the field doing their assessments of the immediate needs and then proceeded with social mapping so they could identify vulnerability. Everywhere throughout the zone, the distinctive white and red cross was a familiar sight – on cars distributing food and water, on First Aid tent stations, on vans filled with volunteers. Secretary-General Tautala ruefully admits, "The first month after the tsunami, I slept here in the office, right next to my desk. I didn't go home…and we had to stop all the phones at night because that's when the calls would come from the other side of the world, and we had to say, *hey, do you know it's our night-time here? You must have just woken up but please, give us a chance to rest!"*

On that first day, it was Lavasa's job to also take care of the 'legality', the 'properness' of the official response. "We had to make sure that while it's an emergency, it doesn't mean that you're going to cut corners and do things illegally. That was one of the main things I had to consider, that we weren't going to have any excuses afterwards…so the key players I was looking at was the office of the Attorney General, the Ministry of Finance and the Ministry of Foreign Affairs. We had to make sure the Attorney General had the instrument of the declaration of a state of emergency and we had to get all the clearances from Finance to see how much money there was to take care of this emergency." In following weeks, this attention to accountability extended to the receipt and distribution of aid materials. The Faleata Fire Station became the collection point for incoming aid. In response to claims of misappropriation of aid, one journalist, Terry Tavita described his visit there. "It's certainly no shindig operation they're running up there, The controls we found out were pretty tight. All goods coming in are recorded by Treasury officials who have a desk at the entrance. So is every item, every delivery that goes out. Each delivery truck is accompanied by a policeman and an audit officer. In the village every item has to be signed by the recipient. Two days afterwards, a follow-up team crosschecks with the recipients if indeed the signed items have been received. It's hard enough for one item to go out unrecorded or go missing on the delivery line, but 'truckloads full of goods being sold to shops? We seriously doubt."

It was Filomena Nelson and the DMO staff who had to deal with the frustrations of managing the day-to-day distribution of relief goods in the months after the tsunami. In the first few days after 29/09, the DMO distributed food and other supplies to everyone they met in the disaster zone. "There were many people who were not affected by the tsunami but they moved inland because they were scared. The government provided assistance to everyone – families that were affected by the wave, families that had lost loved ones but their homes were not touched. We just gave the stuff out to everybody in the camps, because we didn't have the time to confirm *were you really affected? Who's who?*"

There were complaints of course. Some said they did not get much when a *pulenu'u* (village mayor) oversaw the distribution of a container full of aid because 'he doesn't like our family". Other times relief didn't get to the people it was targeted for. A grocery store sent a truck full of food for the Taufua family which was stopped by others and distributed to the rest of the village. At Saleaumua, Lee described the haphazard distribution, "If the goods were delivered by government, and you weren't on the list – you wouldn't get it. If the goods were delivered to the religious sectors, it only went to the members so you didn't get any if you weren't a member of their church. So there was a lot of inequity going on." Mika and Ave told of their troubles with aid . After the tsunami they relocated to the town area which made it difficult at times for them to access the aid deliveries that were going out to the disaster zone. "The Red Cross did their assessment of our family but even by the third day, no help had been received. On Friday they gave us tarpaulin to make a tent. I explained to them that we have moved to Vaigaga, but we go everyday to Lalomanu to look for our children that are missing. I can't sleep at Lalomanu because there is nowhere to sleep, no bedding, no water, no mosquito net. I need many things to help us. Finally, the DMO sent us some gifts – some food and water, three bags of clothing. But the food is only cans of soup and fruit, how about some meat? And the clothing is local stuff. The bags were all full of women's clothes – pastor's wives dresses and hand-sewn things." Exasperated, Mika decided "to make an example". He put on one of the dresses, a voluminous *mu'umu'u* and went in a taxi to the DMO centre. "I said to them, is this the kind of thing you want? These are the clothes you give me! How can I go to town and look for a job or go to church with this dress on?!" Mika returned home with a box of clothing but still he was not happy with the relief distribution.

By Day Two after the tsunami, the DMO was working with other agencies to verify people's needs. Filomena explained, "Sometimes it was hard to reach all the people, the camps were scattered. Under the law we had to work with the village mayors to carry out the distribution, even if he's a crook or what, we have to work with him, that's the gateway. There were also families that had been banished by their villages so we delivered goods to them because they're outside the village so we have to take care of them. We did a list of all the families that were affected but we had a new family every day coming to us wanting to be added to the list and so we had to go out again and check."

There were freight containers pouring into the Faleata Fire Station complex everyday after 29/09. Trucks were running a continual delivery route to the disaster zone, driving an hour and a half there, unloading and then driving another hour and a half back. New Zealand helicopters were flying supplies and people to the zone in less than twenty minutes. Filomena and her staff had much to sort through and some of the container contents had heads shaking. There was a purple ski suit. Several heavy winter coats. Picture frames. A set of mini whisky shot glasses. Knitted woollen scarfs. "We really appreciated all the assistance but some of the things were not suitable for our weather and so we had to hold them back. There are also nutrition policies that we needed to abide by in times of disaster, for example, the infant formula and baby bottles that were donated could not be given out because we try to promote breastfeeding – but everything else, we have given out. The only thing we held back for a while was some of the clothing containers because until people had built their new houses, they had nowhere to store the clothes."

Of their response to the disaster, Filomena said, "The first 48 hours, the relief work was all done by our own people using our own resources. The government's response was very fast and when the international assistance came in - we had already cleared the roads, started reconnecting the water, the search and rescue was underway, the recovery of bodies, the hospitals were following their disaster plan. I was just so grateful that we were able to do all that even though we hadn't covered the whole country with the village preparedness program yet. I'm just glad that we started somewhere and we're learning and we're continuing to improve."

It was not just the Samoa government who responded to the disaster - the entire country rallied together to help. Local companies did not wait to be asked – they stepped forward immediately – *what do you need? Where shall we take it?* Wholesale food supplier 'Frankies', agreed to open its doors and release truckloads of goods without up front payment, without government requisition forms guaranteeing they would get paid. Restaurants were cooking hot food for the wet and weary. Like the Samoa-famous 'Pinati's' which sent many stacks of foil wrapped meals, the silver shine of kindness. Bottled water companies were donating boxes of liquid gold. Bulk clothing distributors filled their delivery trucks and started taking bales of clothes to the inland settlements. Two days after the tsunami, local TVOne held a tsunami telethon appeal – and organizers were flooded with the response. People gave money – over $600,000 ST was raised in less than 48 hours. People gave clothing – truckloads of it. They gave pots and pans, tools and furniture. Schools and church halls were collection points for many village donations. TV3 Samoa helped to organize a musical appeal. Legendary singers Jerome Grey, Dave Parker and a host of musical talent sang for Samoa, raising $26,000 which was given to government to assist with the relief effort.

Others drove out directly to the zone to take their gifts. On the afternoon of 29/09, local lawyer Tima Peteru was one of the many individuals who acted immediately. She filled her car with water, food and clothing and headed to Aleipata to give it to whoever needed it. One volunteer remembers, "As we were visiting families at Vaovai, a bus load of people from Asau had just arrived from Savaii bearing bundles of mats, clothes and root crops for Nu'uausala Touli's family who had lost two grandchildren." Lee Palupe's dentist came to Saleaumua bearing gifts, "really nice stuff, marshmallows, hot dogs, candies and a fancy little transistor radio that could get all the stations so I could know what was going on." Another day, president of the US Veteran's Association, Mark Moors, came with a team to help fix Lee's water tank "and it was so great because I didn't have to carry buckets of water around anymore."

In those first few days after 29/09, it was the bare essentials that were most needed. Everyone was wet. And even if their house was still standing, everything inside it was wet. All the clothes were wet. All the sheets and towels were wet. And the water had not been the cleansing, refreshing kind. Mingled with the ocean was the contents of everyone's septic tank, everyone's pig pens, the accumulated refuse of communal living Many survivors remember the ocean waves with disgust. *The water smelled bad…it was dirty…it had bad things floating in it…it made us sick…it killed the plants….* People needed clean clothing. Bedding. They needed water – to drink, to wash with, to cook with. They needed food. Dry shelter. They needed to know they were not alone.

Thanks to Samoa's close extended family ties, many received their first help from relatives who rushed to gather the essentials. Lee Palupe did not want to leave the wreckage of her home in Saleaumua as scavengers were out and about. She sent a note to her family in town with one of the DMO drivers. *I'm alive. Please, I need a cell phone, batteries and some money. Give it all to this boy and he will get it back to me here.* Sinapo Faleui's sister in Apia came to Satalo with bags of clothes and a hot meal that very night. Tina Niusila's uncle came to the Saleapaga bush that afternoon bringing clothes and bedding for their first night sleeping on tarpaulin under the *talie* trees. Where Tina's four children would listen wide-eyed to the chittering flying foxes. And solemnly regard an owl swooping over head in the mountain sky. A pregnant Nofo Letau lay in hospital that night, crying as she thought of her children sleeping in the forests of Lalomanu. The next day, her husband came with the boys to visit. They all wore clean, dry clothes. Nofo's family from town had brought them mats, tarpaulin and food.

Help came from friends, villages and churches. Still in shock from the loss of her son Quezon, Melody Lesa did not have to worry about any of her family's needs. "At the hospital they immediately gave us clothes. Within 24 hours we had everything we needed, food and stuff. The paediatric staff were so kind. Because I used to work with them so they were like *Oh my God..* and they were giving me this and that, almost every hour."

The Samoan word for gift is '*meaalofa'*. Things of love. Things from love. Ask survivors of the Samoa tsunami about the help they received after their ordeal and they will tell you – there has been so much love that they have been overwhelmed. Ask Faletaulupe Lui of Saleaumua. Talking about the wave doesn't make her cry. Not even the part where they narrowly escaped with their 92 year old mother. No, ask her about the *mealofa* after the tsunami and the tears come. "That first day our church minister shared out food for everyone. Then that night we were at the plantation. It was cold and many people were wet. We were so surprised when the company 'Mr Lavalava' arrived with blankets and clothes. They reached us at about 12 midnight. We were very happy and thankful to the Lord for the love of these people, especially on this first night."

Rita Romeo of Lalomanu, remembers that night well. "We slept in the forest, we had no blankets. It was very cold and I felt so bad for my children. I sat there with tears looking at my kids asleep on the ground and I thought maybe it would have been better if we had died in the tsunami. I can handle it – but my children…The very next day we got help from the government and from generous people…clothing and blankets and food. So many things, *oka*!" Several weeks afterward, there is a twenty foot container beside the Romeo's wooden shelter on the Lalomanu shore. Sent from loving family overseas. Cardboard boxes spill an assortment of meaalofa. Dishes, clothes, books, bags, house ornaments, toys, canned food. There is a pool table under a crooked roof. Where the men are playing. On a break from their work building a stronger, more solid house for Rita's children.

Ask Molesi Pritchard and his wife Silaumea of Saleapaga and they will point to the food piled high at the side of their little open hut. Sacks of rice, boxes of canned mackerel, and packets of biscuits for the children. The rolls of matting, stacks of dishes and plastic trash bags filled with clothing. "The assistance has been great, one hundred percent, it is very good at this time."

Ask the principal of the old Saleapaga Primary School, Solonaima Emau. "I thank God for the assistance we are receiving because we are receiving things from many different people – those that aren't even related to us, not just from our country, but those from outside as well. People needed many things after the tsunami. My clothes box was drifting in the ocean but my clothes were all still good so I went and washed everything and it's all still good to use. But the other people of our village lost a whole lot. That's why I'm thankful to God because of my belongings that were spared because I don't have to want for much, and for all those things donated from other people's kindness."

The *meaalofa* came from churches. Religious worship is an intrinsic part of the Samoan culture, of village life and most villages have more than one church. When they needed it most, organized religion - organized and stepped up to give. At Saleapaga, the Sapelu family found comfort with their pastor. "After the tsunami we came down from the mountain with the pastor and all the *matai* of the families and they gathered us all together with our children and we prayed together. After the prayer we went to the pastor's house that wasn't badly damaged. We made tents and many people stayed there." Pastors from Toamua came to Lepa bearing supplies. Seventy-eight year old Lotolua Niumatapele rested at the inland high school that day. "We only had a cup of tea because at that time no one was hungry, just thinking about what had happened. But God is everything. We gave thanks for the many things people were giving us." Vaelupe Olosepu's church came that day. In town, the women of his congregation had the phone lines buzzing. *We're collecting clothing, sheets and food items. Bring what you can.* Leaders were shopping up a storm, cleaning out shelves of canned food. By early evening, trucks loaded with *meaalofa* were rumbling their way over Le Mafa Pass. "Our church came first to help us and then many others. Even up to this day we are still receiving assistance from those that have been generous and kind to our village."

At Saleaumua, the Reverend Tautasi Fa'alemiga was one of the few whose vehicle had escaped the tsunami unscathed. One observer was moved at the sight of the blue mini-Suzuki "being used as a free shuttle bus by the whole village transporting 3-4 people at a time, food, wood and roofing iron, anything and everything that the small car could carry…making countless trips throughout the day to and from the main road to the inland shelters."

Junior Petelo's family was thankful for their first meal that night after the tsunami in Saleaumua, "We got food from families around us and a three pound can of *pisupo*(corned beef) was given to all the families from the priest – that's what we ate for our evening meal."

The *alofa*, the love, was not limited to Samoans. Visitors caught in the tsunami upheaval were embraced everywhere. Arne Schreiber took a group of German survivors to stay at the Insel Fehmarn Hotel in Apia, wearing only ripped *lavalavas* scavenged from a tsunami zone. "We got there and people just came outside and were giving them clothes and things and these Germans, they were so surprised, I think we are not used to that in Germany. Strangers just walked up to them and gave them clothes and cared about them, it was great."

Nynette Sasse arrived at the Taufua mountain house midday on tsunami morning. She was greeted by Lydia, the owner of Litia Sini Beach Fales. "She walked up to me and said, *don't worry Nynette, we're cooking and the lunch will be ready soon for everyone. The village is all helping to get these people taken care of. Don't worry."* Nynette saw tourists wearing colourful *mu'umu'us* and Sunday best *puletasi*. "These people had climbed up out of the tsunami completely naked. As soon as the villagers saw them, they ran up with their best clothes to dress them. I was so humbled to see how our people took care of the visitors. I was so proud at that moment to be a Samoan!"

In the wave of goodwill that flowed through the disaster zone, there are those 'little things' that people remember as everywhere, people reached out to give what and where they could. Things like, a sandwich. On 29/09, Mosiula Vui and Vanessa Niuwenhuizein saw on television, the crowds gathered outside the morgue. They saw the people that slowly climbed off trucks carrying the bodies of children in their arms, then sat on the cement walkways, waiting. Unwilling to leave their loved ones alone. Mosiula and Vanessa wanted to do something. Anything but sit at home, helpless and hurting for those who had been hurt by the tsunami. They bought boxes of fresh bread. Slabs of butter. Cans of tuna and mackerel. They made sandwiches – stacks of them – and took them to the morgue. So that people with nothing could have a bite to eat. Then the next day, they took their big truck and volunteered with the Red Cross, helping to transport the dead to the hospital.

A handful of nails. That someone left on a brick. Where a white flag was raised in amidst the broken remains of a home in Saleapaga. On the flag an appeal was inscribed, *Need Help: Need H2O tank, food, hammer, nails, axe...Thank you!* Someone passing by responded and quietly gave a handful of nails.

A hot meal. Australian volunteer paramedic Steve Williams was caring for the injured at the National Hospital on that endless Tuesday. He got a call from another pair of volunteers, *Have you had any dinner?* Later that afternoon they came to the hospital. With a homemade meal big enough to feed a Samoan-sized family. Steve remembers, "I'll never forget that you know? They wanted to do something to help and it was just so thoughtful."

A ride to the hospital. One woman remembers the total stranger on the ferry ride from Savaii to Upolu island. Who noticed her distress upon hearing of the tsunami. Her tears of worry for her family in Aleipata who had been taken to the hospital. When the boat docked at Mulifanua, the man offered to take her to the hospital. A forty minute drive. Once there, he then accompanied her inside and was there to catch her when she finally collapsed with the stress of her morning. Francis Craig shrugs, "She was so upset, what else could I do? She needed some help."

Lip balm. Wendy Pearce from Christchurch, NZ lost everything in the waves. "The only thing I had left was the *lavalava* I was wearing." A woman offered to take them to the NZ High Commission. "When I asked her to stop at a shop so we could buy Vaseline for my sunburnt lips, she kindly went in and bought us shampoo, deodorant, moisturizer and a bottle of wine as well – that made us start crying again."

A gas bottle. A veteran Red Cross volunteer was moved to tears when she tried to buy two gas stoves for families hit by the tsunami. "The company wouldn't take any money from the Red Cross. Even better, it donated ten stoves and bottles…[me] and the gas man cried together."

A charge up for a cell phone battery. Tsunami survivor Becky Glew who ran from the waves with her mobile phone was thrilled with the helpful staff at a Digicel cell phone outlet in Apia who offered to charge her battery while they hiked through town in search of new travel documents and a flight booking home to New Zealand. The phone was a lifeline for Becky and her friends who were left with no money and no possessions in a strange land.

A beloved Spiderman toy. A young tsunami survivor was taken from his hospital room for an X-ray. When he returned it was to find that strangers had visited during his absence. A family. With a six year old boy, a cheeky glint in his eyes. Who came with cookies and candy and his favourite toy. To give to the boy "who is hurt and sad from the tsunami…maybe he can play with it and feel better."

A *lavalava.* Given to a young survivor from New Zealand. Who lay in agony from her injuries after the tsunami, uncaring that all her clothing had been ripped away in the waves. "This guy came and took care of her leg first – it was pretty bad. She ended up later needing several operations. He got her leg free and then he took off his own lavalava and covered her. She said she hadn't even realized that she was naked until that moment, and she was emotional remembering his thoughtfulness."

A roll of toilet paper. When people are admitted to the National Hospital they take many of their own supplies. A pillow. A fan. And toilet paper. Stunned tsunami survivors, some of them brought in naked with cuts covering their entire body, were then faced with a lack of the essentials. A woman came to the tsunami ward on Wednesday morning bearing hastily assembled hygiene kits. Frangipani soap and lotion, washcloths, toothbrushes – and toilet paper.

A hot shower. Jeff and Rachel Lynn escaped from Coconut's Resort at Siumu with little more than their lives. That night, friends Rob and Katie Wetzell invited them to stay with them. "That night we had a hot shower. I washed the sand out of Jeff's cuts and bathed him in hydrogen peroxide. Everywhere I looked he had a scratch, scrape or abrasion. We petted cats and dogs then laid down to dream of tsunamis and the events of the day."

Two large pizzas. From local pizza parlour Italiano's. Brought to Wendy and Chris Booth at Seabreeze Resort as they struggled to clean up the mess that was left of their lives. The husband of one of the resort employees worked at Italianos. He spoke little English. After the tsunami, he came with two huge pizzas and a message, *Don't give up. I tell you Chris and Wendy – don't give up.*

A bottle of fragranced Samoan coconut oil. Given to Lee Palupe after the tsunami wiped out her wellness retreat. "All my supplies for massage were gone, so I was happy to get the oil. My friends said to me, *ooh you smell nice for a victim!* It was lovely."

A clean pair of underwear. Steve the Aussie paramedic wanted to help the survivors at the hospital get clean clothes. "People had given them *lavalavas* so I called in favours with some of the other volunteers. I rang one of the girls and I said, *Look, this is going to sound really strange but have you got any spare underwear?! These women here really need some clean underwear*.... And she organized to bring some in for them so at least they had something to put on."

In every tsunami story, there are the 'little things', the small acts of kindness. Thoughtfulness. In amidst the suffering, the dirt and the desolation, the little things that make the misery bearable are there if you look for them.

17
The Hospital

More than three hundred people were injured in Samoa in the 29/09 tsunami. Countless others were classed as 'the walking wounded' – treated, shot with tetanus injections and sent on their way with strict instructions regarding post-tsunami infection. Immediately after the disaster, there were one hundred admissions to the National Hospital with a running total of 134 surgical procedures undertaken. Samoa faced a monumental strain on its health services. And yet, it was a health service that was not completely unprepared. The National Disaster Management Plan approved in November, 2006 called for different government sectors to take measures to individually prepare for disasters. The National Health Service had a disaster plan and in 2008, they had practiced for a tsunami.

Andrew Peteru is the Assistant CEO for Health, watching over the Public Health section. "The department held it's own tsunami drill, we used the Apia waterfront, we had volunteers in the ocean and everything. We put some kids out there and rushed them up to the hospital in cars. We had the fire brigade and the police involved too. We put people all over the place and then responded to phone calls, sending out drivers to bring them all in. It was really busy, people all over town didn't know what the hell was going on! We were telling the 'injured' to scream and yell and everything which they did.. and the similarity with 29/09 was very surreal. But on 29/09 the difference was, there was silence, absolute silence. People came to the hospital and there was stunned silence. And of course the numbers were different, because there were so many dead -but because of the drill, I think it was not as strange or as panicked as it might have been."

The H1N1 pandemic also helped ready the NHS for 29/09. In 2008, an estimated 16,000 people in Samoa were affected with the virus and the National Hospital had a 30 bed turnover filled with the more serious cases. Dr Stanley Dean is the NHS General Manager, "Our disaster planning was mainly targeted for epidemics and cyclones, the possibility of tsunami was never in our mind. The first test of our emergency response plan was during the H1N1 swine flu. A large percentage of the population was affected and we were trying to treat people as quickly as possible and also to stop the movement of people so that we could control the spread of the disease. It was a nationwide problem and we focused everybody on responding to that emergency. When the tsunami occurred, we more or less already knew what to do to deal with it." CEO for Health,Palantina Toelupe agrees, "Our experience with H1N1 got everybody on track with what to do [in an emergency] very quickly. But it goes a long way back to other government sectors because the MNRE and the Disaster Advisory Committee had been very assertive about everyone needing to have our sector plans. So it was very good we had done the H1N1, we had done the drills, and we had an idea of what to do."

What a difference a little preparation makes. Dr Bill Grigg is Clinical Director of Retrieval Co-ordination for South Australia Health. He was on the Australian team that responded after the 2004 Boxing Day tsunami. And the team that flew to assist after the Bali bombings. He arrived in Samoa early Wednesday morning and went straight up to the National Hospital. Matt Anderson, the Australian High Commissioner to Samoa recalls that the first thing Dr Grigg did when he got there - was sit. And watch. And after he had observed the proceedings he told the Clinical Director, Dr Limbo Fiu, "*I wouldn't change anything, not a single thing. Now, that we're here, what do you want us to do? Where do you need us to go?* I thought that was high praise indeed for Samoa's medical service, coming from such an experienced disaster response expert."

At 7:30am, medical staff were getting the call to come in the hospital at Motootua. With details still sketchy, many went in routinely. Dr Teriki remembers his trip to the hospital that morning. "I honestly thought it was just minimal damage. For some reason I just thought, *oh, okay, there's been some injuries*... It was the last thing I ever expected, a tsunami hitting Samoa at that magnitude. So I didn't think it was going to be that serious, I just got ready and went in to work as I would like usual – I didn't rush. I didn't realize until I was actually in the emergency department that I knew how severe it was…these people were being brought in by truckloads."

As she drove in to the hospital, Palantina Toelupe was listening to the first reports of deaths on the radio and her mind was busy with the planning. "*Okay, what are we going to do about getting the injured to town...where are we going to put them all...what if there's another earthquake and damage to the hospital...* At the hospital we left the medical professionals to do their thing, the doctors and nurses were quite organized – we didn't set foot inside the outpatients. We were outside receiving the dead, receiving the injured people. Even my older son and some of his rugby team boys were helping to carry the dead inside. Our management was quick to get here to the office – we had people answering phones communicating with WHO and our other development partners, people organizing food and water and others automatically taking over the assistance with the morgue. We had to organize the nurses' homes for families and relatives, because it was very clear that these people had lost everything and didn't know what to go back to. So while injured were being admitted we were organizing places for their family to sleep, food, supporting them in whatever way we can. But they were determined to go back to their villages, after much crying and all that, they were looking for transport to go back – to keep looking for the rest of their families."

Palantina tells of one particular moment on that busy morning. "There was one incident that really got me. A double cab pickup truck arrived and this really big man got out. He had the saddest of looks, he didn't talk to anyone, he just got out and opened the back door and walked away, didn't even turn around. A woman got out with a young child and then the door on the other side opened and another woman got out with another child. They carried them inside and then they came back and each carried out another child. By the time they finished, they had carried inside eight little children, all under the ages of three or so and every kid that was brought out of that double cab looked like they were going to die. Because you know it's a very long drive here from [the disaster area]. That was the most horrible sight for me because I saw the future. I saw babies and I saw three year olds and I saw all Samoa's future, these kids weren't looking good and it was very, very sad. The older people didn't get me as much, I thought, *okay, it's up to the Lord.* But the younger ones were killing me mentally, really getting to me as a person and as a professional."

When Andrew Peteru got to the hospital he went to help man the phones. Among the many calls from overseas media and worried families was one from a family searching for *an elderly woman with Alzheimers, Madeleine Lewis. She has ID on her, please look out for her. We've lost her...* Andrew went first to a crowded morgue to look for her. "It was silence, deathly silent. Some say, the Samoan word for it is *misialofa*. I think it was shock. There were some people taking photos with their mobile phones and just rows of people sitting outside, with blank stares." No Madeleine Lewis. He checked the fast filling tsunami ward. No Madeleine Lewis. He returned to his post, but an elderly woman with Alzheimers was never far from his thoughts. Later that afternoon he led a team from Public Health to do an assessment of the disaster zones. At each district hospital, each evacuation centre he was asking, *has anyone seen an elderly palagi woman with id necklace, disoriented, Alzheimers, Madeleine Lewis?*

Steve Williams was at the National Hospital that morning. Steve is an Australian paramedic with thirty-six years experience, an AusAID volunteer seconded to the Samoa Fire and Emergency Service Authority. He came to

Samoa in July 2009 to do an 18 month stint as a paramedic trainer. He had gone to the Faleata Fire Station after the earthquake, when he received the call that one of the fire trucks had crashed on the southern coast. With the ambulance on its way to Aleipata to the tsunami zone, Steve rushed to the crash scene in his car. There had been five fire officers on the truck – two had minor injuries, two severe and the fifth, Peter Letiu, was killed. Steve treated the men and accompanied them to the National Hospital, arriving there about 9:30am. Just before the tsunami casualties began flooding in.

Wounded from Poutasi village were the first to arrive. Followed by those from the Aleipata district. As with most disaster trauma centres worldwide, the hospital used a colour code system to denote who needed treatment most urgently. "The ones that could wait we stuck out in the Outpatients waiting room on the beds there. Roughly four patients had one person looking after them. The ones that needed reasonably urgent treatment, got it when they needed it. But the really bad ones that came in were treated straightaway. There was lots of nurses and volunteers there helping out. We had a lot of patients that didn't get admitted to hospital, they just needed a basic check out and quick treatment for cuts and bruises. Then they were taken to hotels and some went to the embassies. Everyone coming in to the hospital though, got their antibiotics, everyone got their tetanus shot which was really good and a lot of them actually got it at the Lalomanu district hospital which made things easier on us at the National Hospital when they arrived."

Steve has definite views about the quality of the Samoan response to the health emergency. "I've worked in major cities in Australia and I don't think that any other hospital would have handled it just as well. The National Hospital here is just brilliant. When you consider the resources that they have and don't have, they did a really great job. I was impressed…I still can't get over how well that hospital worked. The best recommendation I can give the National Hospital is I would take my daughter there. I'd have no qualms."

Steve was asked by the Australian High Commission to "help with the *palagis* at the National Hospital. So as they came in, I recorded their information and tried to help them with whatever they needed. But as the day went on, I basically dealt with all the foreign nationals, it didn't matter whether they were Australian, NZ, German, British or whatever."

Can you imagine what it felt like to be injured amongst strangers in a strange land? With nothing left but the clothes on your back – and some, not even that? "Some of the patients had their clothes ripped off them. One woman was showing us her dress and it was just in shreds. I couldn't believe it – it was literally shredded. You'd look at it and think, *for that to happen, you should be a whole lot worse injured than you are!* Emotionally there were people that were absolutely devastated – parents losing children, people losing close friends, loved ones. Emotionally, a lot of people were just completely distraught."

Many times, Steve acted as stand in family for those who had none. "I was going round the wards just making sure everyone had things they needed. If they wanted to ring home – my phone bill – I'd just put ten tala on the cell and say here, ring. That was my little contribution, if they wanted to ring Germany or wherever I'd say, *you've got ten tala, when it's out that's it!* So they could just ring home and I didn't mind that, it was nice to be able to give them something. If there was something they wanted I'd try and chase it up for them. If they wanted to find out anything from the doctors I could go and chase it up for them. The doctors were great. Here's this crazy *palagi* coming up and asking questions about patients and they'd take a minute or two to give me the info I needed so I could go and explain to the patient."

Reuniting separated family and friends was one of Steve's more rewarding tasks. "It was nice to be able to match up people. There were two friends who had gotten separated, someone came up to me and said, *this guy's outside looking for his friend* and they wouldn't let him in because he wasn't injured. He was just frantic with worry trying to find his friend. So they got me to go look for her and it took me about an hour and a half, you'd think a *palagi* would be easy to find in Samoa, but the waiting room at the outpatients was full, it had beds all the way through it. You go along and you're getting the details for the patients and you're saying, *hmm, I've done you, got you, no, you're a new one* and you'd sort of miss one every now and then. So it took me awhile to find this young woman, I said to her, *don't move, stay right here!* Which was silly because she had a broken leg and couldn't go anywhere anyway! Then I got her friend and brought him in and it totally made his day."

The hardest part of the day for Steve? "Dealing with the morgue, the children. "The adults and injuries to them – it may sound callous, but because of thirty years in my job, I'm able to handle it. Anyone younger than 18 though, it's rough…you've got to let something worry you, because if you get so callous that nothing worries, you're hopeless in your job. So the worst bit for me, was the loss of the children and there were so many children coming through. The morgue staff were very caring with the children, they had their own little spot. It was lovely so see how gently they were all dealt with in the morgue."

Steve recalls a particularly difficult moment outside the morgue that afternoon, "I looked over and here was this guy with a telephoto lens. They had just pulled a baby out of the back of the ambulance, taking him into the morgue, and this guy has got his camera sticking right in there. I went up and told him to get down, and not to bother, probably not as politely as I should have! Basically he said, *I've been told I can take photos anywhere in the hospital!* I said, *I don't care, the bodies need to be given some dignity.* Imagine if it was his kid…you just don't do things like that." The confrontation ended with Steve manhandling the photographer out of there. "We've got emotions going, I've just had one of my friends, Peter Letiu, die on me and you don't really want to see people doing things like that. A little bit of courtesy would have been nice for people."

Another rough moment came that afternoon when Steve met with the mother of the victim he had known well. The mother of his friend in the fire service, Peter Letiu. "That was pretty hard. I went and talked to her for awhile and I found that really rough. But she was a lovely lady, immensely strong." Steve had used Peter's watch after the accident to take observations on the other injured and he was a little anxious about explaining it to Peter's mother. "My watch had broken a week before and I needed one when I was rushing into town on the back of the truck with the boys after the accident. I borrowed Peter's, I was certain he wouldn't mind and he would have been happy for me to do that. So I was using that and I actually gave it to his mother there and then. I was worried about telling her that I sort of 'borrowed' the deceased's watch but when I explained why I'd done it, she was really nice about it."

Steve remembers one survivor in particular, a young woman who had been travelling with a group of friends, including sisters, Rebecca and Petria Martin from Matamata, New Zealand. The sisters were among the nine foreigner casualties and the young woman was seriously injured. She could not be airlifted out without first undergoing critical surgery and she was understandably very afraid. "She was pretty distressed about it. I stayed with her and went with her when she went to theatre, making sure I was there when she came out. It was kind of hard emotionally for me because I have a daughter roughly the same age and looking at this young woman – I was thinking very much of my own daughter and how I would feel if she were in the same situation, I would want someone to look after her, be there to keep an eye on her. She wasn't completely alone though – she had a Samoan family kind of looking after her which was lovely. I don't know how they had met her, but she had this Samoan 'mum' that would come everyday and help bathe her and look after her which was so nice to see."

In the afternoon of tsunami day, Steve met an elderly NZ woman who had come in from Lalomanu hospital. "There was this little old lady who had absolutely no idea what was going on. She had an ID necklace with her name and address on it. She'd lost her family and was quite upset. She wasn't injured though. We gave her name to the NZ High Commission so they could get in touch with her family and then they took her up to the residence to spend the night."

Steve was at the National Hospital until late that Tuesday night. He was there when the Asst Commissioner of FESA, Tipaula Laupu'e came to the outpatients with a suitcase. Tipaula was tired, hungry and dirty. But the FESA team was insistent that they try at least to find the elderly woman. *We've got this bag that belongs to a little old palagi lady, she was evacuated from Saleapaga...* Steve made the connection. "*Hey, I think I know who that is...* it was quite neat because we were able to then organize where the bag had to go, someone ran it up to the woman at the NZ High Comm. house and that's when we found out that everything worked out well for her, all her family were okay."

The epic saga of a suitcase finally had its conclusion. Carried up a steep hill away from a tsunami at Seabreeze Resort, Vavau by helpful staff. Loaded in a fire service ambulance that was flagged down on a debris strewn road. Placed under a tree at Saleapaga, Madeleine Lewis rested on it and looked out over a desolate land. Unwillingly abandoned then rescued by a crew of FESA boys. Brought to a hospital crammed full with despair and then finally, taken to a mountain sanctuary for survivors and restored to its owner.

That night, alone in a dark house filled with tsunami homeless, alone in the wasteland of a memory littered with half-forgotten things, Madeleine Lewis could clutch the one thing that was familiar. Her suitcase.

When people are hurt and afraid, all they want is to get home. As fast as possible. Foreign nationals critically injured in the Samoa tsunami were airlifted out of Samoa. One woman remembers talking to the injured tourists in the hospital the morning after the tsunami. "They were so polite and appreciative of everything we were doing for them. We asked them if they needed anything else and they said, *no, no we're fine...* But then, the first medical team arrived from Australia. They walked into the ward and as soon as they spoke to these tourists, those girls just burst into tears, they said, *please, please can you help us, please get us out of here,we just want to go home.* My heart just went out to them, I cried inside for them."

The first airlift went out on Wednesday afternoon, the last on Friday. Steve helped get the patients ready and moved out to the helicopters that would ferry them to the international airport some distance out of Apia. "The evacuation team was fantastic. Their attitude with the patients was really good – like if it was someone who was a bit large, they'd lift them on the stretcher and they'd all scream and grunt and carry on joking, overemphasise it, get the patient laughing. They actually played around with the patients and tried to make it more enjoyable for them, showing humor in what was a really scary situation for some of them."

The helicopters landed in a small grassy patch between the emergency morgue and the wards. "The last lift was 8:30 at night, pitch black and so the pilot flew in blind and watching them come in was just magnificent. They had to land in this small space and it's not even flat, there's a dip in the ground and they landed in the dip, and I'm going *geeze!* Even in daylight it was pretty impressive and you think back and you're just so impressed with what these pilots did. I watched all the patients go and it was a really nice feeling when the last chopper took off that night, knowing everyone was on their way home to their families."

Dr Ben Matalavea and his team arrived back at the main hospital from the disaster zone late on 29/09. Palantina Toelupe clearly recalls their return. "There's never enough said about the staff. When they returned that evening, a busload of the team that went out to work in the disaster area, it was one of those sights that you really wished you had captured on camera. These doctors out the door looking so tired and then one by one, you see tears and they started breaking down."

Dr Teriki Puni remembers his return home that night. "You get to a point working many years in hospital, that you get used to what you see but that day – it's something we didn't expect, especially in my lifetime. It was a different perspective in terms of seeing death, just a few minutes after that earthquake and then there were that many casualties. You come back and see your kids and your partner and you really, really do appreciate that. It's something that you won't forget for a very long time."

All the injured were moved to the main hospital on that first day. Dr Ben explained, "We didn't leave anybody there at Lalomanu Hospital, we just brought everyone here, even the patients considered mild, there were no patients left. When we got back it was pandemonium. All the management from the Ministry, the NHS, everyone was up at the Outpatients. There were hundreds of people there. The Outpatients itself had been turned into a ward with lots of beds for all the injured ones. Staff were trying to prepare the tsunami ward. I left there around midnight and they did a really good job in sorting out the ward and moving people because I turned up at seven the next morning and all the people had been shifted to the ward."

Dr Ben continued working in the zone over the next two weeks, overseeing the response in the field. Teams provided full medical coverage at the Lalomanu Hospital and also went out into the settlements taking medicine and treatment directly to the injured. By the second week, Dr Ben had twenty doctors from all over the world working with him in the zone – all of them private volunteers, separate from those that came as part of Australia and New Zealand's medical response.

Dr Ben Matalavea remembers a doctor from Brisbane, Australia – "He saw the news that night and hopped on a plane the very next day. He's been to many disaster areas so he's an emergency specialist. He had some leave and so he came with his gear ready. He didn't go through any normal channels of bureaucracy, he just came and met me and gave me his credentials and said *what can I do to help?* I knew he would be perfect for Lalomanu and he worked out there for a week. He explained to me his reasons for coming, *I'm actually from Auckland, NZ. I went to Avondale College and played in the rugby first fifteen team. I was the only palagi in the forward pack, all the rest were Samoans. They were all my friends and when this tsunami happened, all I could think of were my old mates and their families. So I came.*"

Nurses came as well to volunteer. Henrietta Aviga is a Samoan nurse working on Thursday Island in the Torres Strait. She was on holiday in Auckland when her sister called with news of the tsunami. By Wednesday afternoon she was on a plane home, arriving at 7:30 that night. She headed straight for the National Hospital where she worked the night shift in the Outpatient's Department. "We worked through the night with tsunami victims, many with horrendous injuries now twenty-four hours old and infection setting in. Many of them had delayed coming to the hospital because they were still searching for missing family members." One of these patients was Ave Mikaele from Lalomanu. "She had lost her mother, sister-in-law, nieces, nephews and two of her children. Her body was battered and cut by the wave, but she was focused on the fact that the bodies of her two children still had not been found. She was holding herself together remarkably well but I was worried about her."

When Henrietta's shift finished early Thursday morning she drove out to Lalomanu Hospital, first stopping at the Fire Rescue tent. She had been a volunteer fire-fighter in Samoa for five years and so knew the team there well. "Their task was a gruelling one, the smell of decomposing bodies in the tarpaulins was everywhere and the sheer number of the dead was disturbing. They all looked physically and mentally tired and I hoped they would get a chance for some debriefing, some rest – soon."

At Lalomanu hospital, Henrietta worked alongside local and overseas volunteers. There was a team of medical personnel from Australia and the Queensland Ambulance, Fire and Rescue team had pitched their tents beside the hospital providing extra support. The NZ Army and Air Force were based at the nearby primary school. Henrietta worked at the hospital all that day and in the two weeks afterward and like many nurses, found her task involved being a listener, a comforter. "There was so much emotional turmoil about the place, I felt compelled to spend time with patients and their families to try and help with their needs that were more than physical."

"I met many disheartened and dejected people…there was an elderly man who came in with bad injuries who was very tearful as he shared the guilt he felt at not being able to save his grandchildren. A middle-aged woman who was swept away but had caught hold of a tree and held on for dear life. A teenage boy who was riding his bike away from the tsunami, then struck by a car as it washed in the wave – he was thrown into the air and then caught by the tsunami and thrashed around. I remember the fear on a young girl's face when we had to clean out a deep gash that was now infected to her lower leg – after her mother shared with me her ordeal of trying to hold her daughter's head above the water while her leg was caught on something under the water. I spoke with a lady whose leg had been stuck between a car and a tree sustaining a deep laceration that eventually required four operations – she expressed how lucky she felt to be alive. There were those who didn't want to talk. A young man who had lost his mother and brother – he was withdrawn and made no eye contact the whole time we treated him. A young pregnant mother who hadn't felt her baby kick since the tsunami – she was so afraid, there were no words to explain how much."

Later that first week, Henrietta met again with Ave, sitting in the hospital corridor. "She remembered me and told me how she still had not found her two children. She had come today because she had heard a child had been found." The police van pulled up to the hospital and unloaded several bodies, including one small shape wrapped in canvas. Ave was disappointed however. A mistake had been made the day before and Ave's daughter Malo Maryanne had been buried with another woman. This latest child was the other woman's daughter. The difficult task of recovering Malo Maryanne from a shallow grave was carried out and now at last, Ave had one child to weep over. But the searching and waiting would continue for her son Aleki.

Henrietta stayed in Samoa to help for two weeks. She was on duty at the hospital when the country held a national memorial service for the tsunami victims. Henrietta saw on television, a familiar face crying over a tiny coffin draped in white lace. "I saw Ave burying her family, crying openly. I was glad because I felt that would have given her some small sense of relief and maybe help her with her grief."

When Henrietta finally flew back to Thursday Island, she was "an emotional wreck." But grateful for the opportunity to be a part of the medical response to her people's disaster. "The strength of the staff base at Lalomanu was amazing to watch and to be a part of. The local nurses in particular, displayed an unwavering courage to endure their many shifts and see that all work was done. The task we faced was at times overwhelming, but the teamwork was amazing, we shared so much and learned a lot too. It was a privilege to share in that, to be there."

18
Occupy the people with work

After the tsunami, the urgent concern was how to prevent the next disaster – disease - as people living in rough conditions struggled with finding the basic necessities. This was Andrew Peteru's task, to " make sure people have safe water to drink, safe food to eat, that the environment is healthy. That's what we do, we protect and prevent." There was the worry of dengue fever with families living in mosquito havens of forest. Typhoid, dysentery, leptosporosis - as people dealt with basic hygiene when there was no running water. Malnutrition, particularly for children as families struggled with an inconsistent supply of food.

The post-tsunami health and sanitation response team was made up of many professionals from different countries and organizations. Hands that helped tsunami survivors in Indonesia were now here. Minds that had puzzled over the devastation of bush fires in the Aussie outback, bombings in Bali, cyclones in Fiji and more – met together to devise solutions to the aftermath of Samoa's tsunami.

Starting from Wednesday morning and in days after, Andrew's office was jam packed. "We had public health doctors and environmental health experts from Australia, NZ Army engineers, EU sanitation consultants, UN health, everyone was here, but I'm very bossy so I had no problems." Andrew is a warrior for Public Health – tough, uncompromising and dedicated. He has to be – the safety of the food you and I eat at a restaurant, the medicines we are prescribed, the products shops are allowed to sell us, depends on his thoroughness. How well we are protected from a pandemic will rest on Andrew and his team's exactness. The Public Health team knew that far worse horrors would follow 29/09 if the right steps weren't taken.

Phase One for the team was to find out where people had gone. The Army used GPS to track the new settlements. "It took us six days to complete and then we were able to hand it over to the MNRE people and those are the maps that everyone else used." Phase Two was a face-to-face consultation with every single displaced family. It was a vast undertaking and one which some of the support from overseas agencies questioned. "They said, *are you sure?! That's a lot of people!* I told them those people out there are not going to get the message over the radio or television. They won't get the message that way. We Samoan's don't listen, we watch, so we have to go in and do it with them, I'm really sorry but that's what we have to do. And after a few days of teams going out they realized what I meant and it was also good because they were able to experience our people quite up close."

"We sent out teams to all the affected villages and met with each family. You go to them, find out *how many of you, what are your needs, dig a hole here for your toilet, dig a hole here for your water.* It was real hands on, we gave practical instructions, *where's your water? Right, boil this and put it in this clean plastic container and that's your drinking water so don't mix it with anything else. Put this other bucket of water here for your toilet...*the team couldn't move on until they'd done the water training and seen a pit toilet dug, a soap ditch and a hole to bury the families' rubbish in."

Andrew insisted that people were to do their own sanitation work. "I told our teams the families had to do it, they had to dig the holes, make the toilets. It was tricky for some to agree to but I told them – *you are NOT to dig, you have to make these people occupied, occupy them with work.* A lot of survivors were just sitting waiting for things to get done for them maybe because they were still stunned."

Phase Three is still ongoing through 2010 – community surveillance. "We only do it once a week. We go to all the new settlements and just find those in tents and rough shelters and touch base with families to see if they're okay, also touching base with the district hospitals to check if there's a cluster of diarrhoea, typhoid, dengue or whatever – because we have to know those things quickly to see it there's an outbreak And thankfully, there's been no outbreaks of disease. If Apia had been affected we would have had heaps of problems with disease, but this is in the village where everyone went to a new place to settle, so all of the areas that were affected were basically abandoned and then slowly cleaned up." Surveillance includes visits by the nutrition team, checking that mothers are breastfeeding, that basic dietary needs are being met for all ages.

Another concern for the National Health Service – was how to cope with the numbers of the dead. Unlike many Western cultures, death in Samoa, is not the cut off point for one's bond with their loved ones. In Samoa, the dead are carefully washed and dressed by their own family. They are kept as much as possible, in their own homes. Beloved. Choirs will often come through the night before a burial to sing as a family keeps vigil by the side of their dead. Singing to give comfort to the bereaved. Singing to remind the new spirit of the love they leave behind. After the tsunami, people wanted to get their loved ones as quickly as possible. They wanted to see them, hold them, cry over them. They wanted to take them home. But with more than a hundred dead, a quick release from the morgue was not possible. Lucia Hellesoe had to wait three days before she could take her parents bodies. "I wanted to see my parents right away and I was told they couldn't be released because all the bodies were still under investigation. I told them, *that's not right, I'm suffering, I've lost both my parents you don't understand what I'm going through, I want the best for my parents, I want to take them away from here...* You could smell the place from a far distance away and I saw they were putting people on top of each other and it wasn't right, it was very difficult to have to leave my parents there for so long."

On tsunami day, the NHS requested assistance with coolers to help with body storage. Even then, the task proved difficult to manage. Andrew Peteru explained, "When a lot of the bodies came in, they were fine [on the outside] but they had drank a lot of bad stuff, so you cannot stop the deterioration of the body no matter how you store it, so by the time the family saw the body, it would be deteriorating and they would get more upset and want to know what was going on. There were a lot of complaints coming in to the CEO that we had to try and deal with, try to explain to people what the challenges were."

Some of the dead were not found for several days. A few were incomplete. The NHS together with the police issued directives which some Samoan families found difficult to adhere to. *With much respect we advise that bodies be placed in body bags. All bodies released to families are to be accompanied with the advice that the body bags must NOT be opened. Body bags may be placed in coffins. Once bodies are released from the morgues they must be buried immediately.*

A sensitive situation arose in the week after the tsunami when the Australian family of a victim came to get the body of their beloved daughter. She had been buried together with her Taufua family in a communal grave at Lalomanu and needed to be exhumed. Talks were held with the family and permission was given. Approval was sought from the Health Department. Anxious not to cause further trauma to a village and a family still reeling with their grief, the exhumation was carried out in the early hours of the morning. It was difficult to be unobtrusive though when heavy equipment was needed to break open the cement covered gravesite. Thankfully the Australian DVI team was still in the country and could quickly carry out the necessary identification so that a daughter who had died so far away from her parents could be taken home.

Working on the post tsunami recovery was a sharp learning curve for all and Andrew was grateful for the opportunity to work with other international experts."The people who had been at Banda Aceh were amazingly the most quiet people in the team. They sat back in the meetings and never said anything but they would come to me afterwards and say, *good work* and that's when they would give me their advice. Those were the people who I found were the most understanding and the most sympathetic. Even if I said something and they knew better – they wouldn't change it or question it. They would go out and do it that way and then come back to me and say, *okay we tried that, can we try something else now?* Or *hey, that actually worked, we didn't think it would but it did!"*

Andrew can laugh now about one particular lesson learnt, what he calls his "most embarrassing moment." Nurses at Lalomanu Hospital were complaining to him about the smell. *We can't get rid of the dead body smell, it's everywhere, what should we do?* Andrew asked one of the Australians on the team what was the best way to deal with the problem and they replied, *Lime.* Andrew wanted to know, *big or small?* The answer – *small lime.* Eager to help the district hospital staff, Andrew went to the Fugalei Market in Apia and bought boxes of freshly picked limes. He drove out to Lalomanu, his car spilling over with the crisp green fragrance. The next day, one of the hospital staff thanked him, *hey, we had some really good 'vai tipolo' lemonade last night!* Confused, Andrew replied, *okay, but that's not what the limes were for...* Another of the Australians asked him *well what were they for then?* There was a lot of laughter at Andrew's answer and they explained that the lime in question was a powder called holding lime.

"There was a lot that we hadn't encountered before, so we had to 'go down the path that isn't trodden' so it was about doing that and just hoping and praying that there wouldn't be any outbreaks [of disease] because that would be the measure of our work."

Andrew's boss, Palantina Toelupe concludes, "We haven't had any major outbreaks since 29/09 and the credit for that should really go to the people living in the disaster areas. We were only there to help facilitate the application of the technical knowledge, together with colleagues from the Red Cross, Oxfam and other NGO's, but it's really them, the people, who do it everyday. They are the ones who had to live with the challenges and look after themselves in spite of their water and sanitary situation, in spite of their resettlement. Everyone contributed to ensuring that [outbreaks] didn't happen."

December, 2010

Some tsunami stories are running wild.

Did you know that there's a monster at Aleipata?. Or so they tell me. With hushed voices and huge gestures to emphasize how big it is, how fast it runs, how frightening it looks. Apparently, it's a creature that used to live in a cave on the beach. (don't ask me which beach as they can't really say...) It was happily living in this cave, only venturing out at night to get food from the ocean. Big fish, definitely. Dolphins and sharks, probably. Whales, when it got really hungry. It didn't used to bother anybody. But then the tsunami came and ruined it's cave. And the fishing was disrupted. And all the machinery and the mess on the shore drove the monster up into the hills. Into the forests and bushes. The same forests and bushes where survivors are living in shacks and tents. And now the monster is hungry. It's eating stray pigs and chickens now, definitely. And lost cows, probably. And people need to be careful, very careful. The monster is hungry and it doesn't like living in the forest. It misses the ocean. What does it look like? Like a bear, they tell me. A huge, wild, hairy bear. Only much bigger. They tell me I shouldn't be going to the forest in Aleipata to do interviews. No way. Too dangerous. They ask me if I saw all the helicopters that would fly over there all the time after 29/09? They shake their heads knowledgeably – the helicopters were looking for the monster. Only the army didn't want to tell people the truth, in case they got scared. Of the monster.

Did you know that there's a crocodile at Aleipata? A woman calls the newspaper to report a strange creature in her village of Satitoa. She thinks it's a crocodile. A week later, the Samoa Observer has a picture of a little, furry, weasel-like animal. Not a crocodile. A mongoose. Sighted at Satitoa. They think it came from Fiji on a shipping container. Villagers are warned to keep their children away from the mongoose. Special traps are brought from New Zealand. Hunting parties are organized. Success! The mongoose is caught. Everyone hopes it hitched a ride to Samoa as a lonely bachelorette, exiled from all friends and family. And not as a mongoose mum with lots of baby mongooses.

But the monster? Apparently it's still out there. Somewhere.

Running wild like a wild tsunami story.

19
How will we cope?

Samoa did not face its trials alone. Help came from Samoa's closest friends, Australia and New Zealand, cooperating with France under the FRANZ Arrangement to coordinate relief supplies and defence force assets. Both countries initially each made two million dollars available for emergency relief support in the hours after the tsunami – money to help buy food and water on the ground, money for the Red Cross. By the end of the recovery phase, Australia's contribution grew to more than twelve million dollars. On tsunami morning, the Australian High Commissioner Matt Anderson received the official request for assistance from the Samoan Government – a hastily composed, hand-written 'Third Person Note', drafted by the Acting MFAT CEO, Sharon Potoi-Aiafi in the High Commission's conference room. Mr Anderson described this timely request for assistance as "the single most important factor behind the early dispatch of emergency humanitarian assistance" and added that it was "the professionalism of officers from Samoa's MFAT and Ministry of Health in those first few hours that saved countless lives.

The first medical teams from Australia arrived on Wednesday morning, less than 24 hours after the formal request for assistance, along with search and rescue experts the day after – a total of 115 trained personnel who went to work immediately. Air Force medics knelt beside a battered survivor sitting in a wheelchair as he clutched a red plastic bag – all he had left from his Samoa holiday. They stretchered others into the gaping jaws of the RAAF's largest and newest aircraft, the C-17 Globemaster. They went to the inland settlements in search of the wounded, they helped staff battered district hospitals and performed lifesaving surgery on survivors.

Numbers can paint a picture. Over the next week, Australians would conduct 22 aeromedical evacuations, 101 operations, 171 field treatments, 1060 emergency department treatments, and 33 public health village assessments. The Air Force flew in 127 family tents, 1300 water containers, 4000 water buckets, 2,500 blankets, 1000 mosquito nets, 1400 large tarpaulins, 100,000 water purification tablets, assorted gear for cleaning debris and rebuilding – wheelbarrows, overalls, boots and gloves.

New Zealand worked alongside them. At 10 o'clock on tsunami morning, David Dolphin received a request from Samoa's Acting Police Commissioner to arrange for RNZAF aircraft to be sent up to look for survivors and to assess the extent of the damage.

An Orion aircraft deployed soon after and it was able to do an initial sweep of the coastline before nightfall on the first day. Over 250 personnel from the NZ Navy, Army and Air Force were involved in a range of tasks, and among them were 43 medical workers deployed through the NZ Ministry of Health. An NZDF team re-opened the Poutasi Hospital and staffed it for a month afterwards, providing valuable medical care for those in the Falealili district. The huge multi-role vessel HMNZS Canterbury, brought 80 tonnes of general relief materials, 6 tonnes of medical equipment, 7 tonnes of plumbing supplies, 128 tonnes of power poles and *fale* posts, 4 tonnes of electrical equipment. Men and women in black and red shirts unloaded 22,000kg of tents, caskets, chainsaws, tools, shelter boxes, tarps, food and water. 43 NZDF engineers assisted with clearing debris on the island of Manono, including lifting beached fishing boats and returning them to the ocean so fisherman could have their livelihoods back. Other NZ engineers helped to do key assessments of infrastructure affected by the earthquake and tsunami, like hydro-electric power plants, Poutasi Hospital, the Salani bridge and several buildings in Apia.

Search and rescue teams from Australia came to work with Samoa Police, NZ Police dogs, NZDF engineers and divers to complete 550 man-hours of physical search effort over an intensive five-day period. An Australian Federal Police Disaster Victim Identification team arrived on day two after the tsunami to conduct post-mortem examinations leading to the identification of five known Australian nationals as well as one Brazilian, one German, and numerous other Samoan casualties.

The Royal NZ Navy Dive Team repaired and replaced fresh water pipelines to the islands of Namua and Manono off the coast of Upolu. NZ Police radio technicians helped local police establish a police radio network on the south coast. Officials from the NZ Ministry of Civil Defence and Emergency Management came to assist the Samoa DMO team. Structural engineers helped assess the safety of infrastructure and public buildings damaged by the earthquake.

In the finest traditions of the ANZAC spirit, New Zealand Iroquois helicopters, flown to Samoa by the Royal Australian Air Force, were invaluable for transporting people and gear to the zone, quickly and efficiently. They also brought smiles to the faces of Lalomanu Primary school children during a visit by the NZ High Commissioner, Caroline Bilkey to the area. Air Force pilots gave the children rides in the chopper – there was absolute glee as they clambered in and out and posed for pictures sitting in the pilot's seat.

On the 9th of November, 2009, the HMAS 'Tobruk' arrived in Apia and the ship's doctor and medic went to Lalomanu Hospital where they set up a day clinic to monitor the recovery of those affected by the tsunami. LARCS were unloaded. They drove over Le Mafa pass and straight into the ocean at Lalomanu beach to clear debris from the reef. Working with divers and volunteers, they meticulously scoured the lagoon for trash. Little children watched in awe as the truck-boats ferried in tonne after tonne of metal scrap, timber and junk. Tobruk crew responded to a makeshift sign as they drove through Saleaumua village, *Our pre-school needs help.* Men and women in grey camouflage uniforms dripped with sweat in 30 degree humidity as they put walls, doors, a roof – on the little school. Children cheered when they were done, bringing them sweet coconuts to drink.

Amongst the 200 tonnes of emergency supplies, equipment and donations unloaded from the 'Tobruk' was a shiny red fire truck, gifted by the Melbourne and Metropolitan Fire Brigade to replace the one that crashed while responding to the tsunami sirens. Inscribed on the side, *In Memory of Firefighter, Peter Letiu. 2009.* Peter's widow Nele was there at the official handover to the Samoa FESA and Peter's little girl Mile helped to 'launch' the new truck. In June 2010, another two fire trucks came for Samoa aboard the HMAS 'Kanimbla'.

Help also came to Samoa from Germany - $100,000 euros. Announced by Arne Schreiber at the National Disaster Committee meeting held on Wednesday after the tsunami. "I held up my hand and said that Germany is donating this much and it was silent and then suddenly all were clapping… something I didn't expect." Help came from our cousins throughout the Pacific – Wallis and Futuna sent us massive containers loaded with clothing and food. More containers came from the islands of Niue and Tokelau.

Help came from the USA. President Obama released a statement on the 30th of September, 2009 in which he noted, "We stand ready to help our friends in Samoa and the region. Going forward, we will continue to provide the resources necessary to respond to this catastrophe, and we will keep those who have lost so much in our thoughts and prayers." In early October 2009, members of American NGO, 'Samaritan's Purse' flew to Samoa to assess how they could best assist. The organization – founded by the Rev. Franklin Graham – was awarded a grant of $500,000USD by the US Agency for International Development to distribute tsunami relief goods in the affected communities. They partnered with local NGO and church-based groups and identified three hundred families in over twenty-six villages to receive hygiene kits, and recovery packages that contained; mosquito nets, kerosene stoves and lanterns, dishes, buckets, nails, shovels, machetes, hammers and axes. Rev. Nu'uausala of the Apia Protestant Church was the project's main liaison advisor and youth from his congregation worked together to help with the distribution. "It was a privilege to [work with] Samaritan's Purse and to be a part of the countrywide help for tsunami-affected families. God bless the American people for these gifts." Another key part of the project was the gifting of 326 air horns to coastal villages to enable them to be a more effective link in the National Tsunami Warning system. Paul Murphy of 'Samaritan's Purse' described their work in Samoa, "It's been rewarding, from volunteers to local businesses, to the families, everyone has been so grateful. We are having fun, being the Santa Clauses giving out gifts and seeing the smiles on people's faces."

Japan went to the disaster zone with tents and other relief supplies in the first few days after the tsunami – via their local JICA volunteer organization. In the next month Japan spent over $80,000 USD assisting with the waste clearance project in the tsunami zone and then stepped up with funds to rebuild the damaged Saleapaga Primary School. China ensured the roads in the tsunami zone were reconstructed with $300,000 USD worth of bitumen. Italy reached out to Samoa via the WHO, sending vital technical experts to assist with the relief work, paying for mosquito nets, water safety equipment and medical supplies.

Help didn't just come from governments. It came from the International Red Cross and volunteers worldwide. On tsunami day, as her staff rushed to respond to the disaster, it was the Secretary-General's task to worry about… money. " Everyone was running around, trying to find out if all the stories were true, trying to organize to respond and I sat right here at my desk and I was really anxious, I was trying to think what we were going to do because [the Samoa Red Cross] didn't have any money. So I was worrying, *how are we going to react to this…this is such a big disaster, how are we going to cope?!* Especially when my staff and volunteers in the field started calling me that it was true, they were coming across so many dead bodies. But the strength came because I knew that the International Red Cross was there to help. Then they were on the phone already trying to find me, asking, *how are you, do you need help?* The head of the International Red Cross in Fiji, sent their first team that night, arriving in the early hours of Wednesday morning. With the support they needed, Tautala and her growing band of volunteers faced Samoa's disaster head on. Months later, from an office overflowing with helpers, Tautala spoke movingly of the life's blood of the Red Cross – volunteers. "They came right away, on that first day, so many locals were coming here, wanting to help….when things happen, they come. It started from there and it didn't stop. In the following weeks we went up to over four hundred volunteers. A lot of these people came from overseas – these were Samoans, New Zealanders, Australians – people from all different countries. We couldn't believe it. A lot of them paid their own way to come, they just wanted to do something. They would email asking if they can please come and we say, yes of course. I know the feeling of many of the Samoans – you want to do something, even give a cup of water for people that are victims of the tsunami, anything just to help. Another big help was an offer of accommodation for our overseas volunteers from the Christian college, Rhema. They called us, *if you have volunteers travelling here let us know and we can house them here* and that was really nice. There were a lot of ANZ bank staff who came from overseas that wanted to volunteer and so the bank would organize their trip and they would come and help too. Even the airline, Air New Zealand. Whenever the plane crew arrived in Samoa, they didn't want to waste any time or take a break, they wanted to support the work – they would come and go out to the disaster area with a Red Cross team. It's a very strengthening thing to see all these volunteers, to see that people have that sense of belonging and caring for those who have been devastated. It really touched our hearts."

120 years ago, a five feet tall diminutive woman of steel, Clara Barton, faced the first great test of her newly organized American Red Cross when they responded to the Johnstown Flood of 1889 which killed more than two thousand people. For five months after that disaster, Clara "worked round the clock, directing hundreds of volunteers and distributing nearly half a million dollars worth of blankets, clothing, food and cash,stoutly proclaiming that the Red Cross was there to stay as long as there was work to do. *We are always the last to leave the field."* This is an ethos the Samoa Red Cross clearly adheres to – late into 2010, it is the white and red cross which is still clearly visible throughout the villages that are slowly rebuilding their lives. The Red Cross is providing every new tsunami house with a water tank. There are plans to build disaster depots in various rural areas that will be fully stocked with the supplies needed during the next disaster, the next evacuation. Before the tsunami, Tautala had a permanent staff of eight people and now there are twenty-five with hopes for thirty more. Partners from the NZ and Australian Red Cross are assisting with the ongoing tsunami work and helping to train the local team. The activity levels have died down in the hospitals and emergency response centres, but the Red Cross headquarters at Motootua is still a hive of busy energy. *We are always the last ones to leave the field.*

Help for Samoa came from religious organizations based overseas. Islamic Relief USA, based in Alexandria, Virginia is an international relief and development organization that "strives to alleviate the poverty and suffering of the world's poorest people." In response to 29/09, the humanitarian agency partnered with the Church of Jesus Christ of Latter-day Saints (LDS) to send a massive DC-10 plane filled with supplies that included food, clothing, bedding, hygiene kits, mosquito netting, wheelchairs, walkers, crutches and canes. There was enough food on board to feed two thousand people for a month and it wasn't dry, survival-camp-ration-style food either. Visitors to Leua Vi'iga's shelter at Saleapaga, were welcomed with canned peaches and containers of chocolate pudding – just a sample of the food distributed to people throughout the affected areas. The LDS Church provided the goods worth $500,000USD and Islamic Relief paid for the flight costs of another half a million dollars. Reporter Scott Taylor of the 'Mormon Times, Deseret News' met with Tasi Toala, a cargo worker who helped to load the plane at the Salt Lake International Airport. Tasi was scrawling messages on the pallet packaging – hand drawn hearts, his name and email address. "I just want to send my love to all of Samoa." Tasi shared tears and an embrace with an LDS authority, James Humula who said, "You see, this touches not only people there in Samoa, but also people here." The DC-10 arrived in Samoa one week after the tsunami and volunteers worked for twelve hours to unload the plane by hand because the local airport's forklift could only reach 14ft while the plane's cargo doors were 17ft off the ground. A spokesman for Islamic Relief said of their partnership with the LDS church, "As things like this happen throughout the world, we put our resources together."

Help came to Samoa via many individuals. The Samoa Tourism Authority was surprised by the huge influx of visitors in the weeks and months after the tsunami. They coined a new term – the voluntourist. "It is evident that a significant number of tourists that came to Samoa after the tsunami were not here for the regular beach holiday…they have come with the simple objective of wanting to give something back to the community." Some were repeat tourists who had experienced the friendliness of the Samoan people or enjoyed the hospitality of particular resorts and wanted to come help with the rebuilding. Others were first time visitors who came on their own volition – to volunteer with the Red Cross, to work in hospitals, to clean up the wreckage, to help build houses.

Jennifer Villamin, an Office Manager at an engineering company in San Francisco, USA was one of these. Two weeks before the tsunami, she had booked a holiday to Samoa, planning to arrive in November 2009. "When I heard about the tsunami, I freaked out and almost cancelled my trip. Eventually I got over it - it was only a small portion of the island that had gotten hit, 90% of the rest of the country was untouched. I decided I would hook up with the Samoa Red Cross and try to do some volunteer work while I was there."

It was Jennifer's first time in Samoa and after checking in to her motel, she joined many other volunteers like herself who were using their 'holiday' to help in whatever way they could. "It was very easy to volunteer with the Samoa Red Cross, they just asked me to sign in my name on the register, date it, and that was pretty much it! Then we loaded up a truck with supplies and we were on our way to the disaster area."

Jennifer spent a week working with the Red Cross. "We administered First Aid wherever needed, helped to take water to people, refill water tanks and distribute boxes of donated clothing. It was hot as hell and the mosquitoes were eating me alive, but it felt good to be helping people in need. Definitely a very sobering experience to say the least."

Some experiences stand out for Jennifer. "During one of our walk-throughs we came across a village chief and his family who had relocated inland after the tsunami, moving deeper into the forest. They had nothing left of their home and possessions, but they offered us everything they had. There were three of us volunteers in the team – two Americans and one local Samoan. The family fed us a full meal of roasted pig, sweet potatoes, and salted coconut milk and told us to sit and share a meal with them inside their shelter. We tried to refuse since we were supposed to be working but they insisted. Translated through the local volunteer who was with us, the chief then shared his story of how he and his family used to have the biggest house in the village, own a business and live right next to the water with a great livelihood. Now all of that was gone, everything, except for their health and their family. I thought they would be despondent and depressed, and yet they were completely determined to move on and still offered us everything they had by insisting we share a meal with them…I was actually expecting to be surrounded by a lot of depressed and angry survivors, but what I mostly encountered was the complete opposite. Most of the survivors I met seemed to act like everything was business as usual. They were just trying to move on with their lives with whatever resources they had or could find. It was very unexpected for me."

Another Red Cross volunteer was Sean Wooten, aged 29 from West Seattle, in the USA. He heard about Samoa's disaster and immediately knew he wanted to help. Sean's parents had lived in American Samoa for three years in the 1970's and the family still has many friends there. His story, told to reporter Steve Shay, highlights some of the challenges associated with trying to give service. Sean flew five thousand miles to Pago with a first aid kit, tools, and two suitcases filled with clothes to donate to tsunami survivors. He wasn't greeted with open arms though. "I was pulled out of the customs line when I arrived and strip searched. I had my sweatshirt tied around my waist and their K9 patrol dog loved it and was jumping all over. My dog had always slept on it at home! Apparently a week earlier, someone else had flown in carrying drugs saying he was there to help the tsunami victims. They searched all my stuff and then actually apologized. I definitely stood out that's for sure!" From the airport, Sean went to the American Red Cross station – that refused to accept his donations. "They wouldn't take them, they said they'd only take cash. So I went to a school auditorium where FEMA was set up but they turned me away too. I eventually donated my stuff to the American Samoa Community College. I learned that all these agencies do background checks. I should have prepared in advance, but I wanted to just get there as soon as possible." Sean finally found a spot helping out at the Emergency Operations Center (EOC) delivering aid supplies to the affected villages. "I saw whole villages completely destroyed with only foundations remaining. A lot of places I went to made me sad, entire lives, clothes, furniture and appliances completely in one big pile."

A week later, Sean caught the little plane to Apia and was quickly put to work by the Samoa Red Cross. He spent two weeks building cement foundations for water tanks in the new settlements. "Every day I saw homes that were holding funeral ceremonies. I'd drive along the road and see a crowd of people in front of a house with food out, but it wasn't a party." In spite of the hiccups he encountered, Sean would do it all again. "I didn't want to leave. I feel bad for the people and definitely wish I could have done more. They still need a lot of help. There's a lot of debris and trash in many of the affected areas. I'd like to go back and live there. They are open-minded people and not judgemental."

At times, locals were taken aback by offers to help from *palagis*. In Saleaumua, a team of Air New Zealand flight crew came as Red Cross volunteers to help with the cleanup. Lee Palupe chuckles as she recalls the uncomfortable astonishment of the village people, "The Air NZ team had the most beautiful air hostesses, they looked like beauty queens in shorts and sandals and they wanted to do something to help. One of the family chiefs that was here, was shocked - *they can't do this work, they can't come and get their hands dirty with our rubbish, it wouldn't be right!* I told him, they were volunteers but the reaction of our locals was quite taken aback. 'Volunteer' is just such a new word for people in the village, I mean where in the world do strangers come in and pick up your rubbish? So they came in and cleaned all the fence line and the pilots and engineers came and put up the tents for people and I was so happy to see it. And then afterwards they gave the family some money and you should have seen them, they were so excited. It was just amazing to see the transformation of those here who were in despair when those volunteers came."

Some came with other international organizations to help Samoa. Jana Hamilton from New Zealand, is an Oxfam aid worker who arrived in Samoa 48 hours after the tsunami. She travelled to the disaster zone with a team from the Samoa Women in Business Development (WIBDI) as they delivered relief supplies. WIBDI provides agricultural training and business skills support to organic farmers throughout the country, as well as financial assistance through a micro-finance loan scheme. On the Oxfam, NZ website, Jana told of hearing stories of "tragedy, loss and extraordinary heroism."

Oxfam engineers worked with the national response team to supply a safe water supply and meet sanitation needs in the disaster zone. Jana concluded of her time in Samoa, "Samoans have always had a strong sense of community, which comes from their strong sense of culture, and traditions of collective support. The local *matai* (chiefs) are providing leadership at the village level and I can feel the incredible spirit here. For those with family to take them in, they are being well supported. Oxfam is working with all agencies involved to make sure the most vulnerable people are able to recover and begin to rebuild their lives. Once the initial urgent needs have been met, it is clear people will require long-term support for reconstruction of their homes and livelihoods. Being here, in Samoa, heightens my sense of family – and my connection, not just with New Zealand but with all Pacific communities."

David Youngmeyer is a communications manager with UNICEF, New Zealand. He came to Samoa as part of a team looking at education and health impacts of the tsunami and reported on the UNICEF website of his visit. On October 14th, 2009, he flew on a RNZAF helicopter to the small island of Manono off the western coast of Upolu. Manono is home to several hundred people and relatively isolated. It is an hour's drive from Apia and then a 30 minute boat ride. The chopper landed on a concrete slab still peeling with linoleum – all that was left of a small shop that had been completely wiped out by the waves. David's report described how the tsunami had destroyed "up to 40 homes and businesses as well as the only pre-school on the island, leaving some thirty children aged 3-5 years stranded with no early childhood education. Principal Rebecca Mose told us that many of the children are still fearful of another tsunami, with some not wanting to return to school because of its proximity to the sea. There are plans to rebuild but finding the necessary funding is an issue."

David's team were able to gift the school with Early Education and Recreation Kits that contained learning activities, crayons, exercise books and pencils as well as fun games. They were touched by the sight of Manono's children working together to help clean up the debris around the preschool site. "Children of all ages from a Sunday School group were all pitching in to clean up…they looked happy to be doing something positive and helpful. Even the youngest of the children did their bit, sitting on the ground as they picked up little sticks and small pieces of rubble, putting them in woven baskets for disposal. Older children took care of the larger debris and looked out for the safety of the littler ones."

"As we walked the dirt track that circles the island, we talked to families about how they were coping in the wake of the disaster. One mother told us that the tsunami had destroyed the family fishing canoe, meaning that they were no longer eating much fish, an important part of the local diet. Many of the surrounding banana trees had been destroyed, further affecting the normal diet. Water is also an issue with supplies of running water disrupted and family water tanks ruined."

"The NZ military were doing a great job on Manono helping to restore the water supply, rebuild sea walls and provide medical aid. It was really good to see some other Kiwis and know that we are all working together to help provide the assistance that is so badly needed."

Many locally based groups and individuals, organized to receive donations from overseas givers. Kevin Petrini is an ex-Peace Corp volunteer who lives in Samoa with his wife Taialofa and their children. They run a yoga retreat venture called *Raw Shakti*. After the tsunami, they went out to affected villages like many other locals to take food and water. Many friends from overseas were contacting them, *we want to help, what can we do, how can we send help to where it's most needed?* The Raw Shakti website started receiving financial donations and by 7pm on October 3rd, 2009, the Petrinis had delivered to the families of the tsunami in twelve different villages; 2,280 pounds of rice, 2725 pounds of canned fish, 148 cases of water, 720 pounds of oranges, 30 pounds of chicken, 14 cases of soap, 35 cases of milk and more. A note on the Raw Shakti website reminds people that twenty US dollars can buy a 40 pound sack of rice OR 25 pounds of canned fish. An entire case of bottled water is only fifteen dollars. But the Petrini's were not just interested in alleviating the short-term needs. "We wanted to help villages regain their livelihoods, businesses to get re-established so they could start bringing money into the villages for the long term." On October 6th, funds were used to purchase a new motor for a boat on the tiny island of Namua and 1.12 km of pipe to re-connect the island's water supply. The focus now for Raw Shakti is assisting with the rebuilding of the tourist accommodation, Sunset View Beach Fales, on Manono Island.

A group of rugby legends came from Australia with some expert surfers on a 'Wave of Hope' tour to help. Dylan Longbottom, Zahn Foxton, Michael Lowe, Tommy Herschell and Craig Tufin conducted surfing clinics in several villages, hoping "to get the children to jump back in the water and make them love the ocean again." Rugby players like Wendell Sailor, Andrew Johns, Roy Asotasi and Nigel Vagana visited local schools in the tsunami zone, thrilling children of all ages as they showed off their skills and gave them a chance to play a game with their heroes. The group brought with them sports gear and other gifts to distribute that had been donated by many different rugby clubs in Australia. Vagana said, "This trip is not about us. It's about trying to help those people who were affected. It's about coming here and giving something back. We are passionate about this mission and I hope we make some people happy."

The wave of goodwill for Samoa intensified in the following months as people worldwide, reached out to a small group of islands in the Pacific. The *mealofa* took many different forms and spanned all generations. Children and youth in particular were inspired humanitarians taking part in many creative ways to reach out across oceans and cultures.

A seven year old boy in Papamoa, NZ called on kids everywhere to dig deep and give their pocket money to help Samoa tsunami victims. Liam Porter was inspired to start the collection after seeing images of the aftermath on television. He got friend Emily Pol to help him and together they kick-started a fund that raised NZD $92,996.00 which was given to the Red Cross.

Tuatagaloa Joe and Tui Annandale. *Photo - Tuatagaloa Joe*

Tofilau Afatasi and his son at Poutasi. *Photo - Lani Young*

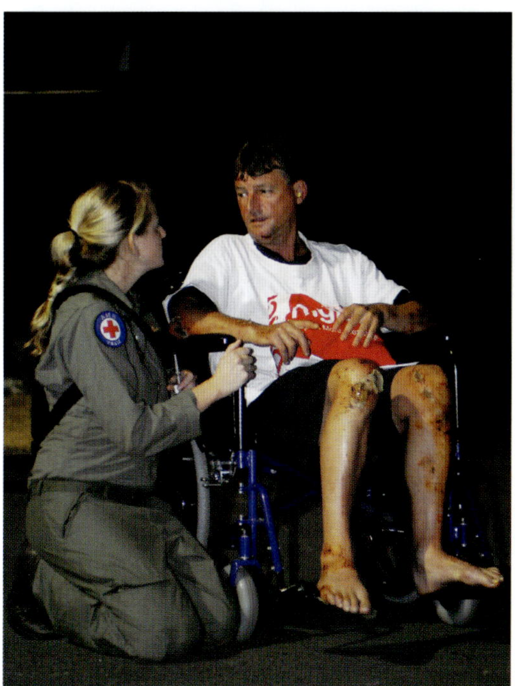

Survivor John Blacker with an Aust medic before he is evacuated home. *Photo - Australian Defence Force*

Moimoi Tumanuvao at Poutasi. *Photo - Lani Young*

The Laufatu sisters, Vaovai. *Photo - Lani Young*

FESA Commissioner Tony Hill (center) with team members at the Faleata Fire Station. *Photo - NZ Defence*

Filomena Nelson is interviewed by media at the Faleata Fire Station. *Photo - Disaster Management Office (DMO)*

Media teams prepare to broadcast, from the remains of the Taufua Restaurant. *Photo - Pele Wendt*

Volunteers busy at a Red Cross station at Lalomanu on 30/09. *Photo - Matt Leal*

The tangled wreckage searchers had to cut through the day after the tsunami. *Photo - Vaughan Simpson*

An injured survivor is carried to a waiting helicopter for evacuation to NZ. *Photo - NZ Defence*

Teams from Australia and NZ worked alongside Samoans to search for the dead. *Photo - NZ Defence*

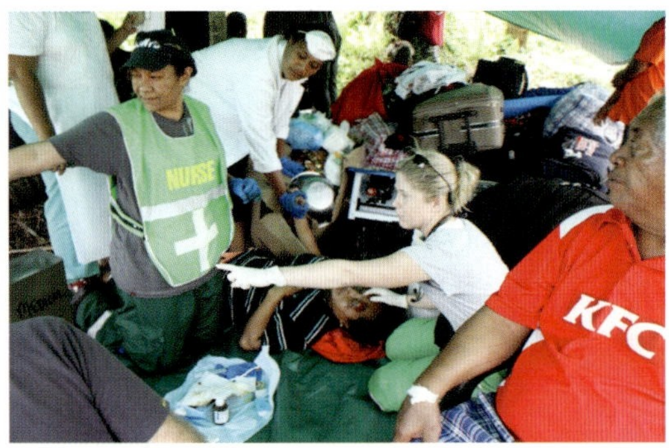

Medical teams treating survivors in the camps after the tsunami. *Photo - Australian Defence Force*

The Maritime Police Wing with their new rescue boat donated by Aust to assist with their search efforts. CO Petona Mauga at far left. *Photo - Australian Defence Force*

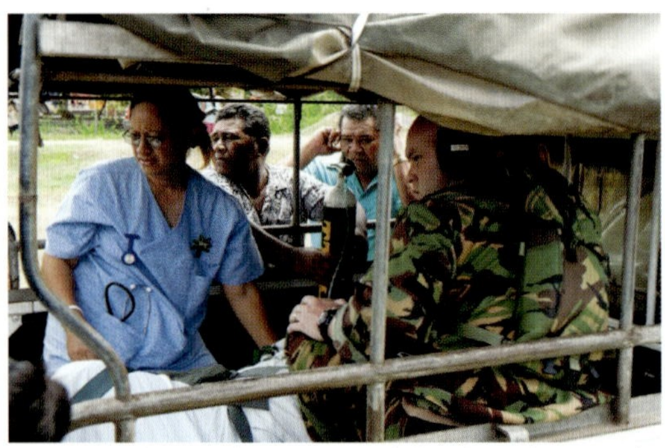

Nurse Henrietta Aviga treats an injured survivor on the back of an NZ Defence truck. *Photo - NZ Defence*

Funerals taking place at Lalomanu beach a few days after the tsunami. *Photo - Pele Wendt*

A local priest blesses the ground before rebuilding takes place. *Photo - Australian Defence Force*

LARCs drive through the disaster zone on their way to cleanup Lalomanu reef. *Photo - Australian Defence Force*

Army engineers install a water tank at an inland resettlement area. *Photo - NZ Defence*

Helicopters like this one on Lalomanu beach, were valuable for transporting supplies and other assistance to the disaster zone. *Photo - Australian Defence Force*

Relief supplies are unloaded from the HMNZS *Canterbury* at the wharf in Apia. *Photo - NZ Defence*

Some of the several tones of debris cleared from the reef at Lalomanu. *Photo - Australian Defence Force*

An Iroquois helicopter returns a fisherman's boat to the sea on the island of Manono. *Photo - NZ Defence*

Paramedic Steve Williams (2nd from left) with some of the FESA team at Faleata Fire Station.
Photo - Australian Defence Force

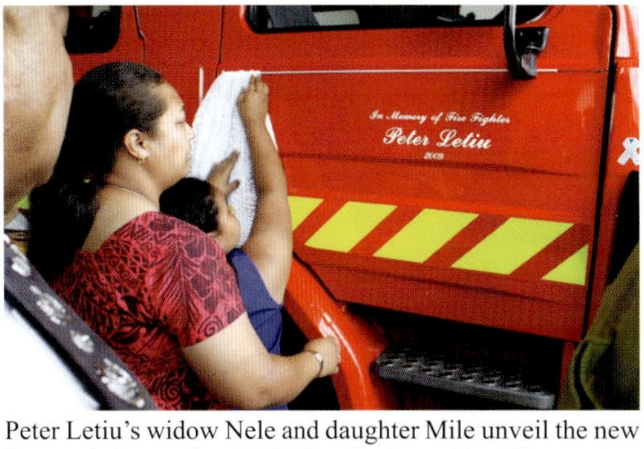

Peter Letiu's widow Nele and daughter Mile unveil the new fire truck donated to FESA by the Australian Fire service.
Photo - Australian Defence Force

Children at Lalomanu Primary school enjoy a visit from the NZ Defence Force. *Photo - NZ Defence*

Nathan Keil (in center) helps give out Christmas gifts during Operation Stocking. *Photo - Mark Moors*

A team of volunteer builders pause in their work at the new Saleapaga settlement. *Photo - Lani Young*

The Taufua family rebuilds after the tsunami.
Photo - Lou Anderson

Liam told reporters, "I want to say thank you to everyone who helped. I'm saving my pocket money now for a family trip to Samoa – that would be cool, to see what we helped with."

Boys at Kings College in Auckland, New Zealand packed over five hundred desks and chairs into a container to ship to tsunami-hit schools where students were now sitting on mats on the floor in open *fales*. Each desk had a personal touch as students attached a letter of friendship to their desk and then added a t-shirt and stationery. Nine year old Nicholas Johnson told NZ reporters, "It's really cool we can give the desks to someone who needs them."

Other NZ schools pooled together with Omanu Primary in Mt Maunganui, to fill a container with school equipment and learning materials, as well as tools for the village of Satitoa to help with the rebuilding. The project was nicknamed, SOS – Support our School Samoa and the white shipping container – decorated by the students themselves with pink hearts and red and yellow flowers – made for a cheerful addition to the Apia wharf when it was unloaded. Organizer Michelle Penn explained, "A lot of love and commitment was put in by the students of Omanu School and their families. We want to offer more than help. We want to build a relationship with the Satitoa Primary School and with the school's principal instead of just bringing the container and leaving. We've been teaching the Omanu students about Samoa and the culture so they will understand why they are donating. We want to keep supporting the children of Satitoa in the future."

Further away in Salt Lake City, Utah in the USA, students at Fairfield Junior High made seventy quilts for the people of the tsunami. Their teacher, Sydnee Seager explained how the students had worked for two weeks to measure, cut, and sew the cheerful collection of colourful fabrics. Teenager Hannah Judd enjoyed the opportunity to do something meaningful. "This gave us a chance to step up and help others." In a tarpaulin shelter in the new Saleapaga village, there hangs a pink and burgundy quilt made with loving hands. It is doubtful whether the students of Fairfield Junior High will ever meet the fragile old woman Fuailuma that lives here, but on a cold mountain night, she will wrap their quilt around her shoulders and feel of their warmth.

Also in Salt Lake City, a group called 'Samoa Relief Wave' organized to collect 1,200 Christmas stockings to ship to Samoa. People were invited to fill a stocking with gifts – stationery, toys, and other treats – which would then be distributed to children in Aleipata. The shipment arrived on the last day of 2009 and the gifts were delivered on New Years 2010. A team of Santa's helpers was made up of Samoa US veterans like "Kiofu" Nansen (USMC), Maae Taala (US Army), "Kit" Faumui (US Army), "Sarge" J. Moors (US

Army), Honourable MP Joe. Keil (USAF), Lee Letiu (USAF) and "Pres" M. Moors (USAF). A support crew of wives, children, grandchildren, and Nathan Keil completed the convoy. Gifts were given to the children of Saleaumua, Mutiatele, Malaela, Satitoa, Ulutogia, Vailoa, Lalomanu, Saleapaga and Lepa. Mark Moors recounted their experience meeting with the youngest of Samoa's tsunami survivors, "It was wonderful to see the young faces light up as they realized this tsunami relief shipment was strictly for them and aimed at bringing some fun back into their lives.Stop after stop we witnessed the same excitement and sparkle in the young eyes as we traversed the newly made, muddy back roads to the inland settlements." It rained that day but the mud didn't stop the delivery. Joe Keil commented, "The happiness of the children and their families was amazing. It was a New Year's day that I'll never forget." As the sun set on the first day of 2010, the team of US Veterans and their support crew made their way home, hungry, thirsty, wet, and exhausted; but completely satisfied in the knowledge that they had helped bring some measure of joy to the lives of many of the tsunami affected children. And with that they report the ***"mission accomplished."***

In Tacoma, Washington, there are two brothers Emmanuel and Shammah Leasiolagi who have never seen their homeland of American Samoa,. The boys' grandmother was killed in the tsunami. "Our grandmother prepares and serves meals at the Senior Center in Pago. She had gone to work early when the wave hit. Some of the stronger workers were able to climb to the roof. When the water began to take our grandmother, the others tried to pull her up but the water was too strong and she was swept out to sea." The boys parents went home for the funeral but back in Tacoma, Emmanuel and Shammah felt helpless. "We are so far away here from our other family in the Pacific. We felt like our hands were tied, we wanted to help." The brothers organized a fundraising drive with their church, *The Love Church,* that filled three shipping containers with building supplies, canned goods and other non-perishables. They helped put on a Samoa-style pig roast *fiafia* dinner as well to raise funds for tsunami relief. A few weeks after 29/09, the Leasiolagi brothers were sending containers of love to a land they have yet to visit.

There are several thousand Pacific Islanders living in Anchorage, Alaska – many with close family ties to Samoa, Tonga and American Samoa. Val Tuiolemotu, who moved to Anchorage in 1999, spoke to the 'Anchorage Daily News' of hearing about the tsunami on 29/09. "I got several text messages that I totally ignored. They were talking about an 8.0 earthquake and I thought it was a hoax because it was so large." Once the news was confirmed however, Val and other Samoan mothers left work 'within minutes', to get their children early from school, to gather together. "When something that devastating happens in the islands – hurricanes, natural disasters – we tend to pull together in prayer." The week after 29/09 was filled with many 'agonizing phone calls and prayer' as the Anchorage Samoan community wept for their homeland. As one Samoan pastor explained, "Somehow, some way, you're either related to somebody or you know somebody who lost a life [in the tsunami]." Anchorage mobilized. Val's church, 'The First Samoan Body of Christ' Church, held a Christian music performance and bake sale to raise funds for tsunami relief. A former NFL player Mao Tosi led a fundraiser at the Northway Mall, collecting over $5000 USD in the first four hours.

A month after the tsunami, three year old Patrick Beautiful "helped raise money and spirits during a wet day at Fogarty Park in Cairns, Australia during a Samoa Tsunami relief concert." The youngster danced cheerfully in front of a large crowd that attended the concert which included hip-hop dance groups, Aboriginal didgeridoo musicians and many others. Cultures came together to dance and sing for Samoa.

Music and dancing also helped to raise funds in the Camberwell Centre Ballroom, Australia on November 2nd, 2009. Feauina Sullivan, Caroline, William and Ken Taitin formed the group 'United Samoans and Friends' (USAF) with the objective of assisting in the relief effort. Pacific Island performers rocked the crowd with their spirited dances and the 2009 Australian Idols, Scott Newnham and Jamilla Ioane sang for Samoa. The concert raised $9,800 AUD which was sent to the Samoa Rotary Club and the Adventist Development and Relief Agency (ADRA). "It was a night of remembrance and celebration of our culture. We send all our love to Samoa and plan to continue raising funds through future events."

Sara Trafford wanted to do her part for the family that had embraced her and her son. "I heard about the tsunami on the Wednesday when it happened, but I didn't know about Lalomanu and the Taufua's losses until the Thursday when Sina Ale's story was on the front page of the NZ Herald. It was like being punched in the stomach and I cried and cried. I was just in shock, I had thought surely the Taufuas would be fine…and it was hard to take in the devastation that had happened there. I immediately decided to fundraise as much as I could." Sara hosted a breakfast at her brother's restaurant with proceeds going to Lalomanu. She wore a floral *lavalava* (sarong), flowers in her ear and collected donations at the local market.

These are only a few examples of the worldwide *mealofa*. Taua Kitiona Seuala says of the assistance, "Some people have called the disaster a blessing. Even the pastors are saying, let's talk about the blessings we have received, don't dwell on the disaster. Some families had really bad living conditions before the tsunami, but now they are living in good houses, they have nice clothing and they have things they didn't have before. I speak mainly about our district and our village. Some poor couples had never seen a microwave before and now they own one. I go visit them now and I feel happy because I see them warm up their leftovers from the night before in a new microwave. Some Samoans moved away from their home many years ago and forgot about their families here, but when the tsunami hit, the brother or sister who had lost contact, picked up the phone and called. They got their airfare together and came over, bringing many gifts for their families in Samoa. We all could see the many donations coming in from many people and many countries overseas – there was a great giving spirit towards those who were affected. And now we see the government comes along and builds families a house so now they're set. So now, we should focus on the blessings."

October, 2009.

There's been a lot of grief in the news lately about tsunami relief NOT getting to where it should. About people NOT getting help. While I'm sure that there are many instances of wonky, misdirected aid efforts and while I know of other instances where some people are receiving - and then lying about it so they can get more....I also see and hear everyday, of tsunami relief that IS getting where it should. To everyone out there who's fundraising, collecting food and clothing, buying building materials, sending money and generally giving blood, sweat and tears to help tsunami Samoa survive and rebuild - this notes for you.

I have seen - a young mother in a wheelchair, her still-healing bandaged leg. So grateful to the plastic surgeon and the medical team from NZ and Aust. that did the skin graft, that ensured she WOULDN'T need an amputation. 'They were so kind. So careful. So nice to me.' she said, voice thick with emotion.

Families living in tents in the rainforest bush. Tents donated by the Red Cross, the NZ Army and countless others. A child regards me with solemn eyes through the netted doorway of his family's tent as mosquitoes buzz and hum. And rain trickles down my back. And the fresh cut bush is thick and heavy around us. A 74 yr old gentleman says to me with tears. 'We don't want to become people of the bush. We must not become people of the bush. There - at Satitoa by the sea - that is where our homes are. That is where our grandparents are buried. That is where we should be.'

Truckloads of wood and other building materials chugging their way over Le Mafa Pass. The first houses being built by Habitat for Humanity and Caritas and so many others.

Huge containers being opened and distributed at various village points along the route. There are crowds of people. Patient and calm. Sitting in the sun waiting while the pulenuu unpack and give out. Boxes of clothes. Canned food. Mosquito nets. Outside one container a team of helpers are building careful stacks of kitchenware....one for you, one for you. More than 30 stacks. A shiny new pot. A kettle. A gleaming frying pan. A handful of plates. Rows and rows of new cookware anticipating their new owners. (Don't underestimate the power of a new pot. I have one miserable broken frying pan and it drives me insane everytime I need to make pancakes.)

A mother of 8 children say thank you for the blankets and clothes they were given on Wednesday the day after the tsunami. At her breast is a chubby one year old with mischief in her eyes who keeps trying to grab the MP3 recorder off her mum and put it in her mouth. 'We spent the night in the bush. We were wet and cold. We had nothing. I didnt care about me but I only wanted to find a way for my children to be warm. Thank God for the clothes and blankets we received.'

If you had any doubts about whether or not your donation was needed. Utilized. Appreciated. If you were wondering whether or not the people of the tsunami still need your help. Whether it's a pot. Some planks of wood. A can of fruit. Your prayers and thoughts.

This note's for you.

20
How many hands does it take to build a house?

After the tsunami, Samoa's communal land structure was revealed in all its beauty. The tsunami affected a total of nineteen villages with approximately 5,274 people made homeless. Many others with intact homes fled their villages and remained unwilling to return permanently for months afterwards. Where did they go? Where are they sleeping, eating and living now? Every family that belongs to a village has traditional rights to land – acres of dense tropical bush – some of it already cleared for extensive plantations. With their villages transformed into vast mass rubbish dumps in a matter of minutes, people moved to their plantations. Some to spindly *faleo'o* (wooden huts) that used to be a planter's refuge, shade during his lunch break. Others using tarpaulin sheets to erect hasty tents. There in the plantations, people had space to call their own. Green bananas to boil for supper. Ripe papaya to satisfy the sweet hunger. Taro to dig for empty bellies. Coconut leaves they could weave into plates and baskets, to make blinds to shelter them from the rain. Everyone needs somewhere to call their own and tsunami survivors in Samoa had that. Accustomed to the after-effects of disasters in other countries, aid workers from overseas nations who came to help Samoa in the following weeks would ask, where are the refugee camps? Where are the homeless? Where are the dispossessed, displaced people? Many tsunami reports noted how the communal land and village structure provided both shelter and assistance. "There were no tent cities of evacuees. All villages included both high and low ground, and the family ties assured that no one was isolated or left alone." Another report, "the customary land system was a very powerful mechanism in the recovery of families. The fact that they had land readily available to them was not only a relief but also a means of survival."

After the tsunami, the resilience of a people who could make 'something from nothing' was revealed. Dr Teriki Puni was one of the medical team that went inland on Wednesday to find those with scratches and cuts that were too afraid and too busy to come to the hospital. Visiting families dispersed through the bush was an experience not easily forgotten. "It was amazing, people just made do with that they had. You saw these very basic structures, just four poles, either that or just some roofing iron, families were very creative in the sense that they had found so many things that they made use of to build their shelters and people were already working, trying to make permanent structures and homes out there."

Once the initial urgent needs had been seen to, the focus for the disaster recovery effort, was building homes. The National Housing Committee led by the Minister for Works, had the task of determining the housing needs and overseeing the rebuilding. With cyclone season threatening, the emphasis was on providing families with a basic, safe and strong shelter, one that could then be added on to later. The design was for a 30x18ft open *fale* with a raised concrete foundation and corrugated iron roof. A square bathroom block would be built at the back. The cost of each tsunami house was approximately eighteen thousand tala. Assessing the names for the housing list was no easy task. Most families before 29/09 lived communally and there could have been more than twenty people living around one main house, sharing bathroom and cooking facilities. The Committee had to wade through tangled webs of familial relationships as they apportioned who should get a new house. Should a house be built only in the name of the *matai*, the chief of the family? What about his four daughters and their husbands, their children? They had all lived together before – should each minifamily now get a tsunami house? The committee met every week in the months after the tsunami and had to carefully consider new cases, new complaints. One week it was the problem of what to do with a family in Saleapaga. A son had asked that his little family be put on the list because, according to the report, the father refused to allow them to share in the tsunami house. Quarrels and conflict. Several other names had to be taken off the list because an inspection had revealed that their homes had NOT been damaged in the tsunami, rather, the families had "smashed down their own houses" so they could hopefully get a new house. For some who had lived in meagre humble dwellings before 29/09 – an eighteen thousand tala house with a flush toilet, running water, and a water tank – was a mansion. For others who had lost more palatial assets, there were complaints about the housing aid offer. "We had a 3-bedroom brick house, a shop, a boat, a car – and the tsunami took it all. Now the government is only offering us a simple eighteen thousand *fale*? It's like a slap in the face really. What about all the millions of dollars of aid that Samoa is getting for this?"

Families were given the option of having their house built for them according to the proposed design, or being given the materials so that they could then build their own home to whatever design they wanted, perhaps combining the materials with resources of their own to build a bigger house more similar to what they had lived in prior to 29/09. Many tsunami families chose to accept the building materials package. Tofilau Afatasi of Poutasi points out his family's new house that stands proudly beside the main road now – an expansive, solid European style building that his family from overseas helped to fund, adding on to the eighteen thousand tala worth of cement, bricks and timber. Mika and Ave of Lalomanu, have moved to a suburb of Apia with their two little boys. In March 2010, tarpaulin shields their building materials from the rain while Mika searches for a team of carpenters that can help him put up their new house. The new Saleapaga surprises the first-time visitor expecting tented misery, as scattered grand brick homes sit nestled amidst wild bush along a narrow dirt road – all are evidence of an extended family that has pooled their finances together to build a home using the government housing aid package as a starter. Another family in Saleapaga shelters from rainy season winds in a house built with the aid package and with labor provided by a flock of men from an inland village, led by their Methodist minister.

Habitat for Humanity was appointed by the Samoa government as the "principal builder" of tsunami homes.The international organization is a non-denominational, Christian, non-profit one that helps more than 60,000 families around the world into homes every year. Their aim is to "help build sustainable communities and eliminate poverty housing worldwide." A Resource Centre was established at Lepa village – a facility for mass production of pre-cut trusses, walls and rebars and a month after 29/09, volunteers started arriving from New Zealand. Between November 2009 and June 2010, 300 tradespeople, 150 handymen and 150 other unskilled volunteers came to Samoa to work on the "Samoa Hope" project.

Habitat for Humanity worked in conjunction with the Catholic charity Caritas Australia which was funding seventy of the tsunami houses. On the sixth of November, 2009, work began in the village of Saleaumua on the first tsunami relief house. The Deputy Prime Minister Misa Telefoni, spoke at the ground-breaking ceremony, "Words cannot describe this loving work and it will be a new era for Saleaumua. I see that Caritas are people of action and not just of words." The patron for Caritas-Samoa is Her Highness Filifilia Tamasese Efi. "This work has given me the opportunity to work with village communities that have been hard hit by the tsunami. The work is uplifting and healing." The first Caritas house completed in Saleaumua belonged to Oselani the fisherman's family. He asks visitors eagerly, *did you see that new house by the road? That's our house, the first new tsunami house in the village!*

Habitat for Humanity relies on volunteers who are willing to pay their own travel costs and then work for free on a project. A number of the Habitat volunteers were NZ-born Samoans – many who had never been to the birth country of their immigrant parents and who were thrilled with the opportunity to learn more about their ancestry. Local Project Manager Lu Maea commented in the Samoa Observer newspaper, "Their immersion in the culture and the discovery of relatives they have not met before is an experience they will never forget."

Other volunteers spoke of the rewards of their Samoan experience. Mr Chapman is an electrician at Cresswell Electrical in Blenheim, NZ. He arrived in Samoa on December 7th to spend two weeks working on the housing project. He shared his story with the 'Blenheim Express' newspaper. "When I first arrived and looked around the rubbish that was still sitting everywhere – it felt like nothing had been really done. It felt like people were too shell-shocked and just couldn't bring themselves to start tidying up." He noticed many changes throughout his two weeks however, including some children beginning to overcome their fear of the water and get in the ocean again. "It was a stinking hot day and I thought, *right, Im going for a swim.* So I walked down through the children and went in the water. I splashed them. Well, talk about having three dozen kids piling on top of you! That was quite exciting for me to see these kids back in the water and enjoying it."

Mr Chapman had initially thought he would be using his electrical expertise but he was put to work building roof frames, something he had little experience in. He laughed, "What I know about power tools and how to use them you could write on the head of a pin. It was a very quick learning experience for me." He told though of the magic of watching people turn on the light switch in their new home. "It was fantastic to see the lights going, [the family] were going to move in that night, even though the concrete was still pretty green from being laid the day before. They were pleased, just absolutely pleased because they were getting a fridge the following week."

He told the *Express* that the most memorable part of his trip was the generosity of the Samoan people. The volunteers were housed in the Lepa church hall and women from the village cooked all their meals and did their laundry. "I left New Zealand feeling like we were going to be the ones giving to them but in actual fact what we got from them was quite incredible. Everything they owned was washed away. It was them giving back to us – even though they didn't really have any stuff to give us."

When it was time for him to leave, the women gave every volunteer a handmade tropical print shirt "that included hand-stitched button holes" and the men cooked a pig on a spit roast as a sign of their appreciation. Mr Chapman arrived home in Blenheim a few days before Christmas. "It was emotionally overwhelming…coming from where I had been, there was nothing but such huge generosity – to be home where we have just got everything. It was probably the best experience I'll ever have. The satisfaction of knowing

I've done something for somebody else…and to me, that was the important thing – doing it because we can."

Another volunteer from the Waikato region, Diana Browning, also shared her story on the HFH website, calling it "the hardest two weeks of my life, physically, emotionally and mentally but also very rewarding knowing that we are helping house families and helping to rebuild their lives." Diana was part of the fourth, twelve member group to go to Samoa. "The first thing I noticed getting off the plane was the heat…*Oh no, what have I got myself into? How am I going to work in this heat?!* Waking up at 4:30 every morning was hard but welcome as it meant I could have my cold shower. Who knew I would look forward to that so much!" Diana helped build houses at Saleaumua. "Being the only female amongst twenty odd guys (a dream for some I know!) was a little scary at first, as I didn't know how they would take to me working on site, but I soon proved I was just one of the boys and didn't mind getting my hands dirty. I managed to do just about everything, from concrete work, to painting, nailing, sawing and even using the skill saw. The boys were great in teaching me and the locals were surprised to see me getting amongst it. They laughed when I tried to carry a big bag of cement – okay, so I can't do everything!" Diana concluded, "Even though the two weeks were hard, the good by far outweighed the bad and the sense of achievement made it very worthwhile. The local Samoan people were fantastic, friendly, accommodating, funny and very humble. There is still a lot of work to be done, but you can see that the little we have done is making a big difference." A few months later, Diana went again to Samoa to do a second stint of building.

Diana is not alone in wanting to return. The CEO for HFH, Pete North, writes, "the volunteers are coming home with reverse culture shock. Many want to go back. The accommodation at 'Hotel Lepa' (mattresses on a concrete floor) is being rated 10 out of 10, the food served by the church ladies of Lepa another 10 out of 10, and the love shown by the people of Samoa is off the charts. Kiwis went to bless and are coming home blessed. We are doing something here far greater than just rebuilding houses, we are in this process facilitating people, discovering afresh what loving one another can do, both in terms of housing output and spiritual transformation."

Many other overseas volunteers came to Samoa of their own accord to build, connecting with extended family or friends in the tsunami zone. David Wimmer from Raglan, NZ went to Saleapaga village for two weeks with another friend, Frank Dreischmeirer – an engineer and joiner from Nelson. The men organized their own trip and stayed with a minister in the village – Pastor Uaea and his family. The Raglan Chronicle reported on their experience "the only two white people staying there with the villagers." They helped with tarpaulin shelters and then worked together with the village men on a seventy square metre fale which was used right away as a medical centre for everyone and as a church on Sundays.

Another fifty tsunami homes were funded by telecommunication company Digicel-Samoa. The company launched an appeal right after 29/09. Many of the international Digicel group of companies contributed to the tsunami relief fund and $1.4 million tala was raised and dedicated to the building of new homes. Samoa CEO Christian Fruean explained, "One of the worst things that happens in a disaster like this is when people are displaced, they don't have that sense of home and belonging. A home gives people some normalcy, so that they can lay their roots down…get their kids back into school, get people back to plantations doing their farms. We wanted to support government's initiative and we worked in partnership with Habitat for Humanity." Tuluiga Utaulu of Lepa, can sit in the shade now, in her new house and look out over a forest that is fast becoming a village cut through with roads and power poles.

The Church of Jesus Christ of Latter Day Saints (LDS) also stepped up to assist with the rebuilding, taking on the cost of another fifty homes. Houses were pre-cut at a factory in Apia and then trucked out to Aleipata where volunteers helped to construct the houses. Two hundred *poumuli* posts were donated by church members in Savaii. The Pesega congregation in Apia volunteered to build three tsunami houses. On the 23rd of January, 2010, Daniel Meredith led a team of forty men that left town at four in the morning with a single-minded objective for that Saturday – to build a house for the Vi'iga family of Saleapaga. Head of the LDS Welfare and Humanitarian Services in Samoa, Dr Felix Wendt, worked as part of Daniel's team and remembers the opportunity to meet with the owner of the fast-materializing home. "When we got there, he was just sitting there, staring out towards the sea, he looked lost. There were two small graves in front of their house. You could feel the sadness." Daniel mobilized the men to attempt the 'impossible' – build a house in a day. "When we arrived we divided into groups to work on the different sections of the house. We were lucky to have several men in our team who own construction companies and so of course they had the expertise we needed to get the job done. The most challenging thing for us was the sun, it was very hot out there." The team included many retirees and some senior missionaries from the United States and Canada. "They were the hardest workers out there – we tried to restrain them from picking up too much but they were picking up barrels of sand and those things are heavy you know! They would pick them up to pour the sand into the cement mixer and then almost fall in themselves! The camaraderie of everyone pitching in really helped, it made the day go quickly." By three o'clock the house was complete and the Vi'iga's needed only to wait for the cement floor to dry before moving in. "I think we boosted their spirits somewhat by being there."

Daniel's team went back twice more to build two more houses for Saleapaga families. "It's a very rewarding feeling to know that you helped to rebuild shattered lives, that you were able to give people a little bit of hope. Here in Apia, I enjoy the comforts of my home, I have all the things I need, we have it good here and to go out there and see so many people lacking the basics like clean water and good shelter – well, it makes us appreciate what we have and want to do something to help."

For every new house that stands in the tsunami zone, there is a story. Of many hands and hearts coming to work together. How many people does it take to build a house? In the case of the Ropati family on the little island of Manono, it takes several communities, schools, churches, international airlines, a government and the finely woven ties of extended family.

Leaitutulia Ropati lives in Canberra, Australia. His mother lives on Manono in Samoa and her house was wiped out on 29/09. Her new three-bedroom home speaks of the love and generosity of many – both near and distant. Mr Ropati thanked them with an advertisement in the Samoa Observer - beginning with the Samoan government for the relief funding used to begin the rebuilding. The LDS Church members on Manono who helped clean up the devastation after the tsunami, clearing away the forlorn remains of the Ropati family home and many others. The Methodist Church congregation and other denominations who brought clothing, food and tools.

The many businesses and organisations in both Samoa and Canberra, who gave generously for the building project. Ropati's son is a senior student at St.Edmunds College in Canberra and the school raised $600AUD for his grandmother's home. Businesses in Canberra gave light with two electrical companies heavily discounting all the necessary supplies. These supplies were carried home by Mr John Wickett, flown on Polynesian Blue Airlines via extra baggage allowance gifted to the rebuilding effort.

Many extended family worked to build the house. They came from other villages. They came from New Zealand and Australia. And all so that an elderly woman living on a tiny offshore island, could have a beautiful new home to call hers.

March, 2010.

Today I tagged along with a group of volunteers as they went to the new Saleapaga inland settlement. A woman named Mareta Craig had organized women from her church to plant a garden for a tsunami family. They took with them cuttings from their gardens, little black packets filled with earth and seedlings, woven baskets overflowing with spiky pineapple shoots. They went with shovels and hoes, trowels and bush knives. They went to sow the seeds of hope.

The men went to build a tsunami house. An open fale with a cement floor, poumuli log posts with a wooden truss and corrugated iron roof. A separate square brick bathroom to the side.

It had rained the night before and the track to the house was mud-filled and gooey. 4WD trucks got stuck and the bricks had to be carried from the dirt road to the newly cut clearing in the bush. Mosquitoes swarmed hungrily as the men carefully laid the blocks for the bathroom, as they wheeled barrows of dirt to fill the foundations for the floor of the fale.The family worked alongside the volunteers. The father hammered nails and chopped logs, pausing only to check on the slow baking of taro in the umu, lunch for the work teams. The mother of the house heftily carried sacks of cement and dug mini-boulders from the backyard to make way for the pineapple grove. During the mid morning break, we drank koko and she told us matter-of-factly about running from the wave, carrying her niece above her head in the swirling water. The other niece drowned. She shrugged, "Her mother just ran, she didn't carry her." We were quiet but she smiled and went back to attacking a stubborn rock that didn't want to budge from its privileged place in the middle of the vegetable garden.

At the close of day everyone was wet, tired and muddy. But there stood a house in the wet forest that had not stood there that morning. There was a garden at the back of it. Cucumber vines curled over black earth. Green cabbage sprouted hopefully. Tomato plants stood cheerful with the promise of red fruit to come. There were little avocado and guava trees. Lemon, lychee and passionfruit. There were flowers too. And row upon row of bristly pineapple.

As we drove away, I looked back at the mother of the house, sitting on a fallen tree in the forest, looking out over the new garden. Pineapple can take eight months to produce fruit. A little over a year after the ocean took her every possession, she will be able to walk outside her house and pick a ripe pineapple. She will carefully cut away the prickly orange skin, through to the yellow sweetness, digging out the pitted points of green. Maybe she will slice it into thick chunks and the children will eat it right away, the juice making sticky faces and fingers. Or maybe she will make pineapple juice, vai fala, with sugar and coconut milk. Her husband will have a glassful of the chilled delight after a sweat-filled morning in the plantation.

As we drove away, I smiled.

Today was a good day.

21
American Samoa

Stand at the furthest tip of Upolu island on a clear day and you can vaguely see the outline of Tutuila in the distance. It takes forty five minutes by plane to get to American Samoa. A little fifteen seater Twin Otter that shakes and hums as it takes off from the airstrip at Fagali'i. On particularly humid days, the pilot will open his door as the plane taxis along the runway for a welcome rush of coolness. Crossing the expanse of ocean that divides the two groups is to be awed. By the expanse of sun dappled blue far below. By the majestic rise of rock as Tutuila comes into view. If you are lucky, you will pick out a line of frolicking porpoises. Or the grey bulk of a whale as it moves purposefully through the water.

You need a visa to enter Tutuila. American Samoa is a territory of the United States and has been since the 17th of April, 1900. Five islands make up the territory. Tutuila is the largest and the location of the capital, Pago Pago. The two island groups may be separated by international red tape, but they are still one, with the same language, culture and traditions. American Samoans play gridiron football and join the US military. Samoans in the western group play rugby and migrate to New Zealand.

Unlike it's western neighbor where the capital Apia was spared along with much of the heavily populated northern coast – American Samoa was hit on all sides. The tsunami was "almost as large on the 'lee' side of the island of Tutuila as it was on the side facing the southwest where the tsunami was generated." One report explained why, "Tsunamis have long wavelengths and so are affected by the depth of the water they traverse from the time they are generated until they inundate land. They slow down where the water is shallower, causing their crests to bend (refract) toward the shallowest areas. As a result, tsunami waves change direction as underwater topography refracts their travel path. This refraction and other possible effects [like] interference of reflected waves and generation of 'edge' waves which travel parallel to shore, resulted in the large wave heights on the north and east coasts of Tutuila."

Wave run-up in American Samoa was higher than in Samoa, reaching a bewildering 16.9 meters in the village of Poloa at the western tip of Tutuila and 12 meters in Fagasa village. Yet, the death toll was much lower. Thirty four people killed in American Samoa. The field survey that Dr Borrero worked on concluded that "given the considerable run-up values and the scenes of devastation that we witnessed, it is remarkable that *only 34* people lost their lives on the island." Several factors contributed to the lower death toll.

First, there is the terrain. Here in American Samoa, there is a wild, rugged beauty. Not seen in Upolu. Maybe in the more remote sections of Savaii. Curved roads hug the cliff side shore. White surf and the ever present roar of the ocean as it pounds against reefs that weave so close to the rocky islands. Flat plains are sparse, so people have had to be far more accommodating of the land than their western cousins. They carve little precarious niches for themselves. Wriggling into minute clefts and terraces. It is a ruggedness that helped save lives during the tsunami of 29/09, as "steep slopes in the immediate back of many villages, [made] evacuation out of harm's way easy and fast." Not only do many people here live on high ground, but there are also extensive roads and access ways up into the hills – unlike in the villages of Lalomanu and Saleapaga on Upolu that had the highest number of deaths and where survivors had to claw their way up steep cliffs, pulling on roots and vines to get them to safety.

Widespread tsunami awareness and community education was another factor that researchers identified as on which explained why the death toll was so much lower here. September was 'Emergency Preparedness Month' "and a number of projects were in place, funded through the *TsunamiReady* program, which included showing videos of the 2004 Indian Ocean tsunami. The village mayor of Amanave for example, had attended a tsunami hazard workshop just a few weeks earlier and used a bullhorn megaphone on 29/09 to alert people to evacuate. Some schools had practiced regular tsunami evacuation drills. Drive along the coast of Tutuila and you can see tsunami hazard signs posted, blue bilingual signs describing the danger and instructing people to evacuate to higher ground upon feeling an earthquake. One report stated, "It is certain that the presence of these signs, which are seen daily by the populations at risk, can eventually instil in them the reflex of self-evacuation."

Talk to the people of American Samoa about the tsunami and one can notice some key differences between them and the people of Samoa. People here do not talk about the wave like it is a living thing. It was NOT a 'beast that jumped out of the sea and ran to devour us'. It was NOT ' a demon…a monster…a roaring, growling animal…a fire that swept across the land.' Perhaps thanks in part to their greater immersion in western culture, American Samoans seemed to find very little about the tsunami that was mystical, otherworldly, or mysterious. Or maybe this was just a reflection of how much more poetic and expressive with language their western cousins are! The waves that swept in to American Samoa were a natural phenomenon, a natural disaster, directly caused by a powerful earthquake. Perhaps this can explain then, the other marked contrast. The absence of widespread, deep-seated fear. Yes, there are those who are afraid of what the ocean can do, but it is not as strikingly apparent as in Samoa's disaster zone. Children in American Samoa are swimming in the ocean at Amanave. People are fishing in Pago harbor. Ask them about apprehensiveness towards the ocean and they will tell you that yes, there is some unease, but "I know that a tsunami is caused by an earthquake so the next time I feel one, I will evacuate inland faster."

Pago Pago: On the Water

There is a stunning natural harbour on the mid-eastern side of Tutuila. A harbour that calls to boats of every kind. Immense rapacious trawlers that plunder the midnight depths and then come in to disgorge their slippery shiny tuna. Starkist. The smell of canned gold is a pungent one, oiling its way into your hair and clothes, becoming one with you and your home so after a time you are no longer offended by its heavy presence. Huge floating rubbish bins return to the deep sea with the refuse. The trail they leave going and coming, a calling card for the most determined of ocean scavengers…Sharks. Nobody should go swimming in Pago Harbor. I see a lone fisherman standing in low tide water up to his waist on half exposed rocks some distance out, and I shudder. Waiting for a creature from 'Jaws' to launch itself from the foul water and expertly rip him in half. The perils of an overactive imagination.

The harbor also calls to others. Massive freight deliverers. Naval vessels. Yachts. Sailing boats of all kinds. Four Asian crew on board a rusting fishing boat, whistle and wave. Friendly and curious. There are sailors of every variety. Come to the American Samoa Yacht Club and you will meet some of them. A tired wooden building, the club sits by the water. A bar, pool table, darts, and on a lucky night, a three man band. A sheltered deck encircles it. Watch your step – there's slats missing in the floor. But the perilous decking and rickety benches are but a minor detail. The view of the harbor is a hushed breath of awe. Silver moonlight trembles softly on velvet water. A gathering of small sail boats commune just out of reach. Solitary masts yearn for a touch of sky.

Americans Kirk and Catherine McGeorge have been docked here for some time now with their six year old son, Stuart. Sailing, living, working here and there all over the world. There is Larry Jones, retired from California, land of sunshine, oranges and buxom babes in pink hotpants. There were two crewing with him when the tsunami came. "but they had to go back home and back to work because they ran out of money." They were an eye-ogling sight on his fifty foot boat. Larry smiles, "I'm enjoying my solitude now." He sits in a front row seat, listening to the music. Alone but not lonely. He is happy to tell his tsunami story.

On the morning of 29/09, Larry's sailboat, *The Tribute* was anchored in thirty-five feet of water. Other yachties had been having some trouble with their anchors, "every time the wind came up their anchors were dragging, so I didn't want to put up with that." The night before the tsunami, Larry had taken extra precautions. "I put out 275 feet of chain on one anchor and then I took the other anchor out and dropped it with about 350 feet of scope." This would make all the difference when the wave came.

After the earthquake, Larry was worried about a tsunami but reassured by the chatter on the VHS "people said they had checked with the Pacific Tsunami Warning Centre and there was no tsunami reported on their website. So we all relaxed and went back to making coffee." Within minutes, everything changed. "The water started going out of the harbor. There was a reef to the south of us and all of a sudden it looked like a waterfall. The water dropped so far that water was pouring out off the reef…we were anchored in 35 feet of water and it just dropped down to about 15 feet. We knew there was going to be a tsunami and it was too late to do anything about it." The safest place for a boat to be during a tsunami – is out in deep water. "when you're out on the ocean, tsunamis don't even exist. They're just an energy wave going through the water and when they get close to shore, that's when it turns into a surge." Larry has sailed through two hurricanes, but they were nothing compared with being stuck in a crowded harbor with a tsunami rushing in. "When you're out in the open ocean in a storm and you have everything under control, everything's lashed down and you shorten the sails so you can handle the conditions, it's scary but it's not impossible, it's not that scary if you're used to it. When I saw the water cascading off the reef which was ordinarily under water, it was brand new territory. I'd never seen anything like that and I had no idea what was coming."

Larry started his motor and pointed the bow towards the approaching surge, revving the engine. "All I remember is a lot of turbulence and a couple of boats went flying by and other boats started bouncing off of us. Our anchors held and most everybody else's didn't. So the boats that were breaking loose were bouncing off us on their way in to go to ground. I was pretty busy trying to keep everything steady and when it kind of settled down, I looked around and there were hardly any boats left. They'd all gone." Most of the boats around Larry ended up smashed inland. But if the first surge in had been rough, things got scarier. "There was a period of calm and then the water turned around and started rushing back out again. When it came back past us, the water was full of incredible stuff – refrigerators, pieces of roof, big chunks of wood, all kinds of stuff, it was just solid debris instead of water, like a moving mass of it."

There were several more surges. “The second and third wasn’t as big as the first one but it was carrying all the solid debris so they were probably the most destructive ones because there wasn’t as much water, just solid debris causing massive damage.”

Larry was desperate to get out of the harbor. “I didn’t want to stay there anymore because we didn’t know if there would be any aftershocks, more tsunamis. My boat wasn’t operational and it scared me because I didn’t know what else was coming… the rest of the day I was very busy trying to repair the damage to the boat. The engine wouldn’t run because it had sucked a whole bunch of garbage up into the intake, the water intake into the engine. So the engine overheated and I had to shut it down.” Larry donned diving gear and went into the water. “Of course I couldn’t see anything because the water was all churned up, it was pretty ugly diving. The surface of the water smelled like diesel fuel, it was ugly work. I had to clear the intake and you wouldn’t believe the amount of stuff that was wrapped around the propeller of the boat, I had to cut it all off.” Even with the engine running though, Larry couldn’t go far because he couldn’t pull in his anchors. “When you pull on an anchor it likes to dig itself deeper. During the surges those anchors dug themselves so far down deep that we couldn’t get them up, they had buried themselves. The boat weighs 41,000 pounds and pulling on that with the speed of the surge that was probably around fifteen knots…well, those anchors were stuck, one was down about eight feet below the surface of the bottom of the harbor.”

Two days after the tsunami, a US Navy frigate came in to the harbor and the crew went out to do community service and assist with the cleanup. “The Navy dive team came and worked for three days clearing anchors and moorings and helping people like me. It took a four man team of divers most of the day to dig up my two anchors and there was no visibility for them at all, just diving in the mud. They actually took shovels into the water and they had to shovel down quite a ways because they were so deeply buried!”

Larry’s tsunami experience has not changed the way he feels about the sea, he seems almost surprised anyone would even ask such a question. A few months after 29/09, he is docked again in Pago Harbor, and all is well with *The Tribute.*

Stephen Bourassa is a yachtie who has been sailing for over forty years. His boat the *Tulak* is Polish for ‘Wanderer.’ He was anchored in Pago harbour the morning of the tsunami not far from Larry Jones, and a faulty engine meant he was helpless against the onslaught of the wave. He shared his story with the Fiji Times. "It was around 7am when the boat started rocking violently. All the other yachties were all panicky on the VHF,, *what is that! What is that!* And then someone said it was an earthquake. My engine was down so I couldn't turn on my motor. So there I was stuck, wildly hoping there wouldn't be a tsunami."

When it happened, it happened fast. "Within minutes the water level in the harbor receded by about nine meters. When I looked around I was absolutely shocked to see the bottom of the harbor. I hurriedly secured my anchor lines but the waves came back in at a speed of about nine knots and broke them so my boat was driven forward by the waves. I stood there gripping the steering silently screaming, *What do I do?!* I was terrified and sensed death was near."

Stephen's boat hurtled over the wharf, clipping between a row of buildings and then rushed down the main street. "I sped between two buildings, hitting power lines down the main street, my mast came down, missing me by a meter. I sailed about 300 meters, I looked out and I saw an elderly couple struggling to stay afloat but the waves took them out again."

The *Tulak* jammed into a pile up of debris and as the first wave receded, Stephen clambered out and began going through the mess looking for survivors. "I heard screaming from people in a high building nearby saying, *Look out, look out!* and when I saw another wave, I scrambled back on board. I didn't know tsunamis had several waves. I was so scared I ran to my cabin and hid there until about 10am when no more waves came. Then I looked out and there were just ruins everywhere. I found the body of the elderly man I had seen earlier."

Stephen left the *Tulak* in search of supplies and came back to find teenagers looting his boat. He chased them away before setting to work trying to repair the damage to his home. "That is a day I will never forget…After that tsunami I feel more vulnerable but I won't give up sailing. I'll do this for the rest of my life, I'm gonna die on this boat…I'm not going to be afraid to enjoy my life just because of a tsunami. I took to sea for adventure, and boy, I sure found it!"

Americans Danny and Joan Wolszewski were "avid sailors with a streak of adventure". They had been sailing their boat 'MAINLY' for several years, beginning in Florida and then through the West Indies before venturing on to the South Pacific. On 29/09, the couple were anchored in Pago harbour when the earthquake struck. Danny went onto the dock to untie the boat so they could head out to the safety of open waters. The tsunami swept him away and his body was recovered later that day. On October 8th, family and friends farewelled Danny as he was buried at sea. Joan shared her memories of the memorial service. "It was really a day Danny would have loved. Beautiful but no muss, no fuss. A local seamstress made the bag required by the EPA. It was white canvas with brown piping on it. My cousins went to the morgue to dress and prepare Danny. I provided a pair of white jeans and a loud shirt and docksiders. Also included was his backpack he has used for years. Bought at a church bazaar, it had Emy written on it in green paint. He always wanted to meet Emy and let her know just how far that back pack had been. If you ever travelled with us I'm sure you knew Emy. We included a few things in the bag as well. Some videos, chocolate, a small flask with 23 year old Ron Zacapa Centenario Rum, his favourite from Guatemala. Also some pictures of the immediate family. The service was beautiful. The Reverend Ekitoa Sopoaga from the Congregational Church in Fagasa officiated at the ceremony and managed to say all the right things. Wade and Cole each spoke of their Dad and the love and lessons they learned from him."

Danny was buried three miles out to sea in 1,000 feet of water. "The water was clear and a very deep blue. As we drifted, we began putting flowers in the water slowly and they trailed away from the boat towards Danny in a meandering path. We had also prepared an empty bottle of the Ron Zacapa Rum with a few details on one of our boat cards. Some cruisers also put their boat cards in too. This was tightly sealed and tossed in the ocean for someone to find someday. We returned to the Malaloa Yacht Basin and traded stories with the cruisers and family over snacks and wine. It was a celebration, not a sad time and Danny would have loved that. Those who had never met Danny felt as though they had, at least in spirit. We found a card in Dan's wallet he carried for years. Wade read the following from the card:

I do my thing, and you do your thing.
I am not in this world to live up to your expectations
And you are not in this world to live up to mine.
You are you and I am I,
And if by chance we find each other, it's beautiful.
If not, it can't be helped

Just to let you know, we found each other and it was beautiful."

Pago Pago: On the Land.

Some people seem to have more than their fair share of challenges. They are survivors. They weather life's storms and in the face of adversity, they seem to bend and then come back stronger. Rose Talalotu is a survivor. Even before a tsunami threw a hardware store on top of her as she ran across a crowded street beside the Pago harbor. Yes, even before that, Rose was a survivor. "I've been through a lot of perfect storms. I've been through two bad marriages. My first marriage I was stabbed and then after the divorce I was in a head-on-collision, a mean car accident. My second husband pulled a gun on me…my kids say I've got nine lives and I guess I've used up a couple of them already. Hopefully I have a few more left!"

Rose was born and raised in Honolulu but she came to American Samoa fifteen years ago to help care for her parents. After they passed away, she stayed on. Rose works at the Bay Hardware Tool Shop and was at their Pago branch on the morning of 29/09. The store is on the ocean side of the main road. Stand at the counter and you can look out over the grimy harbor, the docks piled high with shipping containers. When the quake ended, Rose looked out to the harbour "because I know if there's a big earthquake, then there could be a tsunami. I looked out the window and the tide was still high so I thought, *oh the water's still high so that's alright.*"

Rose made herself a coffee and sat to work on her computer. She had her back to the sea and so did not see the changes taking place behind her. Until Ma'a, her co-worker called out to her, "Hey Rose, come take a look at the harbour!" Two steps and one glance was all she needed to get her moving.

"I walked to the window and looked out. The harbour was empty, all the yachts looked like little toys on the bottom of the ocean. You could see the bottom of the harbour, the dirt and rubbish all scattered there."

Rose yelled at Ma'a to grab the cashbox and run. Instead of taking off straight out the door, she herself went back around the counter to get her purse. She still wonders if maybe those few seconds would have made the difference to what happened next. Rose ran out the door and tried to cross the busy street. It was school hour and buses battled with cars coming in to town to work. Drivers had no eyes to see the water that was fast approaching them. The surge hit the buildings on the harbor line first, many of them crumbling. Then the tsunami swept into the traffic, pushing buses and trucks into a mad jumble of automotive fury. Rose had just made it to the far sidewalk when she saw it. " Out of the corner of my eye I could see the water coming and not long after that, all the buildings and the tool shop, they were right behind me. That's when I got caught in the tsunami. I was just rolling and rolling, for a long time I was under water."

As she tumbled in the water, Rose was being bashed by wood and metal debris. "I could feel everything hitting my head. The water was deep because I couldn't feel the bottom and I couldn't see anything because the water was so dirty and filthy, filthy, filthy…" Rose was taken to the far end of the street. A virtual dam of debris that used to be the hardware store, landed on top of her. She came up gasping for air only to find herself trapped underneath a sloping roof. There was a narrow pocket of space between her head and the black water. She hung on, hoping the water level would not rise any further. In those next few minutes, Rose heard something she will remember for the rest of her life. Someone, trapped in the water beneath her, scrabbling and scraping wildly against the wreckage for escape.

"I could feel that there was somebody under there. I didn't know what to do, the water was so dirty I couldn't see anything…I could feel them creeping and making this scratching noise and then I heard a last breath being released, like a gurgle. I was so scared…I thought *oh my God...*"

Rose knew when the wave changed its mind and began its retreat to the ocean. It tugged and pulled at her as she struggled against its command to go with it. "When the water started going back to the harbor I could feel my hair going back, getting pulled and I was afraid because I didn't know whether all the debris was going to shift and the whole thing would just come down and smash me.."

As quickly as it had come, the water left. Rose fought her way out of the wreckage, crawling towards the sunlight. "I kind of pulled away the lumber and wires and fixtures and electrical wire I was tangled in. I stuck my hand out of the mess and yelled for help. A kid came, Michael, he heard me and he called these other three guys to move all the stuff and pull me out. First they were carrying me up the hill but I wanted to try and walk so I could see if I had any broken bones. I stood up and looked back and there was no store left. I started crying because I just thanked God for my blessings because if I had been in the store I don't think I would have made it. That's when I started feeling the pain…"

Rose had full body bruises, numerous cuts that needed stitching, a severe head injury, broken fingers. Her face swelled up so badly that she was unrecognizable. "I didn't want anyone to touch me because I was in so much pain. And I really wanted to wash because I was so filthy. For days afterwards, I could still smell that stink water on me."

Two days after the tsunami, friends found Rose's handbag that she had grabbed from underneath the counter. The eight dollars cash was gone, but her passport and money cards were still there. She was able to fly out to Hawaii that weekend to get medical care closer to her children and grandchildren.

She laughs as she recalls the look on the Immigration officer's face when she handed him her tsunami-trashed passport. "Oh that thing stank so bad! Everyone could smell my passport…I hated to even put it in my bag it was so awful…"

Rose couldn't sleep beside the sea in those first days after the tsunami "I was staying at my cousin's place and she lives right beside the ocean and I had to go stay at my house inland because I could hear the waves outside and I was so scared just hearing the water." But now, she has taken steps to overcome that apprehension. She is back at work at the newly rebuilt Tool Shop, looking out over the placid harbor. And she is deliberately staying with a friend who lives beside the ocean. "I have to get over my fear. I remind myself that you need an earthquake before a tsunami comes so we have to just watch out for the signs. I like to go fishing and swimming. So I'm working on it. I look out every day at the ocean to make myself get used to it again."

Rose Talalotu is a survivor. What keeps her going? "My kids and grandkids. Nobody wants to lose a parent. I want to see my children and grandchildren grow up. Its mostly for them that I survived, strong mind, strong will. The best way to overcome your fears and your struggles? Just face them. It will make you stronger."

Beside the Tool Shop is a little coffee shop. It serves fresh baked bread and cakes. Delectable treats and pastries, the perfect melt in your mouth complement with a morning coffee. Marge Tafiti worked there. Customers would stop to pick up breakfast on the go and to steal a chat with Marge, a vivacious and sociable woman. Also working at the coffee shop was Marge's sister Moana and Moana's daughter, twenty-one year old Sabrina.

The women were preparing the morning baking when the earthquake shook the store. Marge was the only one who went outside. "I just stood there watching things falling down, waiting to see how long it would go for. It was weird and scary…when it finished I said, *oh well, I think the show's over!*"Back inside the kitchen of the coffee shop, Marge had started on her morning coffee when Moana ran in.

"There's something funny going on with the water in the bay, come look!"

The two women went outside the front of the store and stared in amazement at a fast emptying harbor. Wonder turned to fear as the water began to rush back in. The women stood still for only a moment. But it was a moment filled with choices. Do we know what we will do when death comes for us? When it surges towards us in a boiling mass of muddy water? Will we stand stunned and frozen, unable to react? Will we be able to run for our lives? What will our first thought be? The women yelled for Sabrina who was still in the store, but she was listening to the sounds of her iPod and could not hear them. Moana ran across the street where she would hang onto a steel post when the wave caught up to her. But Marge ran back inside the coffee shop calling for her niece. "She didn't hear me so I ran to the kitchen looking for her. That's when it happened. The water came through the back wall and just blew out the oven. My bread rolls flew across the room and I said *Holy Moly!* Sabrina was there, I grabbed her hand and we tried to run together but it was too late. We were caught right there in the kitchen. We were like pieces of paper getting washed away. We couldn't even move or anything. I could hear crackling, things bubbling up. We were swallowing water and there was no air…"

The wave completely destroyed the coffee shop, taking it like a flimsy pile of cards across the street and down to the bus stop. Marge and Sabrina were separated. Initially Marge struggled, but quickly realized it was futile. "For a while I felt my niece's hand in mine, but then it was gone…I told myself *there's no way you can survive this Marge* so I just relaxed, held my breath and didn't fight it. I just went with the wave. I don't know how long it went for but the next minute, I looked up and saw sunlight. It was really far away, the water was about eight feet deep above me. I pushed up with one foot and went up for air. The first thing that came out of my mouth when my head broke free of the water, was the piece of sausage roll that was still stuck in there, that I'd been eating with my coffee! I blew it out and tried to get some air. I couldn't see Sabrina anywhere. Then I saw everything – all the buildings and rubbish coming back towards me."

The wave was pulling back. It slammed Marge against something heavy and dragged her underwater again. Once again her first instinct was to struggle but then again she went limp, allowing the wave to take her with it as it ran back to the ocean. Marge was taken all the way out to the middle of the bay. "I sat there and I couldn't believe what was happening. I was sitting on the ground in shallow moving water, in the middle of rocks and rubbish. There was water rushing behind my back further out into the center of the harbour. I looked left and I saw a big yacht moving backwards. I looked right and I could see all the buildings, the lumber and everything. I said *hell no!* I stuck my hands down on the ground where the rocks were, trying to hold on to something so the water would stop taking me."

Marge was not the only one fighting the relentless drag back. "I saw an old man going past in the water, sucked from another two storey building and then a young guy still on the shore. He was coming up from under the water and he was struggling and had no clothes left at all. He was looking at me while I was going by in the water and it was like we said to each other - *its okay, just look at me, you can't help me and I can't help you either!"*

Marge was battered and weak but she knew that she had to move, get out of the bay before the wave returned. "I had to think fast at that time…I had no strength, I tried to get up and walk, to climb over the rocks and rubbish to get back up on the shore. I kept falling back in and the water was getting deep again." From somewhere, she found the strength to clamber up out of the bay, over the piles of wreckage. "I finally got up on the land and I sat on the top of a roof and the broken bits of a building. I was vomiting all that dirty water out and breathing really hard. I looked like Rambo. I came out of the bay with my shorts on, all ragged at the edges and my white polo shirt was brown."

The flesh on Marge's leg was hanging ragged where a deep cut pulsed red onto the dirt. Some men came then and helped her to stand. They half carried her across the road and up the hill, telling her another wave was coming. Marge saw many people running around screaming and scrambling through the destruction. She saw some other men carrying her friend Rose up the road. But she still couldn't see her niece Sabrina anywhere.

Marge and Rose were taken to the hospital. "We were the first ones there. We sat there like zombies, just looking at each other… everybody's running around trying to prepare the hospital for all the injured and we sat there waiting for our turn. Nobody paid any attention to us for five hours. Maybe because we were so still and quiet. A lot of people started arriving, they were crying and in pain and me and Rose just sat there and watched. They were bringing people in on Samoan mats because they didn't have enough gurneys or wheelchairs and they were taking them straight into the operating room…people were just lying on the floor all around us. Finally, we said, *okay, we were the first ones here and its kind of starting to really hurt, can we go in now?"*

Marge had surgery to first clean the wound on her leg and then stitch it. She spent three days in hospital. Marge's sister Moana escaped unscathed. But they didn't find her daughter Sabrina's body until the next day.

Marge doesn't know why she didn't die. Or why the young woman who was holding her hand when the wave came, didn't survive. She doesn't know why Sabrina's son - a handsome eight month old baby named Malachi - will now never know his mother because a tsunami took her. But she knows she that she is lucky to be alive.

September 29th was Anthony Wyberski's birthday. His twenty-first. The day a boy becomes a man. The day he is legally and technically, by most reckonings, an adult. Anthony's parents, Ray and Eta, gave him a brand new truck for this most celebratory of days. He drove it in to work early that morning, to the National Park Service office at the Pago Plaza. At home were his parents, his two brothers and the baby of the family, six year old Jasmine. In the two storey house next door were Eta's parents and sisters.

Ray Wyberski is from Pennsylvania in the USA while Eta was born and raised in Pago. They own American Samoa's finest jewellery store – the '*The Treasure Island*'. The main branch is in the Pago Plaza. A visit there is to breathe in luxurious indulgence, You sigh with welcome relief in the air conditioned comfort. Ropes of gold and silver lie draped on black velvet, rings for every finger cry out for attention. The Wyberski home is on the Pago shoreline. It sits slightly raised on the hillside, the sloped driveway graced with several sturdy trucks. In the garage, pride of place belongs to Ray's triumph – a fully restored 1977 supercharged MGB Roadster.

Eta was standing on that driveway saying good bye to her two youngest children before they left for school when the earth shook. "It was stronger than any other earthquake that I've ever felt, we were all so startled. My truck was actually jumping up and down."

Eta's eighty year old mother had long told her children about her experiences with tsunami in American Samoa. "My mom knows of three tsunamis from her past. One of them was so weird because it happened when they were all asleep. They woke up and there were fish everywhere on the ground outside their house. She remembers gathering them and they cooked them and ate them. The last tsunami she knows of was about fifty years ago. There was no earthquake before that one. It wasn't a very big one, it just came up on the road." Because of her mother's stories, because of tsunami awareness from television and school – Eta's reaction was automatic. "Right there and then I stopped everybody from going to school. The first thing that came out of my mouth was, *get out of the cars, get in the house, grab what you need and get back in the car, get grandma, get everyone and put them in the car and lets go up the hill!"*

Ray hurried the family in to the cars while Eta screamed for her sisters in the next house. "We're going to get hit, there's going to be a big tsunami, hurry!" Everyone loaded into the trucks and began to move out, As they pulled out of the drive and onto the shoreline road, the water in the harbor"just flushed out, there was a loud hissing sound like the water was being sucked out and all the boats in the bay there just dropped. When you looked back, the whole harbor was like a valley. It just went down."

The Wyberskis accelerated down the street and turned up the first mountain road that climbed sharply. By then the first wave had ripped into the harbor and was fast approaching the road. They parked the cars a short distance up the steep hill and got out to see what was happening below them. "A bus came past going down into town and I was yelling at him as he was going by that a wave was coming but he didn't hear me. There were kids in that bus. He kept driving down and all of a sudden you could see the bus come back in the wave and the schoolkids are floating around it trying to hold on to stuff. The bus driver got stuck on a mango tree and I was yelling at him to hold on."

Eta's two sons Keith and Nicholas, then did something that months later, the memory of it still brings a rush of panic to their mother's heart. "There were two women in the water. They had been washed off the road in their car and now it was filled with water. They were washed out and the waves were hitting them around and they were trying to hold onto bushes by the road. They were crying for help. My two boys just ran from behind me, from up the hill and jumped in the water."

Eta screamed at them with desperate fury, with the tone of voice a mother uses when you just know you better listen…or else! "What are you doing?! Are you crazy, get out of there, it's strong, that water is sucking people out!" Seventeen year old Keith turned back to reply, "Mom, we can't just leave them there…" Words Eta couldn't argue with. "I told them to be careful, so they were clinging on to trees and then grabbing the women and pulling them up."

The waves did not stay long. They came in with fiery haste and then just as quickly, they rushed out again, as if chasing another engagement. The Wyberski's went back to their house, or what was left of it. "There were cars, buses, containers and everything all piled in the front of our house, all smashed up. The whole house was wet, the inside was upside down. Everything was ruined – it all went into the garbage. The wave had rushed in and broken the front wall, it brought a big tree and a shipping container into the front room. That night we slept outside on the cement driveway."

But it was not the state of their home that concerned Ray and Eta. It was the thought of their eldest child, Anthony, at work in the Pago Plaza. "We didn't even think about the house and the damage or anything. We just thought about our son – everybody here was safe but we had one more that we needed to find…" Ray and Eta went into town. The Plaza building was standing but the lower levels had been blasted through by the wave. It was difficult to make sense of the wreckage though, because of the hordes of people who were scrambling through the ruined stores. Looters. Grabbing anything and everything they could get their hands on. Mainly passersby from other areas unaffected by the waves. Spectators who did not have to worry about searching for missing family members. "It's amazing how you have two sets of people. One care about life and the other have no thought about anybody else, all they were thinking about was what they could steal. That was it, it was like night and day, two sets of people."

One couple had come to check their offices at the Plaza and watched the thievery with horror. "One of the hardest parts of the day was the looting. Watching people at their worst was not something I will soon forget. People were looting stores, vehicles, offices, really anything they could get their hands on. Cars that were overturned were soon missing their tires as thieves came by. Some people were literally scavenging before a woman's body was even removed from a car..."

Into this madness, came Eta and Ray looking for their son. Recognizing them, several people came up to them, "Do you know they're looting your store? They're stealing all your stuff." Eta just waved them away, shoving through the crowd, screaming for her son, *Anthony! Has anybody seen my son?!*

On the internet, there is video footage of the wave sweeping into a car parking lot at the Plaza. Dimly you can see a man halt mid step as he is walking out towards a truck. He sees the water coming, he stops, he backs up, turns and runs. The wave swirls over the cars and through the lot. That man is Anthony Wyberski. One of those trucks that eventually ended up stuck halfway in a house, was his birthday present. Anthony made it safely out of reach of the water. But when the first wave had barely pulled back, he was one of the rare few who ventured outside to try and help those crying out for aid. Beside the Plaza there is a deep ditch where a murky stream runs. Many vehicles were lodged there. Including a bus. Anthony clambered into that ditch to help bring people out. The bus driver was pinned against the seat and could not be freed. The driver told him to *go, go*. The next wave was coming. "He didn't want to leave him. But that wave came again and he was still in the water trying to pull the man out, the water came higher and the man told him to *go, get out of here, go*. So he got out of the bus and when the wave left, he went back there and the driver was dead."

When Ray and Eta finally see their son, he is at sweaty, muddy work, dragging a body out of a car in the stream. "He was climbing out of the stream after getting some people out and as he was climbing out, he looked at us and he started crying. I hugged him, I was crying too, I told him, *you're okay, you're alive, you're okay.*"

Anthony's first words to his dad were asking for forgiveness. "Dad, I'm sorry about my truck, it's all messed up…and the store, I'm sorry I couldn't do anything about the store, I couldn't stop those people from taking stuff."

Ray and Eta held their son close, filthy and wet from the foul waters of the Pago harbor. They looked at the remains of their beautiful jewellery store, at people darting ecstatically down the main road with pockets crammed full of Wyberski gold and silver. Ray shook his head, "To hell with it son, we can always rebuild, we can get another car, another store. You're alive and that's all that matters. Let's go home, never mind, just leave it alone."

A disaster impacts on people in so many different ways. It sweeps through a family,the force of its impact leaving no-one unchanged. After the tsunami, during the long weeks of clean up, Eta's eighty year old mother felt more keenly her age. There was a sense of frustration. "She was getting really depressed afterwards because she's older and she feels like she's helpless, like she can't help us to clean up and fix things. So she was getting really upset with herself and I tried to tell her, *mom, don't worry about it, that's what we're here for, just take your time and do what you can do."*

The youngest in the family. "Our daughter Jasmine, she was the most affected out of everyone. Even when school started again, she didn't want to go. And when we drive in the car, she used to love to sit on the side of the car where you watch the ocean, she used to love to do it, but not now. She won't sit next to the window or look at the sea anymore." Jasmine's teachers have noticed the change in a once lively student. "The school called us in to talk about it. They made her draw pictures of what happened, what she saw. Because she saw the boys jumping in the water to pull out those women, she saw people getting washed away." The recent spate of natural temper tantrums hasn't helped to ease Jasmine's mind either. "A week after the tsunami we had another earthquake and she was totally freaking out. Then this last cyclone Rene that we had – it's just one thing after another and you've got to think about these little kids who've been through the tsunami and seen it and then to have all these other events coming after it. We were at the store getting it ready for the cyclone, hammering up the windows because the wind was so strong and Jasmine was calling us up on the phone all the time, crying, *mom are you okay? Please be careful, please come home soon.* Every little thing and she jumps, even the rain, it started raining really hard and she wanted to know *is something bad going to happen?*"

The tsunami left its muddy debris on Ray and Eta's relationship as each struggled to cope with the loss of their business and the destruction of their home. "It really brings you down. We were each dealing with stuff in our own way. I was upset because of the way he was acting and he was upset because of the way I was acting. We had a really bad argument and it just got to the point where he broke down, I broke down and we just had to sit down and talk about it."

Even the younger boys have struggled. "My boys, it took a while before it all sank in. A week later I caught my younger son crying by the side of the house. He hadn't really thought about what he had done, about saving somebody's life but then a week later, it all hit and he's like *what did I do? I could have died that day, we all could have died that day.* Then you've got to try and bring them out of it and I'm already depressed myself and now my kids are down."And Anthony? It is certain that he will always remember his twenty-first birthday for all the wrong reasons.

Months later, Ray is emphatic. "I never want to go through something like this again. This thing broke me down. I have never cried in my life until when you lose everything. Growing up, my family wasn't rich and when we work and we buy something, we cherish it and we'll keep it for as long as we can. Everything I have I respect it because I earned it and worked hard for it. To us everything was gone. We lost vehicles, two stores, our home damaged. But we could've lost three boys – we're so lucky. I guess something like this – it can be good in a way because it brings you back stronger. You get knocked down and you get back up."

Clutching the hillside in Pago, is Evalani's. A nightclub, restaurant and motel. Further up sits the family home. For those in search of good food and great music, Evalani's is the place to go. And if you love to dance then definitely head over there when the strobe lights pulse to the beat of the earth-shaking sound system. Twenty-nine year old Kathleen Pearson is Evalani's daughter. She is a coiled spring of energy who helps keep the three different businesses running. Kathleen was sleeping in the motel that morning when the earth shook. Her mother and brother were up in the house. But Kathleen's eleven year old daughter was on the front road, waiting for the school bus.

"During the earthquake, the first thing I was thinking about was my family, so I ran up to the house to see if they were okay. They were already sitting up, saying their prayers, so I sat down and joined them and we all prayed together."

After the prayer, Kathleen decided she wanted her daughter with her. She rushed down to the street, just in time to stop her from getting on the school bus that had pulled up. "It was about three seconds before she got on that bus…it's a really scary thought, I've been thinking, *what if, what if* ever since then. What if I hadn't gotten her off the bus in time? Then she would have been on the road when the wave came in…"

Back up in the hillside house, the family turned on the radio to check for tsunami reports. Nothing. Kathleen went down to the motel office to check the internet for news. Kathleen's aunt who lived in the house next door suddenly started screaming. "My aunty was calling my name and I was wondering why she was calling me with that sort of tone and then right when I opened the front door and looked out, the water was just starting to rise up in the harbour. It came really quickly, it was coming to cover the whole of the road and the front of the motel."

Kathleen ran. She grabbed her bag and dashed back up to the house. "I stood in front of the house and watched the wave…as soon as I noticed the water wasn't getting any higher, that it wasn't going to reach us in the house, then I got my camera, climbed out onto the roof and started taking pictures."

Sitting on the roof, Kathleen had a front row seat to the morning's devastation. She saw it all. The roll of the first wave that smashed through the buildings lining the oceanside. She saw the hardware store crumble. She saw cars from the plaza tumbling in the water and light posts swaying drunkenly."The first wave just took everything. It took the warehouse that was across the street, all the cars from the Town Plaza were just floating right here in front of our house.There was a huge sailboat floating down the main road right in front of our house – it looked like it was a car just driving by and that was scary because you don't see those kind of things everyday! It cut the electricity line from our roof to across the street, snapped the whole power line down and I could see the owner still on board. The boat ended up down the street sitting on top of the tennis courts." A woman stood on the roof of a building across the road from Evalani's. The wave had almost completely engulfed it so that it looked like the woman was literally 'walking on water'. "She came out and knelt down right there on the roof and started praying, she had her hands raised up in the air and she was calling out really loud, praying to God for help…I just felt so helpless looking at her there. We couldn't run across there and get her up to higher ground, we couldn't do anything for her."

The first wave climbed no higher though and as it receded, the woman was able to rush inland for safety. Kathleen came down off the roof and started walking up the main street littered with smashed buildings and cars. " I was walking towards the Pago Plaza…my sister called me on my cell phone. They had heard about the earthquake but they didn't know about the tsunami, so she was getting all hysterical when I told her. Then my dad called me, *Kathleen where are you?* I told him I was walking down the street to the Plaza. *Hang up the phone, turn around and run home right now, it's going to come back!* So I hung up the phone and started running all the way back to the motel and right when I hit the motel steps, the second wave starting coming in. I was very lucky that my dad called me at that time to turn around because I wouldn't have known. I didn't know that tsunamis came in more than one wave."

After the big surges were done, people came down off the hills to help, to look at the confusion and chaotic mess. Some came to see what they could get. "There were people in the Plaza just taking stuff and going through the grocery stores looting. Many of them were passers-by who stopped their cars and jumped out to steal whatever they could. But our Pago people, most of them were out there helping everybody, pulling out bodies from the debris and trying to clear the road so the rescue people could get through."

Kathleen's aunty was one of those killed in the tsunami. "She was on the bottom floor of her two storey house and she didn't make it, it was too late for her to run away."

Evalani's needed a lot of work to get up and running again. "There were coconut trees inside our motel rooms…everything was wet. We had to strip the whole bottom floor, patch up the outside walls and replace all the furniture."

Still, Kathleen's final reflections on 29/09 are gratitude-laden. "This has not changed how I feel about the sea. We're a very strong, very religious family. We have faith in the Lord, right after the earthquake, we said our prayers. We were very lucky that the tsunami hit early in the morning and we were not at work yet. My mom has a problem with her leg so it would have been hard to get her up and out of the motel and to walk up the mountain. We're very fortunate our house is high enough that only the motel was damaged. I'm very, very glad that I got my daughter up to the house quickly after the earthquake, if I hadn't…well, it's just something you don't want to think about."

On the Land: Leone

The village of Leone sits beside the ocean looking out at the main road that meanders past. Like coastal villages in western Samoa, houses are crammed close together. There are no fences to shut out your neighbour or to loudly proclaim – this is MINE and that is YOURS. Families have lived here for several generations. Your elderly mother is every child's grandmother. Your neighbor is your cousin, uncle or aunt. Nobody lives alone. Solitude is unheard of. But unlike her western neighbours, the houses here in an American Samoan village are more grand. More urban. There are fine brick homes intermingled with breezy open traditional fales. People here have decks that jut out over the rocks, ceramic tiling, spa baths and an array of electronic gadgets. Many have air conditioning. They surf the world on internet waves. Most homes have at least two cars parked out front. American Samoans love their trucks. Sturdy Fords and solid Chevy's. Imposing Toyota Prius. Glistening Hummers. That pound the air with their amplifiers. Children ride to school on government buses and eat free lunches. Pregnant and nursing mothers get coupons to buy fruits, vegetables and protein, the building blocks of good nutrition.

At the back of Leone village, there is a black pool laced with twisted mangrove trees. The still waters drain lifelessly into the rocky sea. Clinging to its muddy banks is a helter skelter lean to. There is a table – scuffed and dented, salvaged from debris. Upturned buckets for stools. Here, at the back of a slowly rebuilding village, here is where the Tsunami Club meets.

On a typical night, a full moon hangs heavily in a placid night sky. The evening hums with mosquitoes. A cool wind blows in from the ocean. Light spills from half-finished homes where families prepare for sleep. Hopefully, a sleep unfilled with tsunami dreams. But here in this swamp-side shack, eight men are gathered around the unsteady table. Playing cards. They tell you that for five dollars, anyone can join in. Cold beer costs a dollar. But they graciously offer visiting writers from Apia a beer for free.

"We started meeting here after the tsunami. Our houses were broken. Our days were busy with clean up and building new homes. We're all victims here. All of us made it through the tsunami. So we started meeting here every night. Playing cards. Having a few beers....we're the Tsunami Club!"

The mood is light. The laughter is low. The game is serious. The men are many things. They are shopkeepers. Insurance salesmen. Retired public servants. Customs officers who keep the airport safe with K9 dogs. All are fathers. For their village of Leone, all have history. All have love. The men speak of many things. Like when will their paperwork be complete so work can start on their new house from FEMA? The beautiful cement parking for the new church hall that a few of them hand troweled smooth that day. The mythically perfect shot that one of them hit on the golf course that afternoon. The writer who has come from Apia looking for people to interview... *"I told her to come back later and then I took off in my car and hid in town so she wouldn't find me!"* Loud laughter from all.

But there are other things that weigh heavily on this group that sit and play cards beside a black pool in Leone. Things like - the memory of those they could not save.

Speak with fifty-two year old Tafeama'ali'i Ropati Ofa, the one they call 'the General'. Retired from government service, he is a leader. A man of many talents. He paints, he builds, he repairs. People come to him when they need help. A large, broad man with a cheerful grin. Sweaty and shirtless. Paint spattered. The General lived on the Leone shore, beside the gas station. On tsunami day, he was caught by the first wave as he ran with his wife. He was smashed ribs first into the cement pillars of a hall. The water came to his chin, "but my wife is short so I tried to hoist her up and get her up onto the tin building next to the hall." Then another woman collided with his leg, tossing and tumbling in the dirty water. "I reached down and grabbed her and tried to carry her to put her up onto the building but while I was taking her there, I was hit by another old woman, so I tried to take both of them." The General's attempt to rescue the first two women was successful but it is the others that he cannot forget. "I tried to get to a six year old schoolgirl that was trapped on some corrugated iron. I managed to reach her but then there was another wave and I got battered by it." The wave was not merciful. Churned in the debris, the General was unable to hold onto the girl. Her body was later found in the black swamp. "There was another old woman that I tried to save but when I got to her, she had already died. I was too late."

The wave broke several of the General's ribs and cut him maliciously before spitting him out. He was taken to hospital but discharged himself a few hours later. So he could return to the village to help. "I believe that I am a servant of the village, my assistance was greatly needed. I couldn't stay in the hospital when I knew so many people needed help." There are tears as he remembers those he could not save.

Speak with forty-one year old Iuliano, he of the warrior physique. He graciously puts aside his cards to share his story. Tall and athletic, he had woken early that day for his usual morning run. By seven he was driving two of his children to school with his wife when they looked back in time to see the first wave hit the village. Where his other two children were still at home. With his mother-in-law. Two nieces. An aunt. He ran back, strong and sure to fight his way to his house. Thankfully his aunt had already run with the children to the safety of the back hills. The others were huddled on the second floor of their solid brick home. "I wasn't scared for myself, all I worried about was my kids and my family."

Iuliano took the others to the hill before the second wave rushed in. He could have stayed there. All his family was secure. But he went back. Through debris and hungry water. "I ran back to the scene because I saw a lot of people needing help. But there were so many yelling, you just don't know who to go to first." An old woman cried out from the remains of the petrol station by the roadside. "She was saying, *please help me, I want to live.* I told her, *don't worry, we will try our best to get you out of here.*" She was pinned against a pole by an upturned truck. Iuliano and his cousin Cameron tried to move it. Tried to free her. But the vehicle was too heavy and the force of the water too strong. The third wave was approaching. "I looked in her eyes and I told her, *I'll come back for you. I'll come back."*

Iuliano and Cameron were taken with the third wave, washed to the mangrove swamp. Again, he could have stayed there. But again, he went back. Through debris and hungry water. Back to the petrol station where an old woman was still pinned against a pole. Still. And silent.

"I can't get her out of my mind. It really bothers me. I've known her all my life, I grew up here and she was like a grandmother to all us kids in the village…I keep seeing her face, you know? I looked in her eyes and I promised her that I was going to get her out, I told her I would save her…and I couldn't."

What do you say to survivors who struggle with guilt? Like Iuliano? The General? You tell them that there is nothing more that they could have done – what is the strength of one man against a wave that runs at five hundred miles an hour? And slams shipping containers into warehouses?

You tell them that there is honour in the trying. That there were many who did not go back. Who could not and did not try. You tell them that those women died knowing someone cared enough to try.

Maiava Leota is the Manager of the Tafuna Airport. "I was born here in Leone and grew up here in paradise." Maiava has six children and the Leota family – two sisters and their children, a brother, Maiava's mother and 74 year old aunty – all lived together in three houses on the Leone shore.

After the earthquake, Maiava's family continued with their usual morning routine. His wife left for work. His two youngest children still living at home, went to school. One of his sisters took their elderly mother to get some traditional medicine from another village inland. Maiava thought about tsunamis, "but you know us, we are thinking about it but we just don't believe that it can happen." He was standing outside "having a coffee and a smoke" when he noticed the ocean was very low. He pointed it out to his brother Tasi but he shrugged it off. "My brother said it was because the tide was out." Feeling uneasy, Maiava called out to his sister to get her children into the car and drive inland "because we don't know what's going on." He called for a niece to help his aunty out to his truck. " I kept saying to my brother, *the water looks strange, it's like its drifting out* and he kept saying, *it's because the tide is going out*."

People closer to the road started running, screaming. Maiava's brother came running from his shower, wearing only his towel *Maiava, the wave is coming!* Maiava can laugh about it now. "I felt like hitting him then, because he had kept telling me that *no, there's no tsunami…the tide is low!*"

Maiava looked at the fast approaching wave and he knew there was no time to escape. He put his aunty into the car and told his niece to *come quickly and hold tight to the rails of the car!* "I shut the car door and that's when the wave hit and smashed into us. It was so strong that you can feel your legs and your head being beaten, we were fighting with the wave." The ocean took the pickup truck with it and Maiava knew that would be their only means of survival. "The car started to go with the sea and I said to my niece, *just hold on to the car, no matter where it takes you, just hold on.* So we were being dragged with the car, me and my niece were on the outside of the truck and my aunty was still inside." The two were being pummelled on all sides by other debris swirling in the dirty water. Maiava was caught on the wire of a washing line and couldn't keep his hold on the truck. "So I grabbed onto the wire and the car just went, so fast. I couldn't see it anymore and then the water was going higher than the wire, it was strong and my body was being battered, everything was hitting me."

There was a *ulu* (breadfruit) tree nearby, valiantly holding its head up high through the wave. "I tried to swim to the *ulu* and when I got hold of it, I just hung on for awhile, I was so tired so I tried to rest. But the water kept going higher so I had to keep climbing up, holding on to higher branches." Maiava clung to the tree until the wave receded. He climbed down and went to the back of the destroyed houses, looking for the car that had swept his aunty and niece along. "The water was up to my waist then. I was looking for my family. All the houses were broken and flooded. There was no car anywhere, I didn't know what to do next." Just then, a boy standing on the rooftop of a neighbouring house, called out that another wave was coming. "I looked towards the sea and I saw that it was coming again." Maiava rushed back to the breadfruit tree and as he clambered up, he heard a little girl crying. "She was standing on the road looking at the wave. I called out to her to run towards me. She turned and ran but the wave got her. As she was going past me, I stood and got a good hold on the tree and then I reached down and grabbed her hair and the collar of her school uniform and lifted her up. *Oka*, you should have heard her scream when I pulled her up! I put her in the tree beside and told her *you cry while I try to catch my breath and then we'll climb higher.*"

As the water level rose, Maiava and the child climbed higher. "She was crying that her sister and brother had gone in the wave. She wanted her dad, I told her *You'll be alright, Your dad isn't here so I'm your father right now. We will be okay.* So we just clung on to the *ulu* while she cried." Maiava was worried about things knocking down their tree. Two large shipping containers were moving through the water. One clipped the edge of his house and another was deciding whether to make its way towards the breadfruit tree or continue on through the village. "I was absolutely out of my mind looking at everything going on around us. The sea was rolling around and coming in and then turning and going back out, then coming back in again."

When all seemed ended, Maiava carried the girl back down, handing her over to a neighbour who took her to safety at the inland primary school. "I went with my nephew to look for my family, everywhere people were calling for help." Maiava saw a cousin carrying a woman called Terisa. The red that spurted from her clasped hands was the only colour against the muddy brown day. He kept looking for his family. He saw a small crowd trying to help a man called Stacy trapped in a thicket of trees and dammed debris. He helped the men free Stacy and carry him to the road. He fetched water so Stacy could drink and watched as rescuers forced Stacy to vomit so he could rid himself of the accumulated debris forced inside him by his journey in the wave. Maiava kept looking and there was great relief when he found his aunty and nieces, his sister and her children. All safe. "My truck had been taken very far away. The wave flipped the truck and some boys saw my aunty's hand hanging out of the window so they smashed the window and got the old lady out before the second wave hit and then they were washed together by the wave to the priest's house at the back of the village and that's where they were saved."

People were crying as they called out the names of their missing ones. "We learnt that Michelle had been killed and the daughter of the Korean family was missing. We were all looking for the missing people. Two old ladies and two kids were missing so we all kept looking and when it was descending on dark time, we found one of the old ladies in the dead water at the back of our village."

Maiava saw his brother Tasi, carrying their three year old nephew Pani in his arms. Tasi's leg was ripped open with a huge gash, his clothes were in tatters but Pani looked unharmed. Maiava asked Tasi, what happened?

When the ocean came, Tasi grabbed hold of his little nephew. Together they were churned through the madness. Pieces of wood and various household appliances smashed them about but Tasi refused to let go of Pani. Something cut his leg. He refused to let go of Pani. A fallen coconut tree went by and

Tasi caught hold of it. He sheltered the boy against the tree with his body as he gripped the trunk firmly. The tree was dragged back out to the ocean by the receding wave. Many things tried to dislodge Tasi's grip on the tree. On his nephew. He refused to let go of Pani. They were swept inland again with the second wave, pulled out and then brought in again. Tasi hoped for safety on the land but again they were taken out to the ocean. *There was no hope, I knew that. Every time we reached land, the sea kept taking us back out again.* Tasi was tiring. He knew he would not be able to resist the battering strength of the debris much longer. Already, he could feel his grip on the tree weakening. A freezer came by in the water. Floating upright. The lid open so Tasi could see the emptiness inside. All its contents had long been dislodged. The freezer was floating. Like a boat. Tasi made a choice. As the freezer went past, he hefted his nephew inside it – *Pani, I don't know what else to do...Wait there, sit very still. Don't move.* Tasi shut the freezer lid so his nephew would not fall out. Then, with renewed energy, he struck out swimming after the spinning, swirling white box, hoping that they would both end up in the same place on dry land. Tasi does not like to talk of his fight through the angry ocean crowded with Leone's rubbish, fighting forward to the land. Fighting to keep sight of a fast bobbing freezer. Fighting against the panic as time tore on and a little boy gasped for air in the darkness. Tasi was taken to the shore of the stream in Leone. And a little distance away, the freezer came to rest in a matted tangle of branches and rubbish. Praying he was not too late, praying that Pani had not suffocated, Tasi waded out to the freezer and opened it. Pani was sitting very still, waiting – just like his uncle had told him to.Wide, frightened eyes stared up at him, eyes that lit with a smile when they saw him. Quickly, Tasi pulled Pani out and together they stood on a muddy shore and watched as the freezer was caught in a hungry surge and disappeared out to sea.

Tasi didn't speak to anyone for two weeks after the tsunami. Finally, on a moon-filled night, when the evening hums with mosquitoes and a cool wind blows in from the ocean, Tasi finally shared his story with the Leone Tsunami Club. "We made him have a few beers with us so he can talk and then he opened up and told us what happened to him. As he explained his story, we started to cry, when he said, *there was no hope,* when he told us how the ocean kept taking them back out again. And he cried when he told us how the freezer got taken by the wave after he took Pani out and held him."

The week of 29/09, Maiava and the rest of the men in Leone searched for the missing. "It wasn't easy – after five days, anyone who has a good nose can find the areas to look...We tried our best but there was one more boy we never found. Even late at night, kids were still walking around looking for him...that boy, his sister died too and she was found so the family lost two children. I feel for their father. All we can do is tell him *we tried our best* but it's just so sad. I just wish we could have found him."

At the close of December 2009, the father of the little girl that Maiava saved came to see him. He came to say thank you. He brought gifts. A cake, sweet and rich with frosting. He brought his children with him. They talked and laughed and ate. And looked forward to the new year in Leone.

Terisa So'oto Palmer is a daughter of Leone village. She is resplendent in sunshine yellow, carefully accented with chunks of gold. But unlike many beautiful women, there is no coldness about her. She greets others with warmth and friendly ease. Terisa is one of fifteen children. Her seventy-eight year old mother, Fa'atamali'i So'oto was the matriarch of the extended family. "She was the main pole of the family, she held everything together." Terisa lived on the Leone shore with her husband Alfonso and seven year old daughter Tiare. The Palmers had several businesses adjoining their home. A laundromat that whitened whites from as far away as Poloa village. A car audio shop, computer café and a gamer's haven. Terisa's mother lived further inland with another daughter. Fa'atamali'i was not a woman though who liked to be idle, and every morning she was up early for a day of weaving and craft making with a group of other elderly women at a guesthouse *fale* just a few minutes walk away from Terisa's home. Most days, Terisa would stop there on her way to work, taking loaves of hot bread from the Princess Bakery for the old ladies' breakfast.

On the morning of the 29th, Terisa's daughter woke unusually early. "Usually we have to fight to wake her up early and have her get ready for school - you know how it is with kids and getting them up in the morning! But that morning she was up at five because there was a cartoon coming on TV that she really wanted to watch. I don't allow her to watch TV during the week so she was really trying to persuade me. She took a shower and got all dressed without any problem."

Tiare's eagerness to catch an early morning cartoon meant several things. It meant that the Palmer family was all awake and ready for the day when the earthquake occurred. And, even more crucial, it meant that as soon as the shaking stopped, Alfonso could get in the truck and drive away, taking Tiare to school. Away from a house that would not stand when the tsunami came. Away from a village that would endure unspeakable devastation. Leaving Terisa alone at home.

Once they had left, Terisa started ironing her jacket for work. She had the curtains closed and the air conditioning on full blast, which meant she saw and heard nothing of the outside world. Until it could no longer be ignored. "While I was ironing, I felt like somebody was spraying water at the back windows. Somebody was just spraying water and I thought maybe the man next door was cleaning his house with the hose so I opened the curtain to call out to him to be careful, but when I opened the curtain, all I saw was the wave."

Can you see what Terisa saw? A wall of water that foamed at its crest, that kept building up higher and higher. That seemed to take its time, rising with deliberate ease as it advanced on a woman who stood wide-eyed at the window, hands parting the curtains so she could tell her neighbour that he was getting splashes of wet in her house. Can you feel what Terisa felt? As realization ripped through her brain like fire? Of what was happening. And what was about to happen?

"I was standing there and it was going higher and higher above my head, outside of the house. I looked over to the other side, to the road where there's a bridge that connects our village to the other side and I couldn't see the bridge anymore. I realized right there that something was not right. I turned around to run outside, I said, *oh my God...*"

Terisa took two steps, her hand was on the door handle when the wave hit the house. "I heard something like a tree crack and the next thing I saw was all our house stuff coming towards me. Everything hit me and I felt my body being banged against the next wall of the house. I kind of blacked out then."

The wave obliterated the back wall and then smashed through the front wall as well, continuing on, wiping out all of the Palmer's business complex. From there, the rolling mass of debris swept on to crash through several other houses and head to the swamp at the back of Leone. Inside that rolling mass was Terisa. She remembers waking to the frenzied din of shattering buildings, screeching metal and the snapping of timber. All overlaid with screams. People everywhere screaming. "I could hear ladies, kids all yelling *help us, help me, fia ola...I want to live.*" Teri was caught against a metal pole. The force of the wave had ripped away all her clothing. She was completely naked and covered in cuts. The water was pulling back to the ocean and waist deep at that point. She tried to stand upright, reaching towards the pole for security. That's when she noticed her right hand. Four fingers were almost completely severed. They dangled there loosely. "When I grabbed the pole, I felt my fingers falling down like they weren't even still on, so I pulled them back on with my other hand, trying to hold on to them. I stood there holding my fingers and I was crying and asking God *please help me get through this.*"

Assorted clothing floated by. Terisa salvaged a shirt. Not far from her were others trying to escape the wave. A woman called Ana, clung to the roof of a still standing house, calling to her six year old son T.J. "I saw the father throw his son up on the house and he was crying out *mummy, daddy!* His mom caught him and his dad ran and climbed up the breadfruit tree. Ana saw me and told me to climb up. She said, *Teri, get up here, the second wave is coming, it's coming right now, I can see it!"*

Terisa tried. "I stood up and saw all the house debris coming back , there was glass, metal and everything was coming towards me. I tried to run to their house. The only way to get up to the roof though was to climb up on piles of roofing iron that were jammed on the side of the house."

We can do impossible things in desperate situations. Things that will astound even ourselves when we recall them later. Terisa scrambled through scathingly sharp corrugated iron pieces as the next wave came closer. Those on the rooftop, reached for her, screaming for her to *hurry, hurry*. "I didn't have any shoes and I was climbing and I kept slipping back down over the roofing and it was cutting my legs but at this time, the only thing you're thinking of is to survive. Ana and the others helped pull me up and it was just in time."

The tattered group knew they were still not out of harm's way. Would the house they were sheltering on, stand strong as another wave hit? Would the accumulated wreckage of the decimated village prove too powerful a battering ram? It was with a mangled sense of disbelief that they watched as a twenty foot shipping container that had stood in Terisa's front yard, slowly move in the second wave and come to rest. Directly parallel in front of the house they stood on. "Somehow, we don't know what happened, the container it was coming straight and all of a sudden, it just turned and it sat right in front of the house, shielding us from everything. That was a miracle to us, because if not for that container, the wave would have just knocked that house down and us on the roof with it."

They watched a third and fourth wave sweep in, each weaker than the one before. Looking out over the remains of Leone village, Terisa felt a cold certainty. "I said to Ana, *I just know something has happened to my mom…I can feel my mom is gone."* Once the ocean had tired of its catlike play with a hapless mouse, the group climbed down carefully. The others ran further inland, worried about more waves. But not Terisa. "I couldn't walk as fast as they could because my leg was torn so bad from the roofing iron. I sat down and just waited. A policeman came. He helped tie my hand with the cut fingers and also my leg. Then I saw my family, they were coming to look for me."

The first thing Terisa asked was, *Where's Mama?* Nobody knew at that point what had happened to the old lady. The searching for the living and the dead of Leone had begun. Terisa was taken to the hospital. At Tafuna Hospital she saw another man from her village, Toetu Tauiliili. His leg was cut to the bone. There was defeat in his eyes as he told her what she already felt to be true. "The first thing he said to me, was *Teri, I'm sorry, your mom didn't make it. I tried to hang on to your mom's hand, but the wave was too strong."* The man was distraught and Terisa tried to comfort him as she waited her turn with the doctors. Half an hour later, Terisa's sister Va was rushed into the emergency room. She had not been in the tsunami, but the strain of their mother's death was too much for her and she had suffered a massive heart attack. In all the chaos, Terisa neglected to mention to the nurses that she had been caught in her brick house as the tsunami smashed it to pieces. Which meant that only her external wounds got the doctors attention.

Terisa was in surgery for five hours while surgeons worked on her hand and leg. She would keep all her fingers and her leg would be fine, but scarred. By early the next morning though, the hidden damage from the wave was making itself known. Terisa was vomiting blood and struggling to breathe. Tests revealed she had broken all her ribs and was bleeding internally. Her lungs were clogged with accumulated sediment and blood. There were more surgical procedures. More panic, fear and pain. Finally after two weeks, Terisa was able to leave the hospital. She made it to her mother's funeral.

The Palmer family will not rebuild in Leone. Their beachfront property is in a designated 'Z' zone – a high risk flood and tsunami area. But with FEMA's help, they will build a new home further inland.

For Terisa, the tsunami of 29/09 was many things. It was a merciless thief. It took their businesses – "all of a sudden, everything's gone, taken. Your sweat, what you worked all these years for. Nine years of hard work taken away." It robbed her of her peace. "I've always felt that the ocean was a calm, peaceful thing, it used to be a place where we could go sit and relax and to soothe your mind. Maybe in time, it will be a soothing place for me again, but now, if I go to the beach, all I can think about is that the tsunami's coming back." It took her mother, robbing the So'oto family of many more years with a woman who was their strength. Tiare and her cousins will not have a grandmother's gentle wisdom to turn to as they become young adults and one day begin families of their own.

The tsunami was also a harsh teacher. "It made me stronger and more humble. It made me realize that God created us and God can take you anytime He wants. We're only here temporarily, but life is beautiful and we just have to live it to the fullest every day. This tsunami has changed my life completely. Before, I would only go to church when I have time, I was like the black sheep of the family! But this disaster had made me stronger in my faith. I believe there must be a reason why I was supposed to live. When I think back to what happened to me – there's no way that I should have survived. I should have died in that house. I am very fortunate. I don't know how I survived but it was for a reason. And that's what I'm trying to discover right now. What is my calling? To take better care of my family, my village and my people? In the past months now, I have gotten involved in different church and community groups…now I do things out of my true life, out of my heart. Not because I want anybody to like me, but because I truly believe that I could die and yet I haven't done any good deeds for anybody else. This has strengthened my faith and also my relationship with my husband and daughter."

Toetu Tauiliili is the man who tried to carry Terisa Palmer's mother to safety. Ask the fifty-three year old what he hopes his grandchildren will remember about the tsunami of 29/09 and he answers, "I just want them to remember that their grandfather was trying to save someone's life and he got hurt from it. And that if I ever see somebody in the same situation as that person, I would do it again."

Toetu is a not a man that you forget easily. He is tall and well-built, dark-eyed and handsome like the proverbial novels always say of their heroes. He speaks with baritone confidence and assurance. Toetu and his wife had a business in Leone village before the tsunami came. It was a sewing shop and convenience store. They lived at the back of the store and their home dropped down onto the rocky beach. The main road ran directly in front of the store and across the street was a little gas station, two red steel posts skirting the sole gas pump. On one side of their property was Francis Keil's house. He worked at the Post Office. On the other side was the guest house where the old women came to do their weaving every day, sitting cross-legged in a sea of green coconut leaves. The elderly women were well known to Toetu. They would call out companionably to his children as they left for school and most days, Toetu and his wife would take over lunch for them.

Toetu was home alone that morning. After the earthquake, Toetu tried to warn the women of a possible tsunami. "Me and my friends were outside, looking at the ocean and when we saw it go dry. We went over to the old ladies and were trying to tell them they should go and leave but some of them didn't take us seriously because nothing like this had ever happened before. So it was hard to try and convince them, not until they saw the water forming up into a wave – that's when they started running."

Toetu went to run himself, but as he turned, the plight of the elderly women caught him. "I saw two old ladies, they were having a hard time trying to run, so I went back and tried to help them. At that time, the wave was almost to the shoreline. The other lady, Fa'atamali'i had some physical problems and couldn't walk good so I picked her up. I kind of put her on my back and told her to hang on tightly to me. Then I was trying to run with her and that's when the first wave struck me."

The water smashed Toetu and Fa'atamali'i into the gas station, pinning them against the metal poles. The same wave took Toetu's van and lodged it on the roof of the gas station. "I got stuck between the rail and debris and it was all cutting me. I got cut on my stomach from the roofing iron and my leg was broken. I was still hanging on to Fa'atamali'i at that time…" The force of the wave dislodged the gas pump, that began spurting black liquid. "The gas pump was leaking, I was covered with blood, the ocean and the gasoline. We were under the water and I was really numb, my body was feeling so weak. I tried to hold on to the old lady, I tried, but her hand slipped out of mine. It's like she let go, she slipped from my hands."

Toetu was trapped there until the first wave receded and several men of his village helped to free him, carrying him up the road where a car took him to the hospital. Fa'atamali'i was found after the last waves had returned to the ocean. She was still breathing and attempts were made to revive her but it was too late and she slipped away. Toetu's wounds are healing, but there are still other scars you cannot see. You hear their rawness though when he speaks of the elderly woman he tried to save.

"She was like a mother to us. I'm from Leone village, I was born and raised here. I knew this old lady, her husband, her children – we all grew up together. She's like a mother figure to us. Even though we leave and we come back and we're all grown, she still treated us like we were her kids. That was the sad part about it."

There is comfort though, for Toetu, knowing that he risked all to help another. "I had an option, I could have just run – but I would have run knowing that she needed help. If I had done that, I think it would have hounded me for the rest of my life, at least I know that I made an effort, I tried."

On the night of the tsunami, the village was a deserted wasteland. The empty shells of homes cast looming shadows over the piles of debris, the crushed cars and overturned houses. Everyone had sought refuge elsewhere on the island. Everyone that is, except for the Fitiao family. They camped in their ragged house and on the cement block of their front yard they built a bonfire. Wood shelving, books, upturned furniture, all became the fuel that lit the night sky. Ropes of red, tendrils of orange and splutterings of yellow. It was a fire of pleading hope that called to Taitasi Fitiao's six year old daughter Vaijoresa. Taken from her mother's hands by the wave and still not found. "We were the only ones still staying here that night, nobody was here, they all went to their other families, other villages but we were the only ones that stayed here because I wanted my daughter to come home. I told my husband and my kids to make a really big fire so that she can see her way home, so that she wouldn't be scared in the dark."

Taitasi has eight children. Seven boys and one girl, the baby of the family, Vaijoresa. The older children are grown with families of their own and only three children left at home with Taitasi and her husband. Vaijoresa had left for school with her brother on the 6:30am bus. Then the earthquake happened and Taitasi was worried. She started walking up the beachfront road towards the direction of the school. Midkiff Elementary is an inland school that sits on high ground, but after the earthquake, the teachers dismissed the students and told them to go home. Vaijoresa was skipping along the road with a friend, both thrilled with the unexpected holiday. Taitasi saw them coming towards her. "I was so happy to see her because the earthquake had me worried. She was running and laughing. She saw me and ran faster to me. She was trying to tell me, *mom, the school was shaking!* I reached out to hold her hand and then the wave just hit us."

Taitasi had been so focused on Vaijoresa, that she was caught completely off guard. "I didn't even see the wave coming, I didn't. I was just looking at my daughter, thinking how glad I was that she was alright and I was trying to listen to what she was telling me. She was so excited. I held her and the wave hit us."

Bewildered, Taitasi and her daughter were swept in the water, clasped hands their only link. "I had her in my hand, and then we were moving with the wave. I saw these two cars lifted up by the wave and getting smashed together. I got caught right there between them, those two cars. I was still holding onto my daughter's hand, but there was a piece of steel sticking out from one of the cars and it cut through my finger all the way up to my wrist and I guess that's how I lost her."

Taitasi tried but couldn't free herself. She had to watch as the ocean rushed her daughter further and further away. "She was calling out to me, *mom please, please help me!* And I could see her body with the wave rolling, taking her and I couldn't reach her, I couldn't. When we got separated I was just grabbing and the water was so dirty and I was grabbing all over the place trying to get her. Then I was rolling in the water and I felt stuff hitting my body like rocks and trees and I knew right there that she was gone."

People took Taitasi to the hospital but no-one could find Vaijoresa. Taitasi refused to stay in the hospital and came out that night, asking of her family a difficult thing. To make fire in the midst of drowning wet. Hope burned as a family waited for a daughter that did not come home. Vaijoresa was found in the mangrove swamp the next morning, her body tenderly carried from the muddy water . She was still wearing her blue and white pinafore uniform, her schoolbag still on her back. She is buried at the front of the Siiatia home where once a bonfire blazed. Generous hearts gifted the family with a magnificent tombstone. A dark eyed laughing girl smiles from the white marble and burnished granite.

Taitasi's injuries have healed but she is troubled. "I just hate myself, I should have tried harder, I should have done more, I should have saved her. It seemed like somebody just grabbed her and pulled her out of my hands. I hate the ocean, I hate it because it took my daughter away. If I had seen that wave coming I would have known, I would have moved faster to take her away from there but I didn't see it coming. I was just looking at her smile, that's the last time I saw her smile at me. It's really hard to lose someone so suddenly like that – I didn't lose her because she was sick or anything, she was a healthy, happy child. I should have died with her…I cry a lot."

There is a struggle to find peace with the actions of the school that sent her daughter home – in time to meet a tsunami. "If the school hadn't released the kids, then they would have been fine. My daughter would be alive. They were safe up there but the school told them to go home. I'm still confused, I talk to to my son, I talk to the bishop and they tell me I should let it go and forgive, but it's hard."

She has simple words of advice for others, "Don't take anything for granted. If I had known that Vaijoresa was going to leave us this early, I would have kept her with me more and spent all that time with her. Don't take your children for granted, be with them as much as you can. Appreciate all the time you have with them. Everybody is saying to me, that I'm blessed because now I have a guardian angel in heaven – but I would much rather have my daughter here with me."

Robert Toelupe is the man who tried to breathe life into Terisa Palmer's mother. He is also the man who found Paulina Patu clinging to a closet in her drowning house and swam her to safety. Then returned to rescue a boy called Cameron and his wheelchair bound father. Then answered a cry for help from a desperate man called Stacy – trapped in a mangrove swamp. Robert Toelupe is the man who came back the next morning to dive in the murky swamp and recover the body of Vaijoresa Fitiao and several others.

Robert Toelupe is an oddity, an ex-Navy man on an island thick with US Army veterans. A soft-spoken man with kind eyes and an unforgettable smile. He served twenty years in the US Navy. Naval personnel need to know how to survive in the water – swimming, diving and water search-rescue are all things Robert can do with practiced ease. But that is not all - he was a submariner. They are some of the most highly trained people in the Navy and men must pass rigorous tests before qualifying. Tests that include being locked in a simulation control room that fills up with murky rushing water and you have to fight the panic, finding your way through the darkness, hit by debris as you try to escape, finding your way underwater. Quickly. With calm urgency. There is no room in a sub for wild panic. For claustrophia. Or slow reflexes.

Robert is a man who has seen people drown. And those images were clawing through his mind with vicious fingers as he ran down the hill and into Leone village on that ocean filled morning. Robert was running to find his daughter Anita. Heavily, glowingly, pregnant with her third child. Passing scores of frightened people as they ran up the hill, Robert had eyes only for his daughter's house in the far off distance . "I could hear the rumble of the wave, I couldn't see it yet but I could hear it and by now I'm scared, I'm trying to picture what's happening to my daughter." As he ran, Robert was caught by a young woman, Divine Falaniko, who appealed to him with tears, *Uncle, uncle, my mom is still back there, please Uncle!* Robert gave Divine his cell phone and told her to call his wife, to get to safety. He continued on, arriving at the village, just as the first wave was pulling out. "I got to the bottom of the hill and water is already up to my knees, I looked and everything's gone, I looked straight across, down the road and I see roofs and parts of other buildings all gone, floating. I see Toetu Tauiliili's house, normally there's a twenty foot drop from his place to the water, but now the water is midway up the house, it's up to my waist as I go past. I was wading through the water to try and get to our house and I can't see any people. I figured that everybody had gone. But then out of the corner of my eye, beside a big electrical pole, I thought I saw someone move. I went closer and the first thing that came to my mind, was that it was my grandma, but that couldn't be possible because she passed away many years ago. I walked over so I was right beside the electric pole and that's when the next wave hit."

Robert grabbed onto the pole as the water washed over him. In that moment, his Navy years kicked in. "I figured from training that I could hold my breath for at least four minutes, so I hung in there and I waited until the strength of the wave went by, just hoping that I could withstand the time." A wooden plank with jutting nails smashed into Robert's leg as he stood there, wedging him to the post. Still holding his breath, still controlled, he bent to dislodge it. Once free, he then pushed up and out of the water. 'I looked around for my grandma and I couldn't see her. I could see the roof of a house flipping next to me, doing somersaults. I looked and saw my cousin's house was still standing and I remembered her daughter Divine all hysterical about her mom Isalei so I said to myself, *I may as well do some good.*"

Robert swam about eighty meters to the house, dodging floating debris. Just outside of the house, he banged into a submerged car which he clambered over. He was calling, *Isalei! Isalei!* And then from somewhere in the house he heard a very faint voice, *Fia ola...help me!* In the water-filled house, Robert's head bumped against the ceiling as he made his way through the front rooms, feeling his way through. "The water was so dark and so dirty, I was trying to feel my way, over the refrigerator, the oven, there was furniture everywhere. I was trying to feel for a body and I was still calling, *Isalei! Where are you?*"

Another wave hit then, pushing Robert through a wall and into a hallway. Everything was churning in a cramped, dark space. Like a submarine simulation room. Only there is no one ready to push a stop button if you fail the test. Doubts screamed, will this house stand? Will I ever get out of here? He was being hit on all sides by unseen things but still Robert was undeterred. "I looked in a bedroom and I saw the head of a person going under the water so I went to her. It was an old lady holding on to a closet. I don't know how long she had been hanging on there but I couldn't get her hands to open, they were so tightly clenched, it's like they were frozen stiff. I really had to struggle to pull her off."

The old woman was not Isalei. Her name was Paulina. She had been fighting for her life for thirty interminable minutes and her first words to her rescuer were indignant. *Aikae you fool, I've been calling out for help and you keep saying Isalei, Isalei! What about me?! I know who YOU are, you're that boy from Savaii, from the Toelupe family.*

Robert put the tough old woman over his shoulder and tried to swim out of the house, ripping his leg on some exposed rebars as he went past. Outside , he swam towards the road that sloped upward, where a crowd of people stood just out of the water's reach. Standing. Watching. "I was kind of mad because all those people were standing there at the edge and nobody came in to help us. I was getting tired by then and I was kind of feeling the pain in my legs so I called out, *Please, come help us!* But nobody did. We got to the shallow water and I told the woman I was going to put her down and somebody would come and take her further up but she said, *please don't leave me*. So I carried her up all the way until we were out of the water."

With Paulina safe, Robert turned back. He was still focused on finding Isalei. *Isalei! Isalei!* The water was now waist deep, in between waves. As he was wading past a two storey house, he heard another cry for help. Inside was a teenage boy, Cameron, standing at the top of the stairs. "It looked like he was going into shock, he was shaking. He told me that some people already took an old lady to safety, he thinks it was Isalei. I told him now was the best time for him to leave and run to safety before another wave comes."

I can't go. My dad was hit by the wave and I can't leave him.

Cameron's father is in a wheelchair. His son had pulled him from the rushing water and dragged him up the stairs where they huddled on the landing. Quickly, Robert helped to carry the old man downstairs and out into the water. Half-swimming, half wading back up to where the crowd stood on the hill. Standing. Watching.

Another wave roared in. People screamed as the water rushed over the village again. Robert set out swimming - again. People called out to him, *Where are you going?* He replied, "I've got to find my daughter, I need to help Anita."

Again the images flashed. A house collapses. A daughter flails in the debris. Goes under. Gasps for air, Chokes and sputters. Struggles for a hold. Something, anything. Cries for help. Please God? Is anyone out there? Will no one help me? Hope dies. Strength fades. A daughter drowns. A baby that swims in amniotic warmth will never meet his grandfather. No. Robert refused to let that happen. He swam through a drowning Leone village to the faraway bridge that spanned the dead stream flowing from the mangrove swamp. That's when he saw his son-in-law Shaun Salavea, running along the high ground. He called out to him, *Where's Anita?* Shaun replied, *She's okay, she's okay, she went to work early at six o'clock.*

"For once in her life, my daughter had actually got up and left for work early…I don't know how to explain how I felt when I heard that…to know that she was safe, that she was alright…even now when I think of what could have happened to her if she had been at home…but she was safe. And hearing that, it just gave me strength."

With his energy renewed, Robert turned his attention to others trapped in the tsunami. From far away in the swamp he could hear a sonorous car horn blaring. Calling. He swam to the mangroves where a truck was upturned on its nose. Everytime the water surged inland, the horn would go and then stop as the water dropped. He tried the doors but they were locked. He peered through the tinted windows but couldn't make out the occupants. He had to dive under the water to force the sliding window open and push his head into the car. "The water was gushing out of the window and there was nobody in there. The car engine was still running and the water was shorting out wires so that everytime the wave came in, the horn went off. I turned off the car and pulled myself out. As I was wriggling out, the water went out and the car fell over. I fell out and almost got crushed underneath it. I was pretty tired by then, so I thought, *that's enough, three people is enough Robert.*"

Robert's injuries were starting to slow him down. It had been a long morning. He swam with less strength now to another two storey house that edged the swamp and tried to pull himself up. A group of teenagers stood and watched him. *Wow man, we been watching you swim all over the place. You swam over there and got that lady, then you swam over there and got that man and now you're all the way over here...wow!*

Too tired to answer, Robert just sat and nodded. But the day's work was not yet done. The boys told him there was someone else stuck out there in the swamp, someone who they could hear calling for help. "They were all telling me, *there's somebody out there.* I told them I had already checked out the car and it was empty but they insisted they could hear another voice."

Just then, another massive surge pulled in, filling up the area with water again and as they all tensed, a cry came. It limped over tangled trees and crept wearily through mangled wreckage. *Auoi...help...*

The boys were triumphant – *See! We told you! You have to go help them, come on man, you can do it! You're Navy!*

"I was real tired, I told them they had to come with me because I wasn't sure if I had enough strength to carry anyone. They looked real scared when I said that."

No, we can't swim good...no, what if another wave comes and takes us...no, we have to stay here where it's safe.

Robert was patient with their hesitancy, reassuring and calm. He showed the boys how to jump in the water from the second floor. He told them to stay close, to climb up a tree if another wave came. The boys listened and followed him into the swamp in search of the weary voice.

"We found Anita's brother-in-law Stacy, stuck in some trees. He was cut all over his body. He has really long hair but parts of his scalp were ripped off. He had only one section of hair left down the centre of his head because his hair had gotten caught on something when the wave took him and it just ripped right off. Then he was thrown up high into these trees and when the water went down, he was left hanging by his hair on the other side. He's a really big guy and so his weight ripped the other section of his scalp off. He was hurt pretty bad. It took a lot of us to pull his loose from the trees. We sort of floated him out to dry ground."

Stacy had ingested a lot of water during his journey in the waves and his stomach was hugely bloated. "I told him that if he didn't throw up, he would be in really bad shape. *Throw up, you have to throw up.* So he did and all this stuff was coming out of him – there was mud, rocks and pure white sand. I don't know how all of that got in there!"

Others came to help and Robert left Stacy in their care while he made his way back to the front of the village. "It was about nine o'clock when I got back there and there was a small crowd of people gathered around this broken house, they had an old lady lying on a table, everyone was very sad. They said she was dead." Robert asked if the woman had been checked by a doctor. *No.* An emergency medic? *No.* " I told them, well we can't give up then, we have to keep trying until a doctor arrives and tells us to stop." Robert began administering CPR. There was no response, but he refused to give in to the exhaustion that was beginning to numb his hands, his chest. The crowd looked despondent, pitying the man who would not admit defeat. "Then the old lady started throwing up and coughing. You could see her family starting to have hope. A doctor came and checked her with his stethoscope, he told us *we got a heartbeat! Keep going.* Finally, he said it was safe to move her. So they got a mat and put her in it and took her away to the hospital in a car."

Finally Robert allowed himself a moment to rest, to breathe. It was a moment where his family hustled him to the hospital, worried about the cuts and puncture wounds all over his legs and hands. But after he had been treated, Robert insisted on returning to Leone to keep helping with the search. And at five the next morning, he was up, getting his gear together to go back to Leone. His wife Sinu'u was horrified. *What do you think you're doing!?Where are you going?!You're injured, you can't go back there, you're going to get everything infected.* Her disapproval did not slow him.

Morning broke over a broken Leone village. Sunrise cut swathes of light through the misery. There were still people missing. The men of Leone banded together to comb through the mud filled land. Maiava was there. Iuliano and the General were there. And many others. The search effort was organized by Robert and US Army veteran Faamanu Puaina Sauileone. There were thick pools of water that needed to be searched. Robert got his goggles and diving gear from the truck. Sinu'u was not happy. *What are you doing? Where are you going? Why do you have to be the one to go in that water?! What about your cuts?* Robert began systematically swimming through the water, eyes straining to make out the murky shapes all around him. He searched for two hours before he found her. The little girl in a blue and white school uniform. "I had gone over that area several times and I saw what I thought was two branches lying there in the dirt at the bottom so I kept going over them. I came back to do one more sweep and the tide had come in and out by then, so the water was clearer and that's when I saw the blood. Blood in the water. At first I thought it was ink, I couldn't work out what this stuff was. And then I reached down and grabbed what I thought was branches and my hand sort of sank into something soft and that's when I realized. I went down and got her. I was pulling her up, holding her in my arms like I was carrying her, like she was asleep. That's when it happened. Her head sort of turned a bit and I looked at her and her eyes…they were wide open, they were looking at me and I was so startled. I jumped, I screamed under water. I let her go. I came up for air and tried to get calm. People were standing there, they had noticed the blood, they were waiting for me. I went back down and I got her and closed her eyes gently and came back up with her. I came up and called for help and somebody took her from me. Then all of a sudden, all the strength went out of my body and I couldn't stand, I couldn't move. I just lay there and tried to make sense of all this. She was so young, she still had her school uniform on, she wasn't badly injured…I just tried to make sense of it all."

Robert continued with the search teams until night came. He searched the next day and the next. Until all but one was found – a young boy named Columbus Sauiluma. "Everyday I was there. We went in those mangroves, crawling and cutting through the trees, we never found the little boy. I wonder if he was taken out far out to sea."

For six weeks after the tsunami, Robert could not sleep. "Every time I closed my eyes, this little girl kept popping into my head. I couldn't shake her eyes from my mind, every time I went to sleep I would get bothered by these images, I would see this little girl. My whole family knew it, they would ask me in the morning, *dad, did your friend visit you again?*" Robert's daughter came to tell him that the little girl's mother, Taitasi, wanted to meet him. She wanted to talk to the man who had found her child. Robert refused. He was uncomfortable, what would he say? What would he do? How could he face her? When her dead daughter's eyes were haunting his dreams?

Then one day, Robert and his wife were driving past Taitasi's house. "I saw her standing by her daughter's grave and I couldn't put it off any longer. We pulled over and I went and talked to her. She thanked me for finding her daughter, for giving her back to them. I told her, I wish it had been the other way, that I could have found her alive, I didn't want to find her dead. I told her I was sorry."

Taitasi had something to give him. A bundle of papers she had found in a box while cleaning up a tsunami wracked house. Papers that had Robert's family name on them. Papers that contained vital genealogical information about Robert's family tree and their ancestral rights to a block of land he had been fighting for in court for over ten years. "She told me that when they brought her daughter back, she had asked for the name of the person who had found her. We both didn't realise but we're actually related. When she read the names on the papers she found, she recognised them and wanted to give them to me."

That night, Robert fell asleep reading about his family land titles. It was a sleep without dreams. A sleep without the staring eyes of a little girl found in a swamp. Sinu'u shook him awake the next morning after his first good night in nearly two months "After that day, after I met the mother, I've never seen this little girl in my dreams again. And because of those papers, we are making good headway with our land case. My wife said, maybe that's the reason why I kept seeing that little girl."

Robert is relieved to have found some closure. He is grateful for his Navy training. "In situations like that, I think the training kicks in. Mentally I was prepared to deal with whatever the waves were like, I was prepared for what the waves could do. And then, afterwards, because of my search-rescue work, I knew that time was of the essence. Sometimes us Samoans we tend to react with humour when things are bad maybe because we haven't had enough disaster situations like this one, we take it lightly and it makes you mad. I yelled at some of our *matai*, our chiefs because I thought they could have done more. And during the searching, I yelled at some of my relatives who were just standing around. That first day some guys had set fire to one whole side of the mangrove and it was really hard to breathe in there and we had to cut the search short. The very next day, we found one of the missing women not far from that spot and when we looked at her, it looked like she had crawled for awhile, trying to get out. Time is of the essence. In Samoa we live very relaxed, We don't see extreme hardship, so to some people, the whole thing was like a big fun adventure…but you want them to understand, *hey, we're trying to find these people, we're trying to help them before it's too late…"*

Robert Toelupe is the man who kept going back into a tsunami to save people. People call him "the brave man who ran into the wave" but he disputes that. "No, I didn't go in because I was brave. I went in because I was afraid. So afraid. Louder than the sound of the waves, was the thumping of my heart beating in my chest, and ringing in my ears – that's how scared I was. I don't know how to explain the fear I felt, knowing that my daughter and my grandchild were in danger. I've seen people drowning and just remembering what goes on with a person when they're drowning, that scared me. I felt so scared for my daughter and I felt so scared for everyone in the village. The fear gets to a point where you're not worried about safety. I wasn't prepared to see my children drown, the whole time I was praying I wouldn't see that happen to my daughter. I ran in because I was afraid."

On the Land: Amanave

Amanave is another coastal village that had received earthquake and tsunami awareness training in the past. Many like forty year old Vaea, remembered the workshops well, "We did training in our village with the Marine Wildlife Office about things like tsunami and warning signs so we knew what to look out for."

Once the violence of the earthquake marked it as 'different' from others, the high chief of the village came out with a loud speaker, instructing everyone to head for higher ground. *Don't take the children to school. Prepare to move to the mountain inland. Everyone prepare to evacuate!*

Still, it was difficult to believe that there was anything to be worried about. The grey sea lay quietly, the gently rippling surface seeming to ask a wide-eyed, innocent faced question – *Who me?! What did I do?* And so, while some bustled their young ones into big brooding trucks to drive inland, still others, chose to do nothing. To wait and see how the day would play out.

"We sat under the *niu* (coconut tree), me and the minister and we were just talking about how strong the earthquake was and if anything was going to happen. We just sat there and looked out to sea and it was just coming in gently like usual…"

Then several things happened at once. Vaea saw the ocean recede "You could see the bottom of the sea, the ground…it was so ugly because you could see everything that was usually hidden under the water. I was surprised because I thought the tide had suddenly gone out so fast…I was thinking how ugly our village looked without the ocean in front anymore…"

As a little crowd of spectators stood staring, puzzled at the seafloor, the minister's wife ran from the two storey house beside the church. *There's a warning on the radio that everyone needs to evacuate because there will be a big wave in fifteen minutes!* The words were tumbling excitedly over themselves and into the morning air - as the water began rushing back in over the reef.

Everyone ran. Vaea ran to her house for the children. "I didn't care about anything, any household goods or belongings. The only things I grabbed were my kids. I told my older daughter to grab the little one and to run to the plantation. I could see the first wave coming, it took our *faleo'o* (hut) by the beach, it was tossing it about. I ran and shut the front door, I didn't want to look at it coming. I grabbed my other daughter and we ran out the back."

The waves swept in more powerful on their second and third onslaught. "When the second wave came we started to sprint, we all ran so fast. We ran back further and as the wave came, it was like something was chasing us and the sound it made was like an engine growling. We got high up on the hill and standing there we could hear the sound of the third wave was different – because by then it was full of all the rubble from our houses."

Vaea's neighbor, fifty-three year old Mine Tupou remembers that breathless run very well. She ruefully admits she had no idea where to head for when people yelled for everyone to run to the plantation. "My husband tells me every Saturday, *let's go up and work on the plantation* and I tell him *you go first and start*. And I never go. And so when the tsunami came, I didn't even know how to get to the plantation! Because I never go there to do work with my husband…"

Mine Tupou is diabetic and as she ran barefoot over sharp stones, she was worrying about the damage to her feet. Months later, her tsunami experience has left her with some important reminders. "I will never forget trying to run for my life. There is an elderly woman that lives next to us, Fa'aope. She's more than eighty but she doesn't look old, she is still fit and strong because she is very active. She does everything for herself, she's always cleaning her yard and she walks everywhere. She's not like other old ladies who always say *I'm so tired.* I was surprised to see her strong and well after the tsunami. I thought her kids must have taken her in a car or something but she went with her walking stick up the hill. They carried her part of the way and she walked the rest. I was so amazed. That is why, now when I eat, I think about it first and I'm watching what I eat because we don't know what will happen in the future and what we will need to do to survive. I am being more careful with my food and I'm going with my husband to the plantation, that way I will be fit to run next time.. "

When the waves finished, the people of Amanave village prayed. "When it started to calm down, that's when we sat down and prayed to the Lord to thank Him for letting us live. Even though there were no houses left, but we thanked God that no-one in our village had died from the young to the old. There were deaths in neighbouring villages but not here, because our *matai* did the warning early."

On the Land: Poloa

At the very tip of Tutuila is the village of Poloa. It is a narrow strip of coast at the foot of a steep cliff. There is only room for one car coming or going down into the village. The most distinctive feature about Poloa is the noise. It is difficult to hear yourself think here. The surf crashes angrily, repeatedly on the jagged line of rocks a few feet from the shore. The roaring never stops. It is not an ocean for children to play in.

Poloa is a small village. A cluster of solid homes, a church, a school. Thanks to American dollars though, it is a far from meagre place. The school is new. A stunning steel beamed building with criss cross rafters. It had a reading room, a cluster of classrooms, a bathroom block. Curved steel chairs and desks. Thirty-three year old Simao Masoa worked there as a teacher's aide. She had been living in Hawaii but decided to return home two years ago to be with her parents, her grandmother, two sisters and their children.

Simao was asleep when the earth shook. "I thought somebody was playing a joke on me shaking my bed. It didn't even dawn on me that it was an earthquake. But then it kept getting harder so I jumped out of bed."

Simao's family ran outside. Her grandmother was distressed as the earth continued to rock. "She was panicking, she was thinking that because the earthquake was going on so long that maybe the earth was going to crack open. I was telling my dad to watch the ocean for any changes, but my grandma was saying, *no, don't look at the sea, look down on the ground!"*

As the earth seemed to still finally, lines of children from the elementary school started streaming past the front of their house. The teachers were evacuating them to where the road curved up the mountain. Simao's father called out to them.

"Hey, where are you kids going?"

"We're going home. No more school because of the earthquake."

For reasons he himself cannot explain, Simao's father urged the children to move faster.

"Don't walk – you should run. Go on, run up the road fast!"

Encouraged, the children let loose with whoops of glee and started running and shouting as they danced along the sand scattered street. The teachers who were bringing up the rear, called out to contain them and gave Simao's dad disapproving looks as they chased after their young charges.

"I don't know what made my Dad tell them to run, but it gave them pretty good time for them to get out of there because we didn't know it but we were counting seconds at that time..."

Simao and her family had been standing outside for about ten minutes when it happened. Something that *never* happened in Poloa. Silence. Total and complete silence. The ocean that never stilled – stopped roaring. It was an eerie, unearthly quiet.

"Everything stopped. Everything. There were no waves. No noise. Everything went quiet."

The ocean dropped. And because of their stones throw nearness to the reef, Simao clearly saw the coral rock shelf as it was revealed. "The water didn't suck back. It just dropped down." Simao yelled for her dad. He was quick to respond. "He turned around and said *oh my God, get inside the truck right now.* Our truck was parked in the garage. I ran and got the car keys and shouted to my grandma, *let's go!*"

As they were climbing into the big truck, Simao remembered their neighbour. "Next door to us lives a paralysed man, he had a stroke some years ago and can't walk. I saw his son outside the house and I called for him to get his dad, *get your dad, get him out of the house now!*"

The father is a large man. His son dragged him outside in a blanket, Simao and her dad helped to lift him into the vehicle's back passenger seat beside grandma. Simao was driving with her father beside her. Her mother sat in the back of the truck with Simao's sister. They reversed frantically and began speeding along the road that ran parallel with the fast returning sea. As they drove, Simao's mother was screaming out to every house – *Galu lolo! Tsunami! Run, run!* They stopped beside the Reverend's house when they saw his car still in the driveway. *Reverend, get out, get out, the wave is coming!* They stopped at another house that had not evacuated but the woman refused to leave. "My mom kept telling her to get in the car, to come with us because the wave was coming, but she said *no, you guys go ahead, there's nothing happening, don't worry, I'll be fine.* So finally my dad told me to step on the gas."

They met two other cars going in the wrong direction and Simao honked her horn, yelling for them to back up and head for safety. They could see the ocean as it began piling up and over the seawall. "The ocean was climbing up as it built – it wasn't like a wave, it was a rush of water coming in all at one time, it got to the seawall and it just climbed up and rushed over it." At the end of the road, Simao accelerated to turn up on the mountain and as the car jerked, Simao's mother fell out onto the road.

"I heard my sister bang on the roof of the truck, yelling to me *mom fell out!* I looked in the back and the wave was coming so fast. I stopped the truck, got out and ran to get my mom. She stood up and she kept telling us to keep going, don't wait for her. But I got her back into the truck and accelerated again up the hill. The road was full of the school kids that had passed in front of our house, they weren't even halfway up. I had to honk at them to run faster, I don't think they knew how quick the water was coming after us."

At the top of the hill, Simao pulled over and the family looked back at the ocean. They couldn't see around the corner into the village but the power poles beside them were being pulled towards the direction of the ocean as further down and out of sight, something was tugging on the wires. Something that was consuming their homes. Standing there, Simao remembers a strange sight. That of the wave rebounding and heading back out – towards Upolu in the far off distance. "I saw three ripples in the ocean going back towards Aleipata. Three ripples that built from here in Poloa. And the ripples were full of debris, all this rubbish that was being taken with it. And I realized that was our village. All that stuff going out in the ocean ripples, was our houses and our stuff and our cars and everything."

It didn't take long to wipe out Poloa. It is after all, deceptively easy for nature to erase man when she feels like it. Simao came down the mountain road with her father before even half an hour had elapsed. "There was nothing left, not even a wall…" At the end of the village only the elementary school building was standing. Without sides or innards, an empty hulking shell. Months afterwards, one can still find the tattered remnants of primary school readers, scattered on the shore rocks.

One woman was killed by the tsunami in Poloa Village. Perise Sula, who did not want to leave her house when Simao and her family called her to join them. Thanks to past earthquake drills and swift response to the calls to evacuate, there were not more deaths. But there will be no more Poloa village down here where the crashing surf never sleeps. It has been designated a V-zone, unsafe for rebuilding, the tsunami risk too high. Children in yellow and blue pinafores will no longer skip to school along a sun drenched shore.

How did the Masoa family know to react as quickly as they did that morning? "We lived our whole life growing up in Hawaii where there were a lot of trainings done about tsunami. Also, we would have regular evacuation drills at the elementary school, so I guess that's why we knew what to look for and what we had to do."

December 2009.

On the plane ride home to Apia, I think of many things. I think of dolphins at Fagasa. And the unbearably loud surf at Poloa. I think of 18 foot hammerhead sharks caught in Pago Harbour. But most of all, I think of Leone. I think of a father who runs into a tsunami fearing for the life of his daughter. Who does not find her but then returns again and again in search of others. I think of a mother sitting beside a fire of hope in a decimated village, watching, waiting for a daughter to come home. I think of the Tsunami Club. I think of heroism - defined as exceptional courage in the face of danger. We usually think of heroes as being those who 'save someones life'. But there is still another kind of courage. Another kind of hero. Those who didn't save the life – the elderly woman, the little girl. But in the face of insurmountable obstacles they still tried desperately to overcome. Returning again and again as the hungry sea fought against them. Another kind of courage. Those who 'want to die' because of their grief and loss, but who choose everyday to keep living.

Yes, there are heroes in American Samoa. I pay tribute to them and hope that with time – they will find peace.

22
Tonga

The earthquake that shook Samoa on Tuesday the 29th of September also rocked Tonga – but on Wednesday the 30th of September - thanks to the International Dateline which runs directly between the two island groups, putting Tonga always a day ahead of Samoa. The Kingdom of Tonga consists of 169 islands that are clustered into three main groups – Tongatapu, Ha'apai and Vava'u. The Pacific Tsunami reached two of Tonga's islands within five minutes of the earthquake – the two closest to Samoa – Tafahi and Niuatoputapu.

Thirty minutes after takeoff from Faleolo Airport in Samoa, you can see from the window on the starboard side of the plane – two islands, 205 km directly south of Savaii. One is a sharp cone, a mountainous peak jutting out of the blue, Tafahi. The other is a low lying, 4 km sq island, Niuatoputapu, *Forbidden Island.* Legend has it that centuries ago when the Tongans 'invaded' Samoa, Niuatoputapu was their launching point. The island is closer to the Samoan group than it is to the rest of Tonga and the local dialect there has many words similar to those of their Samoan cousins. The rest of Tonga sometimes refer to the people from Niuatoputapu as Samoans – and thanks to our sometimes turbulent history – such a reference is not considered a complimentary one!

Reaching Niuatoputapu is no easy task. From the main Tongan island of Vavau, one can catch a ride there on a Twin Otter – but be prepared to stay there for a week. A boat makes the three day journey from Tongatapu once a month. There are approximately 850 people living on Niuatoputapu and fewer than a hundred on the more inhospitable island of Tafahi. The main sources of income for those living on both islands include mat weaving, fishing and remittances received from relatives residing overseas. The majority of Niuatoputapu inhabitants live near a broad lagoon on the northern side of the island while the few on Tafahi reside inland on the cliffs in a village only accessible by foot stairs.

The Tongan word for tsunami is '*peau kula'* which directly translates to *red wave.* There is an oral tradition among some elders in Tonga of tsunami big enough to cover an island at one time and researchers are currently studying boulder deposits on the islands of Tafahi and Vavau that are more than 25 meters above sea level and 450 meters from the shoreline. Allan

Morton of the Central Arizona University writes that the tsunamis which may have carried in these giant rocks "were probably generated by local events such as volcanic eruptions, earthquakes and submarine landslides produced by activity in the Tongan subduction zone. Local traditions say the boulders were placed on Tonga by the ancient god Maui." Morton goes on to recount the earliest written account of a Tonga tsunami, discovered in the personal history of a Tongan who was in Vavau in 1919. "According to his story, a large 'tidal wave' struck the shoreline east of Ha'alaufulu and climbed nearly halfway up the cliff while fish, sharks and sea life of all kinds were thrown on the tree tops. From his description, it appears as though the run-up could have been as high as 20 meters or more. A small village of beach houses was destroyed but no one was killed or injured."

On 30/09 the *peau kula* hit Tafahi first which acted as a breaker somewhat for Niuatoputapu. The wave was immense, eclipsing 22 meters. Many survivors on Niuatoputapu told of feeling a second earthquake after the initial 8.3 shake and some scientists speculate that could have been the impact of the tsunami hitting Tafahi first. The wave raced on to Niuatoputapu, completely overrunning its northern tip, Hikuniu point and wiping out an entire forest of trees, leaving them uprooted, stripped of bark and scattered amongst coral boulders. Beaches on the southern shores were covered for distances of 600 meters and flow depths of 10 meters. Thankfully these coasts were uninhabited. Niuatoputapu's three villages were not spared – the ocean ran in for 260 meters at Falehau, 143 meters at Vaipoa and 585 meters in Hihifo. The island had a total of 255 private residences of which 79 were totally destroyed and 56 partially damaged. Most public utilities and government buildings on the island were wiped out as well as water and sanitation systems. Every fishing boat on both Niuatoputapu and Tafahi was smashed.

Eight people on Niuatoputapu were killed. On the nearby tiny islet of Hunganga, the caretaker of the Palm Island resort had no chance as the flat, low lying islet was completely covered by the waves, taking the death toll to nine. A local television crew travelled to Niuatoputapu in the month after the tsunami to film the stories of survivors. People told of three distinct waves. The first sent people running for the hills but it was not the killer. No, the third wave is the one which stood "taller than the tallest coconut tree", smashed through cement and brick buildings and tumbled heavy duty trucks like featherweights. Fisherman Ipeni Vaikata was just putting his boat out to sea when the tsunami came. "I was standing in the ocean beside my fishing boat when it first hit. I held on to a rope and the wave took me inland with the boat. I hung on tightly. I looked back and saw another wave coming. It looked like white sand, but the next wave after it was so huge and black."

Father of two boys, 49 year old Malakai Mini of Falehau, "I looked at the sea and it was different. It was bubbling like it was boiling and then I saw something swollen coming out of the bubbles. It was a wave but not like any wave I have ever seen before. I ran into the house where my wife was cooking the breakfast. I turned off the stove and called to my family to run. I carried my five year old and my wife took my three year old. I grabbed my bush knife and we ran. That's all we took. The wave destroyed our home but at least we had our children and a bush knife."

In Hihifo, Lomanu Losalu saw a wave that was higher than the trees. "The *peau kula* took me. I was very afraid. I came up out of the water and tried to swim. I saw the Catholic church with the picture in the window of Mary. I called to Mary to save me. Then I swam with the wave and I was saved – Mary saved me."

Also from Hihifo, 54 year old Maleko Onesi was fleeing the ocean with his family in a pickup truck when the first wave caught up with them, pushing the vehicle along in a swirl of water. "My son called out from the back that his brother had fallen out. I got out and started going through the water looking for my son. I was very frightened that he was lost. Another wave came. There was so much rubbish in the water. Then my missing son called out to me – he was hanging on to the roof of the truck. We were all saved."

68 year old Uluaki Losalo is in mourning for her mother. "My mother was 87 years old. She is not strong. When the *peau kula* came we put her in the neighbour's truck so she could ride to safety while we ran to the bush. The wave caught the truck and my mother died."

Fokikovi Lefai remembers her escape from the wave as if it were yesterday. "I will never forget that day. We were so scared. The wave was black. We ran and we thought we had run far enough so we stopped and looked back but the wave was still coming, so we ran some more. It just kept coming and the sound of it was so terrible. Now, when I hear loud noises, I am afraid. I am trying to forget the *peau kula* but I can't."

At 3:15pm on tsunami day, the National Emergency Management Office (NEMO) of Tonga chartered a special flight to fly over Niuatoputapu to carry out an initial assessment of the damage. Shortly afterwards, relief supplies from NEMO and the Tonga Red Cross were dispatched aboard the Tonga Defence Service patrol boat, Voea Neiafu, arriving at Niuatoputapu eighteen hours later. The hospital on the island had been destroyed in the tsunami and a makeshift clinic had been set up in the LDS chapel. The nine dead had already been buried side-by-side in the local cemetery. The Tongan government arranged for a charter flight to take four critically injured survivors to Tongatapu. Kalala Lefai, aged 58 from Hihifo was one of them. She shared her story with Linny Folau of the 'Matangitonga News'. Kalala had been at home with her 60 year old husband Sosefo and their two grandchildren, aged one and two years old. Sosefo had suffered a stroke some years before and so was not mobile. When the wave came, the family caught a ride in a neighbour's car to try and get to safety inland. "As we were fleeing the vehicle broke down and some of the passengers who were cramped inside got out and ran but we remained. Then the car started again and we made another dash for it but then the car broke down again. I did not get out of the car because my husband was limping and there were our two young grandchildren that we were holding, so I stayed still because I was not going to leave them."

The wave careened into the car and rolled it. Kalala was thrown out of the vehicle. "The last thing I remembered was being thrashed around under water together with the debris of timber, trees and rooftops and I thought this was it, I am going to die," But the wave deposited Kalala in a tree which she clung to. The rest of the family was not so fortunate.

"I heard my husband shout for help but I just couldn't budge because I was injured. How I wish to God I had an almighty strength so I could rescue him." When the wave retreated, Kalala was rescued from the tree but the bodies of her husband and both grandchildren were not recovered until much later that day.

Another survivor at the same hospital was 65 year old Sulifa Losalu. Her 69 year old husband Heneli died while trying to help her escape the red wave. The mother of eight children had just returned home from early morning Mass when the earthquake struck. Heneli yelled to Sulifa to *get out of the house!* She ran from their home but then remembered her little statue of Mary and turned back to get it along with her handbag. It took less than five minutes – but it was minutes too many. "I heard a strong roaring sound like a machine and Heneli shouting to me, *run, there's a huge wave coming!* My husband waited for me and as I ran out of the house, I saw the wave which was above the coconut trees, coming towards us. It was a horrifying sight. I ran and my husband kept yelling from behind me to run faster. I looked back at him and that was the last I saw of him. The wave struck him and then me. I was underwater, swallowing water and being thrown around like a piece of stick, hitting debris and trees that came in my way. I just remembered praying to Mary to please help me."

Sulifa was flung onto a rooftop and somehow managed to hold on and emerge from the water so she could get air to breathe. Her son Osika found her there and took her to safety. "He carried me down and the ocean was still up to our waist but the waves had subsided. It was not until 4pm that my husband was found dead, kneeling down, at another area. I was shattered. I feel this aching hole in my heart after losing my husband because I know he could have made it – but he waited for me to come out of the house before running. I love him so much and being apart from him is devastating." Sulifa was determined to return to Niuatoputapu as soon as her injuries healed. "Although there is nothing there and no house left, I want to be with my husband who is buried there."

It was not easy for relief supplies to reach Niuatoputapu because of its remote location. On the 11th of November, 2009 the HMAS *Tobruk* travelled to the island via Apia, arriving in the early hours of Friday 13th of November. The naval vessel carried vital supplies for the people of Niuatoputapu from Australia. On board was the well-known Samoan filmmaker Galumalemana Steve Percival who had been commissioned to make a documentary on the 29/09 disaster. Galumalemana met with the Tongan government representative on the island, Sione Peau Haukinima and then went on to interview eighteen survivors about their experiences as well as several scientists there as part of the team studying the tsunami. Galumalemana wrote of his visit, "Most of the women on the island are involved in craft activities mainly weaving of pandanus. Different kinds of woven mats and *ta'ovala* (ceremonial wear) are ordered by stores in Nukualofa and then sold overseas, usually on a commission basis. Men assist the women with this work by preparing the pandanus leaves. Men are also the subsistence farmers and fishers…there has been a decrease in fishing activity since 29/09 as many still fear the sea."

Thankfully, Niuatoputapu food crops were largely untouched by the tsunami as they are primarily situated some distance inland. However, the water supply was severely damaged. Hihifo and Falehau's main piped supplies were cut for a week while more than 50% of their water catchments were washed away in the tsunami. Tonga has a land ownership structure similar to that of Samoa and so survivors were able to find refuge inland and families have begun rebuilding. However, Niuatoputapu's road to recovery will be a long and challenging one. One government report on the situation concluded –

"At times nature and Acts of God deliver us blows from which we cannot easily recover. In the face of these trials , the way forward will depend primarily on the resilience, strength and ingenuity of the Tongan people and the wisdom of their leaders and rulers. International agencies, NGOs and governments can prove to be valuable friends and allies in such times, in their roles of supporting the Tongan government and population."

February 2010.

I cannot sleep. There is a beast that crashes through my dreams. It clambers up out of the sea, breathing smoke and roaring with demented fury. I could handle it if it were only chasing me. But always in my dreams, I am with two of my children. The baby. The one with the most piercing scream.And the eight year old. The one with the most gentle heart. Who cries if a teacher raises their voice at someone else in the class. The ocean comes and I am running. Carrying the baby. Trying to pull the hand of the eight year old. And we are going too slow. And they are both crying and looking to me for hope. And I can't run fast enough. I'm not fit enough. I'm not strong enough. I am a bad mother. And we are all going to die...wake up, wake up. Quickly, wake up.

Another night. Another dream. I receive an excited phone call. "Come quick, they are pulling survivors from the harbour waters. They thought they were dead but in the ambulance, wet and soggy people are waking up. It's a miracle! Come to the tsunami ward at the hospital and interview them!" I go. Yes, it's true. The tsunami ward is full. Families are smiling. There is desperate happiness choking the air. But it all starts to go horribly wrong. The ocean survivors turn feral. They snarl. They bite. Families scream. And it turns into one of those sickening B-grade horror movies where zombies are running around ripping chunks of flesh out of terrified people... Wake up, wake up. Quickly, wake up.

Another night. Another dream. There isn't an ocean in sight. Or any corpses. What a relief. I am standing on a mountain. Looking out over a myriad of houses, roads and shops. All seems well in paradise. I am cautiously optimistic. Could I be having that most elusive of things...a GOOD dream? Ha. What a ridiculous notion. If my brain has exhausted the tsunami scenarios, there are plenty of other natural disaster options. Behind me the top of the mountain explodes and a river of lava runs past me. Around me. Seething red and orange consuming the green land. Even asleep I am tinged with disbelief – what...first tsunamis, now volcanic eruptions, what's next? A tornado ripping down Apia's main street? A plague of locusts? A blizzard? I scream but no one can hear me. I watch. The lava flow covers the houses, the roads, the shops. People are drowning in molten fire. It is so hot that my useless tears hiss and steam in a heartbeat of heat. Children are crying... Wake up, wake up. Quickly, wake up.

In the light of day, dreams seem powerless. Relive them, recount them and they even sound absurd. But when night comes and you finally fall asleep at two in the morning because you can't avoid exhaustion a moment longer, then the dreams hold you captive once again. And as you turn to run from yet another tsunami or from yet another swarm of decaying zombies – you ask yourself – if this is what MY nights are like, what's sleeping like for a REAL tsunami survivor?

23
The broken hearts and broken spirits.

"Most victims have received the food, the clothes they need and that's great, but how do you mend the broken hearts? How do you mend the broken spirits?"

This is the question Samoa, American Samoa and Tonga must answer. This is what the counselors, the psychologists, the social workers and the community support networks grapple with. A year after the tsunami, houses will be rebuilt, debris will be cleared and the ocean will be calm but there will be other tsunami remnants, not as easily seen. Like grief, guilt, anger, hopelessness and fear.

There were many who took on this challenge after Samoa's tsunami. Elena Peteru is a counsellor working at the University of the South Pacific campus in Apia. She was the first counselling professional called in to assist with the survivors in the National Hospital the day after the tsunami. Local survivors had their extended families and pastors on hand to offer support while those visitors and tourists caught in the tsunami did not. On that Wednesday, Elena was given a list of names. The list included couples who had lost children, friends who had lost several of their travelling companions, and people who were badly injured. "I was there mainly to let them know that there was somebody there, a local, that was looking out for them, working together with their own embassy people, just trying to help, to ease what they were going through, maybe bring them some comfort. People asked about help with sending emails, getting photos taken, phone numbers and asking for just basic things like clean drinking water and toiletries. There were a few who weren't ready for anyone to offer support, and when I sensed that, it would be a matter of moving away and saying, *I'll be back.* One woman was really upset with the media coming through, she was yelling…another man kept calling out that the loss of his child was all his fault. I took time to visit with our own people as well, some had quite horrific situations. A couple who had lost their three children and another couple who had lost two children and the husband's parents as well. At times, there was a sense of futility – when a man asks you, *what can **you** do, I've lost my son...*what do you say to that?" Elena worked with survivors through to the following Monday when a response team was officially organized and she became one of the co-ordinators of the tsunami field work.

A Psycho-Social Response team for Samoa, was set up on the 5th of October, 2009 under the National Health Service with technical assistance from the NZ MOH who sent a team of 14 counselling professionals. A range of local government ministries were involved as well as different NGO's. Outreach teams were organized, made up of a registered Mental Health professional (psychiatrist, psychologist, mental health nurse), social workers and counsellors – this included, local church ministers, nuns, theological students and their wives, grief counselling and community support workers. Members of the PSR teams were provided with training to 'enhance their capacity to respond to the needs of the affected people.' Additional training was undertaken by the Piula Theological College and the Malua Theological College to ensure their faculty and students were well-equipped to help in the disaster field. The Catholic Archdiocese of Samoa also participated in training for the catechists, Moamoa Theological College, members of several Catholic Samoa organizations and other volunteers including CARITAS Samoa.

Again, Samoa's close communal spirit was a vital tool in the work as pastors, women's committees and village councils were called upon to assist with the PSR. An effort was made to ensure that each team had a local representative for continuation and sustainability of services. Outreach teams met daily before going out into the field. The aim of the initiative was to meet with every single family affected by the tsunami to assess their needs – emotional as well as physical.

One counselling information sheet for those impacted by the tsunami explained, "Many people will experience acute stress. This is a normal reaction to an abnormal situation. Most people will manage and when social needs are met, distress will subside over time. However, there are some people that will have more profound reactions, particularly later on sometime after the event, and will require assistance. Those who have lost loved ones, or have been seriously injured, will need particular support and care." The handout goes on to detail some of the normal reactions – emotional ones like crying spells, sadness and apathy. Cognitive reactions like nightmares, poor concentration, self-blame, disorientation, confusion. A range of physical reactions; difficulty sleeping, upset stomach, tension, 'jumpy', shakiness, loss of appetite. Interpersonal relations might be affected with people experiencing distrust, withdrawal, conflict and irritability. For many, anger and blame may be common and some could become angry at God."

The initial PSR report was released on the 29th of October, 2009 and certainly, many of the above reactions were noted. The report represents the massive amounts of time and effort dedicated to the response by many different parties, giving both summaries and individual specifics. A selection of their conclusions, offers insight into what people were coping with directly after 29/09. The family haven organization, 'Mapusaga o Aiga' was assigned to three villages in Aleipata. In the village of Satitoa, 105 people were affected by the tsunami out of a population of 603 with 13 deaths. In Malaela where no coastal houses were left standing, 169 people out of 181 were directly impacted with seven deaths. Ulutogia, 169 people affected out of 226. There were no casualties but 85% of the village was left with no homes and the majority of families were camped at the inland primary school. The 'Mapusaga o Aiga' team noted that strong leadership was coming from the church ministers, like the Rev. Mose Lata who was working closely with the mayors and women's committee. Grief hung heavily over both villages with some of the dead still missing. The team heard many stories of acts of heroism and bravery including that of a disabled man saving the lives of his grandchildren. A concern was noted that parents were busy with rebuilding and neglecting children.

The Ministry of Women, Community and Social Development (MWCSD) covered the villages of Matafaa, Sapunaoa and Iliili. Twenty-three families were affected by the tsunami with water being the key priority and the team reported they would be providing the technical assistance required to write a proposal from the villages to procure funding to support a water project. No specialized counselling was required. Families' plantations were not affected and relief supplies had been distributed. Sapunaoa needed help with building a new *Fale Komiti* (Women's Committee House) which was the center of all village-based activities and programs for women and children. There was a concern for families living in overcrowded conditions while they awaited rebuilding of their homes. "One family has eighteen people living under one roof, another has seventeen, one has twenty people."

The Samoa Umbrella for Non-governmental Organisation (SUNGO) assessed six villages. They met children who could "only sleep in mother or father's arms" because of their fear, a teenager with foot wounds who needed shoes, another who was having nightmares every night. In Salani where twenty people are crowded into a home that originally housed eight, a nine year old cried as she told of running uphill with a group of schoolmates, staying there until 5pm that evening. A twenty-eight year old mother would need additional follow up for extra reassurance after her experience of being caught underwater with her baby in the wave "left her with great fear of a repetition." Those who needed it were referred for additional follow-up and counselling.

Others in the PSR team targeted the children of the tsunami. The children of Saleapaga for example, did not go back to their coastside school after 29/09. One *matai* explained, "The children are scared to go to the school there by the sea so we are sending them to Lepa school which is far inland. We are going to shift the school up here in the mountains because the kids don't want to go back, they are terrified because they can't get out of their minds what happened, what they saw. Because many of them were in the school when the tsunami happened and they ran behind the school fence and went up the bank and looked at it. My other children saw the children who were caught in the tsunami and died. They saw the strength of it." 'Faataua le Ola' organized workshops for different schools in the disaster area where children took part in art therapy and fun interactive games that were 'designed to relieve tension and promote the safe expression of emotions without reference to the specific traumatic experience." Solonaima Emau, Principal of Saleapaga Primary School observed, "Many children who have not been attending school since the tsunami… have enjoyed the [workshop] so much that I am hopeful it will encourage them to return to school in the temporary facilities in our new village." The principal of Vailoa School, was thrilled to see the children's response to the workshop, "I have not seen them enjoying themselves so much in class since the tsunami. During the [workshop] they had so much fun, with laughter, singing and games that I'm sure it has lifted their spirits and will improve their sense of wellbeing at school."

Of particular concern for educators and the PSR team alike – were how senior students would cope with their national and regional examinations which were scheduled for less than two months after 29/09. Pelesala Togafau teaches maths at Aleipata Secondary School, "I didn't want them to sit exams, because even if you're not affected with a physical injury, but the mind of a child – you can't see what is going through it. The students have been badly affected and on top of that the exams will further frustrate them, not to mention that we are surrounded by the damage. If you sit the exam and then you look around you at all the houses that were damaged, you will remember that day and then you won't concentrate on the exam but you will be thinking about the tsunami and what it did to the village." A select PSR team in conjunction with the Ministry of Education worked with local volunteers to run a special tutoring program to help students study and prepare for their exams.

Elena Peteru worked to co-ordinate the PSR team from the Apia centre and so met with many of the more seriously traumatized survivors, those who needed to be referred to the National Hospital for further support. "The key need was for grief counselling. The main issues that people needed help with was more for the death of their own child or a niece or nephew, loss of a loved one and then also loss of property. Some of them all their belongings had been lost in the tsunami. There were many struggling with nightmares and with feelings of hopelessness."

What was it like to be on the PSR team, to visit with survivors out in the field? Malia Manuleleua is a Samoan counsellor and writer based in Auckland, New Zealand. In the week after 29/09, she "slept, ate, and breathed tsunami." She describes her eventual decision to come to Samoa to work on the Psycho-Social Response team. "It was hard to sleep. Everywhere people were talking about [the tsunami]. We were a thousand miles away but Samoans everywhere in the world were united in their grief and shock over such a terrible tragedy. I was on the phone on a daily basis with a friend who was coordinating volunteers for the PSR team. She asked me, *why aren't you here?* I replied, *yes, what the hell am I doing here alright*...It wasn't enough for me to stay in New Zealand and watch it from a square box from the comfort of the sitting room. I had to see things for myself, feel and experience it and do my bit in the relief efforts. I was on the next available flight out to Samoa with 30kgs of donated lollies, clothes, medicine and anything else that would fit into my suitcase. Thanks to the airline PolyBlue who covered the extra 10 kgs, many were able to benefit from the goods in my bulky suitcase. It was a typical preparation for at trip to Samoa but for very sad and unique circumstances."

Malia spoke eloquently of her experience in the field,"In the villages, despite the circumstances, families welcomed us into their tents or makeshift fales, offered us water and food and still managed to give us a warm, dignified smile. It was a humbling experience. When our truck was stuck in the sand at Tiavea, the village boys pushed it to firmer ground. When driving back to the main road, up the rough access track, our trucks right rear wheel hung off a ditch a metre deep. Elderly women, men and small children in the tents came to our aid and helped to lift and push the truck back onto the road – amazingly, we achieved this on one attempt. These acts of kindness by locals in the tsunami-affected areas were commonplace. We were walking among angels. We were blessed with heavy showers from the heavens that cooled us from the heat, cleansed the dirt from our faces and quenched our thirst till the next day. We met a puppy called Sunami, talked to children, mothers, fathers, a blind woman, a disabled person, a fisherman, a planter, a teacher, carpenter, a shop-keeper, taxi driver, a beach fale operator, ministers and their wives, chiefs, ninety year old great-grandmothers, a pre-school teacher and many more. Each had their own remarkable story to tell – stories of survival and loss, of incredible acts of kindness, of bravery. Some emotionally, physically exhausted, dazed and lost, some philosophical and strong, grateful to be alive and a gentle acceptance that it is God's will and that life must go on."

There was one case in particular that stood out for Malia. A father, Samuelu (not his real name) who had lost his wife and two year old daughter. He sat with the counselling team and told them about his little girl. *After a long day at the plantation I would come home and she would welcome me back with a kiss and a hug, we would play together...*

"I see a sparkle in his eye for a second which quickly dies out. His daughter was obviously 'daddy's girl'. There is uncomfortable silence. What can you say that will take away his pain? He looks toward the ocean as if he's just seeing through it. He says in a quiet voice, *It is God's will.* His three sons play aimlessly on the other end of the fale. He seems to be in a zombie-like state of stress, shock and grief. He looks as if he hasn't slept in days. You can see that he's just 'keeping it together'. I wish he would just cry. As we offer our condolences, support and words of comfort – they seem so empty. I feel like we have imposed on his pain. I'm not sure whether to hug him, to give him money, something so that my pain would go away too? But what can you give? What words can you say? Nothing can compare to the loss of a child, a mother. I struggled with these conflicting thoughts and emotions. I gave him some money anyway. It was hard for me to come this far and not to be able to give something. He looked at me as if to say, *what's this for? Can it bring back my daughter, my wife?*

As we drive away, he remains where we left him, seated outside his fale, where a big fallen tree lies blocking the entrance. His eyes are focused on the sparkling blue ocean fifty meters away. I wave to him but he doesn't wave back. His full concentration is on the sea. I look back and see a young man with a broken heart and a broken spirit. I am forever haunted by that image. I sometimes wonder how he is, how his children are and I pray that he is recovering well. Later, we get information from Samuelu's sister that she can hear him crying every night. We put him on our referral list for a follow-up visit. But that doesn't stop the thoughts of whether we have done enough."

The PSR report estimated that "about 90% of the affected population has received some form of counselling...the vast majority of these were counselled at the primary community level and did not need referral to tertiary hospital level. [But] the extent of the mental, spiritual, emotional and psychological impact of the affected population is a big challenge. We are currently in the early recovery phase, it is anticipated [the tsunami] is likely to have further negative effects on those affected and the rest of the population, if appropriate mid to long-term psycho-social interventions are not put into place now."

One of those who worked closely with the body recovery effort would agree. "There were some crappy things out there. You block it out, I've just about blocked it out completely. I had one session with the counsellor, I think there should have been more but after going through what we went through, that was the least of my worries and I'm not going to stick my hand up and say *give me more help* when there's worse off, there's bloody dead people out there. They should have done more [counselling] and I think that they should still be doing a damn sight more for the people out there in the villages. What they went through is just incredible. I think it will catch up with people through the years."

Dr Ben Matalavea has been carrying out ongoing medical work at the Lalomanu hospital, post tsunami. "I come across many people that lost their loved ones, their children and they come in to the hospital for follow up treatment but they're also eager to tell you their story and you have to sit down and listen. They want to come and tell you what's happened to them and if you can help out in any way, you do try. I listen and try to talk to them and I feel I've done something. But it's not easy." Dr Ben has seen firsthand, some of the traumatic effects on individuals and families. "There was a seventeen year old who was brought in who drank paraquat, now this kid is going to die. He was mad because his grandmother told him off for something and so he drank weed killer, so I don't know if it's a direct result of the tsunami, but I think that the pressure that everybody's under out there is huge and the young people don't have that coping mechanism to deal with it, so they resort to this sort of thing. There was also a man who came in to treat his hand that was badly cut and it was because he hit his thirteen year old son in the mouth and broke off his son's front teeth. So he came here and sought treatment for his hand and I said, *but where is your son? He needs treatment!* So later that day, the mother brought in the boy to get help and the mother was having a go at him, complaining that he doesn't listen to his parents. I thought, *oh my goodness!* I had to tell them, her son's teeth were very badly damaged and he would need specialist dental treatment and yet the husband came in for just a few cuts to his hand! I think these are some of the social problems that are happening out there. I have a feeling it has something to do with this pressure, people coping with what happened to them."

For some who are trying to cope, blocking out the tsunami memories has been their best defence. Sally Taiva'asele was in hospital for three months. "There were a lot of pastors and doctors that were there in the hospital trying to help people whose minds were affected by the tsunami. I don't think about it too much, I try to forget about it, because if I dwell on it and keep thinking about it, my mind might also get affected. There was another woman in the hospital who to this day is not right in her mind. She was treated and discharged for her injuries but she would walk around swearing at the doctors and yelling. For me – I don't think about it."

Si'a Eli is a fifty-seven year old fisherman from Poutasi. "It was only through the love of God that people were saved. I am used to the waves of the sea because I go fishing but this was something too strong. To me, I don't want to remember the disaster. I call it a curse, it was like a fire that came and destroyed everything. I just want to forget."

Notuse Semo was in intensive care after the tsunami. She is emphatic in her refusal to relive the event that caused her injuries. "I don't want to talk about it because I don't want to remember, I don't want to think about what happened. When I was in the hospital, I told my family that I would never return to Saleapaga because I didn't want to remember what happened. My brothers came from overseas and built us a new home up here inland far away from where it all happened. When I saw it, I'm thinking now that I will stay here. But I don't want to talk about the tsunami."

For others, talking about their experiences and sharing their pain with others has been healing.

Jared and Netta Schwalger have found strength in each other. "After the tsunami, our relationship became even closer. I know when some people lose their kids, they stop talking to each other and they want to get separated because of what happened because one blames the other for what happened – but us, no. It's because we talk to each other, we talk about our kids and we talk about our parents and about our whole family, about how it was. We talk about the funny days – sometimes it brings tears to our eyes but we always cheer ourselves up." In the weeks after 29/09, Antonetta was interviewed by five different television stations – networks from Australia, New Zealand and Wales. "Like I always say to the psychiatrist, the more I talk about it, the more I feel better, the more I get stronger. We have had a lot of comforters, friends and family coming to support us which has been a big help for me and Jared. Even though talking about what happened brings me tears, it makes me stronger."

Brent and Diane Cederman from Raglan, NZ were grateful the tsunami had been a shared experience. "Our therapy for months after was just to sit and talk to each other and have a couple of glasses of wine, that was our counselling between each other. So we were probably better off than most people because having the two of us go through the tsunami together, we could share that."

In American Samoa, the Leone Tsunami Club has been a lifeline of support for the men who gather there to play cards. Men are stereotypically not known for their willingness to bare their souls to each other, talking about 'feelings' is usually relegated to women. But gathering to play cards and to share a few beers has opened the doors for the sharing of tsunami experiences and sprinkles of humour make the horrors endurable. "After you have a beer, you start to get talkative and relaxed and then you can talk.We talk and each of us tell our story and find the funny side of it, that's what we do every night. Some of the stories make us cry but when we listen together, it makes it better. So that's why we are quickly healed I think, because we are tight. We talk every night. And when our friends come to our meetings for the first time and each of us tells a story about the tsunami and we are laughing but they – they are crying because it's so sad."

Sometimes, as people try to cope - there is anger. Faaumu Laufatu from Vaovai has noticed the difference in her three year old son Kenneth. "He's fine physically but he's not the same. I try to talk to my son about the tsunami and about his experience when he was drowning and then when he was found. But it's hard. He cries a lot and when he isn't crying, he gets angry easily. He wasn't like that before. He gets angry a lot and when he does, he wants to hit people."

One father is targeting his anger at those responsible for what he sees as the unfair distribution of aid. "The other village has water tanks for their tsunami families but there's hardly any tanks for families in our village… and we were given only $5,000 for each of our kids lost – that's it, a mere $10,000. It's like a slap in the face, it's like you're not important and who the hell are you? I feel like the government is using us tsunami victims as a way to get money from overseas. I feel as if I'm the one that's standing in the war zone and the Prime Minister is hiding behind my back with a gun and I'm being used as a human shield to grab aid from people. It's like the puppet and the puppet master. We are fighting against the government to get our share of what really is supposed to be ours."

For many survivors, there are nightmares. Rebecca Hamilton of Gisborne, NZ recalls her months after 29/09, "I had trouble sleeping as images and memories of the event would float in my head. The sound of the tsunami coming closer and closer would wake me and I would sit bolt upright in bed dripping in sweat. Everything still seems so clear, the smell, the pain and grief of people's faces and most of all the horrible noise of what sounded like a train roaring towards the shore."

Maua Toa from Lalomanu cannot rid his mind of the memory of trying to hold his son above the water as they clung to a breadfruit tree. "When I'm doing my everyday chores, it's like watching a program on television in my mind, like a video, You just stick in the tape and start it from the beginning from when the earthquake hit. It's like I'm watching it happen on a screen, it's so real. Even when I'm sleeping, suddenly I will scream and my wife will hit me in the face and wake me up. I will get up and wonder what's happening and she will ask me, *why are you screaming?* She doesn't know because she wasn't there and she didn't see it. I tell my wife that I don't know when it will stop, these memories, these dreams. Not until I am dead and buried will I forget a disaster like this, that's how I feel."

One obvious impact of the disaster – is fear of the sea. Or just a heightened respect for it. When you live on an island, it is difficult to escape the sight of the ocean, let alone its distant roar. Many people have relocated permanently to inland areas while others struggle to sleep with the sound of the ocean. Beach fale resort operators are trying to meet the challenge of rebuilding their businesses. The CEO for the Samoa Hotel Association, Nynette Sasse is working closely with the SHA members. "Many of the operators are living up in the mountains and have gone back to farming for their day to day living. They're looking for the financial assistance they need through the tourism tsunami rebuilding program, that will be their lifeline to start rebuilding their beach fales. But a lot of them are too scared of the water, they don't want to come back. It's just a shame seeing that beautiful area down the south coast looking like a ghost town. It's kind of sad every time I drive through there, it was such a vibrant area. Hopefully things will come right in the long run. Give it time, it's only human, time will heal and they'll want to come back down."

The livelihoods of many fishermen have also been compromised. Dr Ben tells of his meeting with some of them, "We were out in the field in another village a couple of weeks after the tsunami and there were these ten fishermen that had been out at sea when the tsunami came. I was quite surprised that they had survived their journey in the wave. They were all congregating at this one house. We checked them all out and told them they were fine but they were scared to go anywhere. They were afraid to go back by the sea, so they were huddled up there in the bush and these are full-grown men. It was like a kind of shock that they were in." Three months after the tsunami, Oselani from Saleaumua still has not gone fishing. The man who used to enjoy the thrill of scaling rough surf in a frail wooden canoe, now leans forward to quietly confess that he is afraid of the ocean. He shrugs, "I'm sure I will go back again one day, but I don't think I will ever feel the same way about it like I used to. Not now. Not since I have seen what the sea can do." One elder commented sadly, "There is an entire generation of our people now, that are going to be afraid of the sea, children that are not going to even consider the idea of becoming fishermen."

During the cyclone season, Samoa was rocked by several near-misses as strong tempests decided whether or not to hit the country. Strong winds during Cyclone Rene whipped the ocean to a frenzy on the Aleipata coast and it was a cause of much anxiety. Agriculture Minister Taua Kitiona spoke to the Samoa Observer. "I am here at Lalomanu and the waves are getting higher and higher. The wind is very strong… I still feel fear the sea, even though it has been five months since the tsunami."

Rachel Cant from Cornwall in the UK who came to Samoa "for the surf" now writes, "My uppermost fear now, is still the ocean. I am better than what I was. For the first few weeks after 29/09, I could not sleep and when I did, I would have nightmares and flashbacks of that day. It still affects me emotionally today. The ocean scares me, especially at night if I can hear it while I sleep."

Maiava Leota from Leone, American Samoa agrees. "Right now I feel that the sea is the enemy for us. At night when the waves are really big and loud, you worry. When the waves break, your mind jumps. Your heart jumps. All we do at night is try and block our ears and try to sleep."

Ngaire Woodward plans to return to Lalomanu for the anniversary of the tsunami. "We still have to have our swim there as we didn't go into the sea of our own choosing…this is what will kill the demons. I have been in the sea two or three times since 29/09 but I am still wary. The first time I went in my son walked me in and talked me through it until I stopped shaking."

It is not just the tsunami survivors who have been deeply affected by 29/09. The helpers, the rescuers, the medical workers, all have had their resolve tested. For many who had to put aside their emotional responses to the tragedies around them, their release came later. For Elena Peteru, working with the overseas survivors, it was when the last patient was airlifted out. "At the time, it was the adrenaline that kept you going. I went in to do a job and I didn't really – maybe subconsciously – I didn't allow my feelings to get involved. It was only when I saw the helicopter that came in to take them out, it was only then, I waited there until the last group left – it was only then that I felt the sadness of the whole situation." Elena is reflective about the impact on her fellow PSR workers. "There was a lot of burnout. I'm pretty sure there's a lot of emotional stuff that has not yet been tapped into for the workers who actually went out there in the field and assisted. I don't know if the [counselling] workers have been looked after themselves…doctors and nurses were offered counselling support but I think everyone forgot that even the counselors or support people in the psycho-social team – at the end of the day, they too need that support." Elena cannot swim and that ties in to her own personal tsunami fears. "The fact that I can't swim is a nightmare because then [in a tsunami] you've got zero chance of surviving. So one of my goals now is to learn how so that I can kind of see a chance of survival if we ever get hit with a tsunami again."

For Dr Ben, it was two months after 29/09. "I was driving to Lalomanu the other day and I was thinking about why I hadn't felt something yet, and then suddenly it all hit me and I was crying, because I couldn't believe all that had happened. I think for everyone, there has to be a point that it will hit you and you've got to let the emotions out."

For some, it is later trauma that triggers the emotions of 29/09. One key member of the team at the morgue can recount without any emotion – the work they did after the tsunami with 143 victims. But when he tells you about the horrific bus crash at Solosolo that killed eight people a few weeks after the tsunami, that is when he breaks down and weeps in a sterile room.

Andrew Peteru remembers a driver who worked at their office, who was there with them at Lalomanu hospital on the eve of tsunami day, "A nurse came up to me and said, *well, they've just found a headless body down there.* So I told the police that were at the hospital about it and they didn't want to go. So I asked our driver to go down and get the body and transport it to town. I can't imagine what that must have been like for him. He worked very hard the next few days with transporting bodies and then he didn't last long here at the Ministry afterwards. He quit shortly after." Several PSR counselors expressed their concerns for those who assisted with the dead. "We spoke to many who had transported bodies from the villages to the hospitals. Remember many of them had to drive a truck load of the dead all the long way to Apia, late at night. They were told to drive slowly… so there they were on long stretches of deserted mountain road by themselves.They had very real fears of *aitu* or the spirits of the dead they had transported. They were definitely in need of counseling and support."

Vaughan Simpson made sure his people had access to some debriefing after their week of recovering the dead. "Our workforce at Ca'bella have been together for years…counselling was offered to everybody. We made it available to everybody, we encouraged it. We insisted they do it. Counselling was several weeks later and we would take four or five guys from the worksite down to the counselor at a time and just keep doing that. A lot of them weren't keen to do it, counseling isn't something that you do normally! But I think the ones that were right in the middle of [the work] knew that they needed it." How has everything impacted on him? "You try not to discuss it when you come home, you just try and be strong, you don't discuss it off the scene. It's made me appreciate my kids, my family, more – definitely because something like [the tsunami] could happen to anybody. I haven't changed my mind towards the sea. You can't live in fear of the ocean. I wouldn't go swimming at the beach – but I wouldn't do that anyway because I know how many sharks are in there! But now, I wouldn't take the kids back to the beach. Because I've seen what a tsunami can do, you couldn't run fast enough to get away from that."

There are many support and relief workers who's only chance to talk about their contained emotions has been when a writer comes to ask them for their story. Their responses catch them off guard, spilling over the secure

The first wave in Pago. *Photo - Kat Pearson*

Streets and buildings under water in Pago.
Photo - Kat Pearson

Pago drowns in the tsunami. *Photo - Kat Pearson*

A yacht sails down the street in front of the Evalani Hotel.
Photo - Kat Pearson

Kat Pearson stands in front a devastated building in Pago.
Photo - Kat Pearson

Ray and Etta Wyberski, American Samoa.
Photo - Lani Young

Rose Talalotu, American Samoa. *Photo - Lani Young*

Toetu Tauiliili with his son in Leone, American Samoa.
Photo - Lani Young

Robert Toelupe stands in front of the remains of his family home in Leone, American Samoa. *Photo - Robert Toelupe*

Robert's daughter Anita holds her son in Leone after the tsunami. *Photo - Robert Toelupe*

Faataui Fitiao and his wife Taitasi with a photo of their daughter Vaijoresa.

The Leone Tsunami Survivors Club, American Samoa. *Photo - Lani Young*

Terisa Palmer, American Samoa. *Photo - Lani Young*

A devastated Niuatoputapu. *Photo - Australian Defence Force*

Crew from the Tobruk walk through a village on Niuatoputapu. *Photo - Australian Defence Force*

National Burial Service - Tafaigata Cemetery, Thursday 8th October 2009

Photos by Ron Kubik

The Lieselore Rock memorial service at Tafaigata Cemetery. *Photo - Arne Schreiber*

The Hon. German Consul Arne Schreiber (2nd from left) and others pay their respects at Lieselore's memorial service. *Photo - Arne Schreiber*

walls they have built to keep them in check. They are embarrassed, “I’m sorry for crying, for carrying on like a little girl.” They talk about their sense of helplessness at not being able to do more, save more, help more. They talk about nightmares and sadness that creeps in when they least expect it.. At work, at home. But most of all, they talk about how much more they appreciate their families, their children. How humbled they are by the kindness and support of so many – friends and strangers alike.

Grief and loss are the most cutting of tsunami trauma. Well-meaning people tell Netta and Jared Schwalger that *you’re young, you can have more children…* “It’s not the same, it will never be the same. Everyday when we wake up, it’s like death. When I think of the kids and especially when I’m alone, I always think of them and I picture in my head everything that we always did together with my daughter and my son. I always picture my son Teancum being naughty and running around, it always brings tears to my eyes and I whisper to myself, telling them how much I love them, how much I miss them, how we’ll see them again. Maybe not now, not soon, but later. We have moved on slowly. It was hard when I realized I had lost my kids. It was hard listening to children’s voices, hard seeing babies, mothers holding their babies and it was really hard to get used to that. I see other people with their kids and it’s hard but I have to accept it.”

Mika and Ave know just how the Schwalgers feel. “We spend many days weeping and mourning. It’s really hard for us. When people talk to us, we can put our weeping aside for a little while. But when it’s just us and our children, like when we have our evening prayers, when we eat, when we’re lying there together and talking, I can never forget. I cannot forget because each of my children had their own personalities, we had a long time together. I knew them, I spent a lot of time with each of them through hardships when they were growing up so how can I forget them? Like the other night, we were eating dinner and my wife was crying because there are some foods that we eat that our children enjoyed, so we would eat and look at that food that my children loved, then I would lose my appetite and I would cry. How can I eat when that person that liked that food is gone?”

For some, grief has been channelled into renewal and relief work. Stacked high in the Australian High Commission office in Apia are boxes of reading books, marked for the Lalomanu Primary School library which has yet to be built. An ocean away, in Canberra, the Salavert-Wykes family have been fundraising and buying books in memory of their six year old daughter

Clea. "Clea loved reading. She borrowed many books from her school library and had created her own library at home with her own bookshelf. Along with reading, Clea also loved swimming. She loved the beach and she was having a wonderful holiday in Samoa. We consider that a library near the beach in Lalomanu where Clea perished is a fitting memorial to the lively, intelligent, beautiful and loving child Clea was. The Lalomanu Primary school has about 150 students…this library will be an opportunity for the school to help expand the world of reading for Lalomanu's children…" Soon, very soon, clusters of children in orange and yellow pinafores,will gather round shelves lined with adventure, magical make-believe and the wonder of discovery. Imaginations will take flight and dreams will soar, in Clea's Lalomanu library.

Andrew White returned to Raglan, NZ and set up the 'Lalomanu Village Samoa Trust' to honour the memory of his wife Mary Ann and "to recognize the assistance and compassion he received from the survivors of Lalomanu." Brent Cederman is one of the trustees. By April 2010, the Trust had raised over one hundred thousand NZ dollars and Brent returned to Samoa with his wife to see how funds could best be utilized to help Lalomanu. The Trust will be carrying out several projects in the village including the construction of an effective escape path up the hill. The proposed pathway and access steps will be lit with solar lighting and of a robust, easily maintained design. "This project not only has the potential to save lives but it would also help promote economic development by way of tourism, through giving tourists more confidence to stay at Lalomanu, knowing it was safe and well prepared for another disaster." If ever the sea again roars into Lalomanu, people will not have to scramble up a rocky hillside, clinging to tree roots and vines to pull them up. And visitors to the area will be able to take a scenic nature walk up into the hills, pausing to enjoy the panoramic view of the ocean horizon. The Cedermans also visited the Lalomanu Primary School which has four teachers and caters for 150 students. The LVS Trust will help the school by purchasing necessary resources like computers and a photocopier. Mary Ann had "been a part of everything that makes our Raglan community strong" – in life she had been a part of the local St John Ambulance, Plunket, the Raglan Area school, and active in kindergarten and PTA committees. Thanks to the LVS Trust, Mary Ann will now be a part of what makes the Lalomanu community strong.

In May 2010, Stephanie Hodgins-May - daughter of tsunami victim Vivien Hodgins – returned to Samoa and to the Lalomanu beach where her mother had died. Stephenie came to determine how best to use funds of $20,000AUD raised through the 'Vivien Hodgins Foundation' which was established shortly after 29/09 by her family. On this trip, Stephenie donated seeds, tools and seed planters through the local 'Women in Business' organization that aimed at encouraging independence for women. There will be gardens grown because of Vivien Hodgins. Stephenie spoke to 'Courier' journalist Kim Stephens about her return to Lalomanu, "I feel more emotionally ready than previously. I am feeling anxious of course, and when I head south and see all the debris, I imagine it will be quite confronting because I haven't really been able to bring myself to look at the pictures yet. But I'm staying on the same beach that Mum did. I don't want it to be that Mum went on holiday to some foreign island and never came back, I want to establish a connection. I originally thought it would be about closure, but it's not closing something, it's re-establishing something." Vivien had been a teacher at Mt. Clear College in Ballarat for 33 years and many generations of students had benefited from her dedication to the Performing Arts. The Foundation has also established a scholarship that provides assistance to eligible students to pursue their career in this area and in 2010, Yseult Mantelli was the first recipient of the scholarship. There will be many more songs sung and art performances that inspire us because of Vivien Hodgins.

The Martin family of Matamata, New Zealand, lost two daughters in the tsunami. Twenty-four year old Rebecca who was a teacher at Rototuna Primary School in Hamilton, and twenty-two year old Petria, a team leader at the Matamata Sports Centre. The sisters had been on holiday with their cousin Jodi McGlashan and friend Olivia Loeffen. At the funeral where the sisters lay in caskets of their favourite colours – green for Petria and purple for Rebecca – their father Kerry Martin asked, "How can something so exciting as a holiday in Samoa go so wrong? How can our girls have been on a beach on the one day a tsunami came and swept them away? They would have fought hard but stood no chance." In memory of the sisters, the Martin family and friends contributed funds to complete two of the new homes at Lalomanu, built by Habitat for Humanity, NZ. In May 2010, a plaque inscribed with their names was unveiled and two families who had been living in the bush in tented shelter, were able to move in to their new homes.

Looking beyond 2010 the question remains, *how do you mend the broken hearts and broken spirits?*

24
We remember the dead

Any visitor to Samoa will tell you that we like to bury our dead close to us. Graves are a familiar sight outside many homes as you drive through villages. They have little roofs to shelter them from the rain and elaborate headstones with photographs and Biblical inscriptions. They are adorned with plastic wreaths and ribbons and at Christmas time, many are lit with flickering lights. Family graves are kept impeccably clean, weeded and scrubbed. But during the tsunami, even the dead were not safe from the sea, as graves were destroyed and some had their contents churned up in the waves. After 29/09, many families had to move their ancestral dead further inland for reburial. And for families with loved ones killed by the tsunami there was some indecision of where to bury them – on desolate land wracked by the ocean? Or in the plantation bush next to a tarpaulin shelter? After 29/09, funerals of every kind were taking place all throughout the disaster zone. A three year old daughter was carefully wrapped in a green mat and buried as soon as she was found on Day Two after the tsunami. The Taufua family buried a group of their loved ones in a single grave at Lalomanu and the accumulated grief of so many mourners was a choking, overwhelming thing. At Jerry and Metita Schwalger's funeral service in Apia, their five year old grandson Niel sang for his Nana and Papa. A bandaged, still-dazed mother went straight from the hospital to the burial of her three children. She spoke to the television crew from TV3 New Zealand. *God gave me these three children to love for a short while and now He has taken them back. I'm thankful I could be their mother, that I could have them for this short time.*

Mindful of the pressures many grieving families were under, the Samoa Government offered to carry all the funeral costs and have tsunami victims buried at a mass grave site at the Tafaigata Cemetery. Vaosa Epa explained how the proposal came about, "After a couple of meetings of Cabinet, the Prime Minister suggested we have a national burial for the victims – but we had to move with caution, we wanted to ensure there would be dignity, and there had to be consent and approval from the families themselves to have their people buried away from the usual places." It was agreed that a monument would also be built at the Tafaigata site, in time for the one year anniversary of 29/09.

There were only a few who chose to accept the government's offer. Many families buried their dead quickly, choosing to keep their loved ones close to them. The new inland settlements all have little cemeteries in front of the slowly rising houses. At the back of Malaela, drive up a ragged road through mangrove swamp and there is a cluster of fresh graves sitting in amidst trees and shoulder-high grass. Jared and Netta Schwalger buried Abish and Teancum on family land in town. With no parents, no home, and no children to return to in Malaela, the Schwalgers have moved to New Zealand to live. Jared's sister Lucia watches over the family graves.

There were some people who did not like the government's proposal. One asked, "After everything that's happened to us and yet the government still has the nerve to ask people if they could bury their family in a communal burial place at Tafaigata. What's the first thing that pops into your head when someone mentions *Tafaigata?* It's where the prison is and it's the trash dumping area! How could they want to bury our loved ones next to those places?"

Still others were just hesitant about a concept that was very new for Samoa. As Vaosa explained, "There were people who didn't support the idea, they had concerns. Firstly because, it's a foreign idea and when the word went out about it being a 'mass grave', I think some people were discouraged because they have seen things on television in other countries where people are just thrown into a big grave. So we tried to explain that no, everyone was going to be buried in their own grave, but that it was going to be one big area designated specifically just for the tsunami victims. After the burial, there were a number of families who then expressed their regret that they buried their loved ones elsewhere. After seeing the dignity of the occasion, how it all turned out, they said, *we really should have brought our family here.*"

On Thursday the 8th of October, 2009, Samoa held a National Memorial Service at the Apia Park Stadium for the victims of the tsunami. The country stood still at midday. The flag flew at half mast and the park stands were wreathed in black and white with red ribbon bows. Huge bouquets of red ginger flowers swayed gently in the afternoon breeze. His Highness the Head of State was there along with dignitaries from many nations. Local government and business leaders were there. Families and friends of the dead were there, hands clasping framed photographs of their loved ones. Mika and Ave were there. Some families brought banners proclaiming their losses. *Taufua Family, 13 Loved Ones, Four Generations.Thank you for all the support.* Search and rescue workers were there. A row of black uniformed police were there with search dogs sitting quietly at their feet.

Dyna trucks draped with fine mats drove into the stadium carrying the coffins that were draped in white lace. Ave's mother Malo was one of them. Four of the coffins were little wooden boxes. Children. Mika and Ave's daughter Malo Maryanne was one of them. Men dressed in white and black with yellow POLICEMAN jackets carefully carried the coffins from the trucks to the stage where they were placed in a row. Eight men strained to lift the large coffins while only four policemen were needed to carry the children.

The service was led by the Reverend Oka Fauolo. Dressed in white, he slowly walked with a silver cane to the podium and spoke with gentle authority. All could feel of his strength. *Comfort ye my people says your God. Speak tenderly to Jerusalem, cry aloud to her, now that her warfare is ended.*The Samoa Police band played their sonorous hymns and the gathered congregation sang familiar words. Broadcasted live to the nation, others sang with them, from the National Hospital where nurses clustered around the front desk to listen and bandaged survivors leaned forward in their beds to add their voices to the refrain. From villages all over the country – from the shade of a spreading tamarind tree at Vavau that looked out over a shattered resort. From a feeble wooden shelter in the new Saleapaga settlement. From the Lalomanu Hospital that had first received so many of the dead. From an eerily quiet Poutasi. And the sloping hills on Manono island. The country united in song.

Other ministers spoke after Reverend Fauolo. *Thanks be to God for He has been our dwelling place in all generations. A pillar of cloud and fire when the day is dim and the night is dark. He is our refuge and strength. Therefore we shall not fear. Though the earth should change, though the mountains shake in the heart of the sea, though the waters roar and foam, though the mountains tremble – thanks and praise be to God for all His goodness to us.*

The Prime Minister spoke to the nation. It seemed a struggle for the usually confident speaker to make it through his address. *The depth of grief and desolation felt by everyone who lost loved ones is beyond words to describe and we can only grieve with them.* He acknowledged Samoa's friends, particularly the hands from Australia and New Zealand that had helped to carry the tsunami burdens. The hearts that had been moved to give.

The memorial service culminated in the reading of a list of names of the tsunami dead. As each name was read out, selected schoolchildren would walk forward to lay a wreath beside the stage. The names of foreigners killed in the tsunami were not read out due to privacy laws. The names of those Samoans still missing were not read out either, leaving a total of 129 names. It was a lengthy process. But there was complete silence in the stadium for the duration. Vaosa Epa had the task of reading that list. "I was standing there and the whole stadium was charged with emotion, you could feel the enormity of it. My heart was full of very strong emotions. I was really questioning myself, *am I able to carry this?* I really wanted to be able to do justice to this duty, to honour those whose names I was reading. I had to pray, *Lord, please help me keep my emotions in check.* Because as I was reading the names, I would look up at the people crying and then I would look down at the coffins lying in front of me and you see the students coming up to lay the wreath as each name was being read. I wouldn't have been able to carry out that duty, if it wasn't for the grace of God. If it wasn't for His strength that kept me going, I tell you, I would have broken down so many times."

The National Memorial service was also broadcast live in New Zealand on Auckland's Radio 531 station and thousands of the Samoan community and friends living there were tuned in. Samoan Member of Parliament, Hans Joe Keil was there and he was one of many moved to tears as they listened to a nation mourn as one. "The solemnity of that service was felt all the way to New Zealand. As the names were read out, I could feel and hear the emotion in Vaosa Epa's voice and I prayed that she would be strong enough and have the courage to see it through."

Vaosa began, *With the utmost respect, I stand to read the names of those loved ones – the fathers, the mothers, the children who died as a result of the national disaster. As each name is read out, a wreath will be laid in their honor by senior leaders of various secondary schools.* Malo Maryanne's name was the third one to be read out. A teenager laid a wreath for the little girl who had woken her mother to pray with her on the night before the tsunami. One after the other, students placed purple and white wreaths at the front of the stadium congregation, took two steps back and bowed their heads for a moment before returning to their places. Nine wreaths were also laid in remembrance of the foreign casualties. Vaosa concluded, *Today we unite as one, we grieve as one.*

Oceans away in New York city, a similar memorial service was being held with Samoans living in the United States. Organized by Feturi Elisaia, Samoa's Ambassador to the US, representative to the United Nations and the father of Vaosa Epa's children. Feturi had asked their daughter Aurora, a poet, to compose something that could be read at the New York service. As Vaosa prayed for strength to 'do her part', her daughter's words were speaking to grieving hearts in America. *We are people of the sea...sons and daughters of land, sun and waves...soon we will bury our people en masse. Families lie together side by side, Never before have we done so, never more do we want to...We are resilient and strong and while breath is still in us, we will still strive forward, to rebuild again. We have always been people of the sea, but we try now to turn our backs on her; she has hurt us beyond words...for now we will comfort and be comforted, but maybe someday not so distant, we will be once more – people of the sea.*

After the reading of the names, Reverend Fauolo concluded the service. The burden of a nation's spiritual comfort weighed heavily upon him but he spoke with authority and certainty. *In this world we have tribulation. Let not our faith be shaken...but be of good cheer, the Lord said, for I have overcome the world.* The stadium of mourners stood as one as the coffins were returned to their places on the shrouded trucks. A procession of cars followed the trucks as they wended their way through Apia and on to the Tafaigata Cemetery. White uniformed policeman stood sentry on the streets as they went by. Tamarind trees hung their heavy heads over Beach Road and crowds of people waited patiently in the shade as first the police escort and then the Head of State drove through. The trucks had to slow as the crowds moved forward to place flowers on the coffins. The whole route to Tafaigata was lined with mourners. At Lotopa, a young mother stood with her children all dressed in their Sunday best. Two little girls wearing lacy white, held baskets of frangipani petals and their mother carefully lifted them so they could scatter the gently fragranced confetti over the coffins. At Siusega, a grandmother held stalks of red ginger flowers and stood with her sons, waiting for the funeral procession.They stood on each side of her, propping her with steady arms as she swayed a little in the afternoon heat. People cried. People waved. People threw flowers and branches on the road as the trucks went by. People held signs, *God bless the tsunami victims.*

At Tafaigata Cemetery, a chasm had been cut into the earth. The sides were lined with red and blue fabric. Garlands of yellow alamander and green fern were threaded through the security railing. A concrete stairway led to the eleven waiting graves – cemented, tiled vaults. There was a tented pavilion for the dignitaries and clergy to sit under. Others sat cross-legged around the gravesite. A police honour guard carried the coffins to each vault. The children's wooden boxes sat forlornly in cement resting places that were way too big for them. Each gravesite had its name marked.

*Lafitaga Suafai Sasulu
Born: 16/ 12/ 2007Died: 29/09/09
Saleapaga

*Malo Fanene Uili
Born: 17/7/1939 Died:29/09/09
Lalomanu

*Malo M. Mamea Mikaele
Born: 27/8/2006 Died: 29/09/09
Lalomanu

*Tusipepa Hanipale
Born: 1967 Died: 29/09/09
Saleaumua

*Lolani Ikenaso Tevaga
Born: 31/5/2003 Died: 29/09/09
Saleapaga

*Niusilani Mania
Born: 24/8/1979 Died: 29/09/09
Saleapaga

*Miriama Faletoi Satui
Born: 24/8/1979 Died: 29/09/09
Satitoa

*Vaimoana F. Lupeamanu
Born: 30/9/1939 Died: 29/09/09
Poutasi

*Filisi Tavita Lupeamanu
Born: 2008 Died: 29/09/09

*Ranui Simanu
Satitoa.

When every coffin was in place, Reverend Fauolo gave a brief prayer and it was time for the families to farewell their dead. Mourners crowded into the gravesite, filling it with love. They brought wreaths and garlands. Lengths of lace. They hugged framed portraits of their family, the smiling pictures of how the dead used to be. They brought their children to say goodbye. A woman lowered her toddler into one of the vaults as she herself bent to kiss the wood coffin. Sternly striking in a black suit, Mika stared down at Malo Maryanne's grave while Ave wept and clutched at the lace coverings. The late afternoon drew on. The dignitaries slowly departed but the families remained as the whispers of evening were heard. Cicadas stirred. Doves called as the first stars slipped from their hiding places. People sat on the warm earth in little groups around the burial site. Those who had labored long to prepare everything for the memorial could at last sit, take a breath, reflect. "It was beautiful, families didn't want to leave, they were milling about, sitting by the graves, looking at their loved ones. I hope the memorial helped provide closure for families and for the nation as well...You look back on it all and you realize, that we can only depend and lean on God to take us through these things, and to take us forward from here."

On the 20th of March, 2010, almost six months after the tsunami, one more victim was laid to rest at the Tafaigata Cemetery – sixty-seven year old German woman, Lieselore Rock. She had arrived in Samoa on the 23rd of September and was staying at Faofao Beach *fales* at Saleapaga when the tsunami hit. Her body was recovered by Commissioner Tony Hill's team and could not be officially identified until the 12th of October, 2009. A week after the tsunami, German Honorary Consul, Arne Schreiber had received a phone call from Ms Rock's children in Germany, asking for any word of their mother. Arne had gone to the airport to check the flight that Lieselore was supposed to have flown out on but his trip was a futile one. The National Hospital contacted Arne with the news that the Australian DVI team had one more unidentified Caucasian victim and it was possible that she could be the missing Ms Rock. Everyone worked together to obtain the necessary information so formal identification could be done and on a Monday morning, Arne went to the morgue. "There was a horrible smell...I had to sign some papers and then they said, *that's it, this case is now all yours.* And I thought, *thank you very much, what do I do now?!*"

Lieselore's will specified that she wished to be cremated and her ashes scattered wherever she had died. Samoa has no crematorium but Lieselore's travel insurance paid for her transport to New Zealand where her final wishes could be carried out. Back in Samoa, the Ligaliga Funeral Parlour was not comfortable with the rather 'un-Samoan' practice of scattering people's ashes. Arne explains,"The family asked us to scatter the ashes in the air, either in New Zealand or somewhere in Samoa but we – me and the funeral people – looked at each other and we said, *we can't do that.*"

Permission was obtained to bury Lieselore at Tafaigata Cemetery. A grave was prepared for the small white box of ashes and a quiet memorial service was held. Her Highness, Masiofo Filifilia Tamasese was there, along with members of the German community in Samoa. Commissioner Tony Hill was there. Arne Schreiber concludes, "She was travelling alone, just arrived here and went straight out to Saleapaga. It's amazing how people go around the world alone, just backpacking and going places without fear. I think for her to find her final resting place in Samoa, I think it's magic. Because I feel like her roots in Germany were not so deep, and I'm sure in the short time she was here, she would have felt the closeness of family life here, the friendliness of the Samoan people. I think she would be quite happy with this decision, and the burial place looks out to the ocean, it's very beautiful."

When the tsunami monument is unveiled on the 29th of September, 2010, Lieselore Rock's name will be among the list of Samoa's dead. The woman who came the furthest distance to die in the Pacific Tsunami.

There are two victims of the 29/09 tsunami who were *not* killed in Samoa – but did return home for their final burial. Thirty-six year old Kapisi Sima from Fogatuli, Savaii was working for COS Samoa Packing in American Samoa in 2009. He had worked the graveyard shift at the cannery and was returning home when the waves "washed away the bus he was travelling on, killing him and others." Kapisi's father, Afualo Sima went to American Samoa, desperate to take his son home. Kapisi's employer could only provide $500 USD towards funeral expenses. Apulu Veronika Mortenson is a Samoa Red Cross volunteer living in American Samoa. Afualo came to her for help and she set the wheels in motion for his son's final trip home to Fogatuli. She wrote to the Secretary-General in Apia. "I am desperately seeking help on behalf of this family, from our government in transporting the body of one of our own, via Polynesian Airlines, for a proper burial in our homeland." Tautala took the matter to the National Disaster Council and the Prime Minister led the Council's emphatic response that of course, the Samoa Government must assist. The Samoa Red Cross and Ministry of Health would arrange the details and government would cover all the costs. All was prepared to fly Kapisi home on the 19th of October.

On the 13th of October, Apulu Veronika was contacted by another Samoan family regarding their son who had died in the American Samoa – thirty-eight year old Pita Tanielu from the village of Tuana'i. Pita worked as a bus driver in Pago Pago and he was the father of four children. Cabinet approval was also given for his transport. American Samoa Congressman Faleomavaega Eni Hunkin paid for the two men's caskets and they were both flown home so their families could bury them. So they could have their sons close to them. Apulu Veronika concludes, "Albeit has been a trying time for all of us, we ought to be comforted and encouraged in knowing God has taken the lead in all that we are doing since day one…the decision of our Samoa Cabinet is fantastic! It is amazing how we all pull together as one people during difficult times."

At the close of 2009, five people from the Samoa tsunami remain missing. All are children.

Aleki Vai Mikaele	*1yr*	*Lalomanu*
Alfie Cunliffe	*2yrs*	*New Zealand/UK*
Fa'atamali'i Shen Yi,	*3yrs*	*Saleapaga*
Savelio Logotaeao,	*3 yrs*	*Saleapaga*
Jamie Suafai,	*4 yrs*	*Saleapaga*

On the 22nd of October, the people of Saleapaga held a memorial service on the beach for their three children still lost. There was a brisk wind and the waves lapping the shore were not quiet ones. Family and friends lined a beach still littered with tsunami leftovers. Reverend Uaea Isaraelu from the Congregational Christian Church led the service after which, a wooden raft with three flower wreaths was set afloat. Two men pushed the raft unsteadily through the waves, pushing it out to the uncaring sea.

Fatima Logotaeao spoke of her daughter Savelio. "She would sing all the songs from the movie 'Mamma Mia' and dance too. She would make us all laugh with her singing. She followed me everywhere. I couldn't do anything without her following after me. She wanted to go with me everywhere. It is very hard to stop thinking about her."

Ioka Shen Yi is the mother of four children and Fa'atamalii was number three in the family. "She was a very clever girl, she loved to talk on the phone to her dad whenever he went to Apia. She wanted to call him and talk to him all the time." Ioka's husband added, "My girl would point to her eyes and my eyes, and say – I'm Chinese like you, ay dad? We are Chinese!?"

Another mother said nothing. She stood and sobbed as the reef waves caught the little raft. Flowers scattered in the surf, some carried back to the desolate shore on the lilting waves.

25
After. Thoughts

"The earthquake of 29 September 2009, generated a tsunami which reached the Samoa islands in fifteen to twenty minutes and the Tongan islands of Niuatoputapu and Tafahi in under five minutes. The definitive death toll is 34 on American Samoa, 143 in independent Samoa and 9 on Niuatoputapu with a total of 189 persons killed. Economic losses are estimated conservatively at USD 200 million. These figures make the Samoa tsunami the deadliest documented tsunami in the region." 2009 Field Survey of the Samoa Tsunami.

"This was a disaster of great magnitude. We need to believe in the love of God for saving us because there's nothing else you can do in these kinds of situations. It's a very important lesson that has come out of this tsunami… the wave has come, it has ended, blood has been spilt and lives lost. People were taken by the sea, even the tourists died. I will never forget this, even my children won't forget because their mother was taken away." To'omalatai Lauvai, Vailoa.

"How can we remain unaffected? If you look at seven members of our family were lost, ten, twenty members of our extended family…all those who died, we are all related. Lalomanu alone lost forty-three people and they are all related. It's not easy, even up to now when I think about it, I cry inside. Some of those that died had a huge role in the family, especially the elderly, they had an important role in the family and in the village. We cannot forget. Sometimes, it's like a movie playing in my mind. I'm in the car and I drive and drive and it just rewinds in my mind what happened on that day and when that happens, I am heartbroken." Taua Kitiona Seuala, Minister for Agriculture and Fisheries, Samoa.

"I was hurt by the tsunami, but I got over that because I've always had God on my side, He has always been with me and my family. He's the one that's given us the strength to continue and through Him we return praise, even on bad days. You have to put aside the problems and move on. I never had any training in tourism but I cared for my parents, I know that caring for people comes naturally for me." Faafetai Taufua, Taufua Beach Fales, Lalomanu.

"The people of American Samoa and Samoa did the right thing after they felt the early morning earthquake – they self-evacuated from the coast to higher ground in the 10 to 15 minutes before the tsunami arrived. Some fled as soon as the ground stopped shaking; others fled when they saw the sea withdraw from the shore. Had residents not evacuated, the fatalities would likely have been in the thousands instead of fewer than 200. This underscores that education saves lives. Recent tsunami education based on knowledge obtained after the Indian Ocean tsunami of 2004 – was critical in preventing greater loss of life." Bruce Jaffe, USGS Science team on the Samoa tsunami.

"Today if a truck drives past and the ground shakes, I find myself looking around with a bit of panic at the faces of people around me. When I'm driving and a foul smell comes in the window, I get a flashback to those days searching. When I'm driving out to Aleipata now, I can't help but stop and think back to what happened. It affected so many people, it brought so many people together, to work and do unimaginable things. To do what we had to do in those conditions has been for me, one of the most amazing aspects of the entire sad situation." Vaughan Simpson, Volunteer Emergency Services, Ca'bella Construction, Apia.

"I think when you go through a traumatic experience like this, the only people that really understand how you feel are the people that were there and you always have a bond with them. Usually when you go on holiday, you meet people, have a chat and then probably never stay in contact again. But this [tsunami] was such a massive experience that you feel – I've got fifty new friends who I didn't have – and we're all connected now." Diane Cederman, Raglan, NZ. Guest at Taufua Beach Fales.

"This event has been a learning experience, it's strengthened me as a public servant and it's made me that more determined to do what I have to do while I'm in office now for the country and for the people that I serve here in government, with what little knowledge that I have. For me as a person, this event has made me realise how in this life, anything can happen between now and the next hour, so you have to embrace what you have now. Love your family, love your children, continue just to draw on each other while you have the time, because you never know when you're going to be taken away suddenly, you never know what's going to happen the next hour or even the next day." Vaosa Epa, Secretary to Cabinet, Samoa.

"Starting from 7:30 that morning, only half an hour after the tsunami, I started receiving phone calls from overseas media. They wanted to know about the deaths, the destruction. They wanted the blood and gore details. The calls continued all the next day. They were not happy with my answers. They wanted to hear about the bodies and I did not know about anything like that. Then when the earthquake happened in Indonesia – nothing. Nobody was interested anymore in Samoa." Arne Schreiber, German Honorary Consul for Samoa.

"I think the tsunami did not only affect the people of Aleipata and Falealili district, it actually affected the whole psyche of the nation. I think we were all in shock. The outpouring of grief was very evident and at times, there was a lot of people in our country who were in a daze for a long time afterwards…I was amazed by the outpouring of emotion not only within the country, but from family and friends from overseas and a lot of countries who came to Samoa's aid. It was so encouraging to see that while we stand alone as a sovereign nation, we have a lot of support, a lot of friends and a lot of love from many countries." Pepe Christian Fruean, CEO Digicel Samoa.

"I seriously believe there was a reason for this tsunami, it was a warning. These are signs of Biblical times. The flood occurred in the Bible and fires and major things occurred. This has to be a sign. We can talk scientific and say the plates shifted here and there and the heat has to come up through some volcano, well whatever, so be it, but it's a miraculous thing. It's a miraculous thing that we are all alive and we're born in this world and we live in paradise, maybe we've abused it a little too much and taken it for granted. We need to go back and look at how and where we live and how not to abuse it." Letiu Lee Palupe, Saleaumua.

"I have seen, I have felt and I now know the concept of a natural disaster. The biggest challenge for me was seeing my own people suffer and the disbelief that this would happen in Samoa. This has given me greater perspective about what is important in my life. I have learnt that family are precious and we need to treasure time spent with them. I am thankful to have been able to serve at home during a time of great need and will always remember the strength of our Samoan people. Their resilience and courage has been an example to me. The opportunity to actually be there and work with other health professionals was priceless. To witness the spirit of the local and global Samoan family in moving so quickly to assist and provide for tsunami victims was uplifting. It was a privilege to be there." Henrietta Aviga, Volunteer Nurse, Thursday Island, Australia.

"The Samoa tsunami is yet another example of a breaking news story that took place far from the watchful eyes of the media, yet was witnessed firsthand by many ordinary people who felt the need to document and share what they saw. For most this was unlikely to be about finding fame or fortune, but rather about sharing what they saw with friends, telling how they'd been affected and asking for information or advice on what to do next. So rather than submitting their images and videos to mainstream media organizations, they post them online on Facebook, Twitpic, or wherever their friends are likely to see them...we're going to see a lot more of this in the future – people using social media to give their own accounts of what happened, and others being drawn into the story, in a more personal way..." Robin Hamman, Head of Social Media, Headshift, U.K. '*Telegraph.co.uk*'.

"I'm so grateful that I was here when it happened. I don't want to ever go through it again but I'm pretty proud that I was here to see and witness how the Samoan people dealt with such a monstrosity of a disaster and they did – they were really good." Steve Williams, Volunteer Paramedic, Australia.

"Recommendation: Build on traditional village structure to promote resiliency. Incorporate tsunami safety into village pride to sustain community efforts. Consider the one year anniversary of the tsunami as an opportunity for villages to develop evacuation routes and practice drills." EERI Special Earthquake Report on the Samoa tsunami.

"There were so many people who came to Samoa to volunteer. We found that the help was quite overwhelming and that really for me it lifted the spirit. How can we be tired when all these people have come so far to try and help us? It was really something. For me, that's the thing that most touched me." Dr Ben Matalavea, National Hospital, Samoa.

"Two months after the tsunami, Mike and I went to Raglan. We met other tsunami survivors...we stayed with the Cedermans and Andy White. We ate and we drank. We shared our memories and we shared our stories. We know we now have something special. A special connection not only with each other but also with the beautiful country of Samoa. We spoke about going back sometime soon. We spoke about ways to continue to help the people of Lalomanu. This place, in particular, will always have that special part in my heart. And yes. We will be back...there is a plan forming already." Sally Adye, School teacher, Gisborne,New Zealand.

"When I was little, I would see our Samoan people working together, sharing and helping each other. But it was different on the 29th of September, that's not what I saw, people didn't help, they came to take things from you, they didn't care. I asked myself – what's happened to Samoa?" Lina Bell, General Manager, Coconuts Resort, Samoa.

"We couldn't have done it without international assistance – those people that came in and worked on the ground. Never mind about the 'big wigs' who say 'we helped, we did this and that.' Im talking about those who worked *with* us. Their careers are moving from one country to another in response to disasters. The people who helped us here – they're now in Haiti helping with the earthquake response, that's all they do. From here some had to leave quickly to go to the Indonesia earthquake disaster. That gave me strength that there is this international support. So when the Haiti thing happened, I felt for the Public Health director of the country, you're thinking about all your counterparts and all these kinds of people when you watch TV and you see what's happening over there. You know exactly who those people are and what they're doing. For me, it was really amazing, I'm very grateful for their strength." Andrew Peteru, Asst. CEO Health, Samoa.

"This has definitely changed how I look at things. Each time I go surfing now I have an exit strategy, know where I am, know what's the best way to go, tell people where I'm going, and make sure that the people with me are aware of the risks too. The tsunami hasn't put me off the sea – look at Joe Annandale at Sinalei Resort, they're rebuilding down by the ocean , it hasn't put him off…I don't think we need to wait for a warning system, an earthquake itself is a warning signal that basically alarms everyone – if you're on the coast then you've got to get out and if you're in the water, then you have to get out to deeper water." Darren Greenwell, Surfer, Samoa.

"During the previous tsunami drills, people would complain *ah what is this? Tsunami will never happen, why can't the government go and do something else, they're wasting taxpayer's money!* Our boys would help enforce the evacuation and during the drills when they went to the market, people didn't want to go and leave their stuff, they said they would rather prefer to stay and die there – but when we had the second warning after the September tsunami, there was nobody left at the market. Hopefully future generations will be more alert than ours and hopefully they will take drills and exercises seriously." CO Petone Mauga, Nafanua Patrol Boat, Samoa.

"When we were interviewed for this job, we were asked whether we would be ready to respond to an emergency at any time – even those who work in the office – the Commissioner is always stressing that we need to be trained and ready, we need to be operational. This has woken us all up to be prepared, because everyday Commissioner Hill is always on our case about this, but it wasn't until the tsunami that we finally woke up and realized that we're not here just to put out fires or wait for fires to come, anything can happen any day at any anytime. We have to be prepared, not only to act but also make sure our families are prepared. Because when the emergency happens, we can't go home to take care of our families, we have to stay and respond to the call." Peka Salapo, Fire and Emergency Services Authority, Samoa.

"After the tsunami we stayed in Apia for a few days, trying to pick up the pieces...as we talked to others who had lost family and friends and watched the death toll rise on television, my only question was – if God existed, then why would he have done this to such a religious country?" Thomas Gogola, Surfer, United Kingdom.

"Never take life for granted, tragedy can strike at any time and we can waste too much time worrying about the wrong things…I've learnt that in the midst of all the adversity, violence and hate in this world – there's so much love out there. The outpouring of care and support that's come from all over the place, all over the world has been overwhelming, fantastic. That in itself is the inspiration for us to keep going. I think if you do right by people, give people a helping hand when they need it, they remember, and it does come back to you." Sose Annandale, General Manager, Sinalei Reef Resort, Samoa.

"Our people are very resilient – I think they are able to move forward because of their spiritual grounding, because the extended family plays such an important role so that if you're suffering, you're not suffering alone, you've got the support of family, then community and then country. On that level I feel that Samoa will move forward." Elena Peteru, Counsellor, Psycho-Social Response Team, Samoa.

"So many lives changed on the 29th of September. As a nation, Samoa has endured so much…and the world has stood beside this little country in the middle of the Pacific Ocean. These are definitely days to remember. Mistakes will be made, we are only human. But let us not forget the bountiful good that has been done and will no doubt continue. On my last visit to Vaovai village, a matai farewelled us with a familiar Biblical verse, '*O oe o le Isaraelu moni…*' to express his deep gratitude for all those who assisted and gave so generously in his time of need – from all over Samoa, from all four corners of the world. The work by all has been heart-wrenching and yet so uplifting. We came to give but what we gave was nothing compared to what we received." Malia Manuleleua, Counsellor, Psycho-Social Response Team, New Zealand.

"On Christmas day we went to the graveyard to wish my parents happy Christmas. This is the first time not talking to my mum and dad and going to visit them during the holiday. I feel so sad looking at other families. They're enjoying looking at their parents while I miss that. I miss a mother's love and a father's love. Although I wasn't there during the tsunami, but I still feel that part of me was there because I lost my dad and my mum and I feel sorry for my brother, he lost both his parents and his kids as well. I wish there was something I could have done to prevent it." Lucia Schwalger Hellesoe, Lotopa.

"Honestly I have to say that I have lost my faith in our people. I have lost my faith that some Samoans deserve a better chance in life, a better opportunity. After everything that's happened, I try to make sense of it – why would people do such things? Why would people steal from people who have lost loved ones? They even stole the earrings from my daughter's dead body. I think it's because they never had those things, like mag wheels, pots and pans, spoons… We're suffering because we lost our loved ones, we lost every single thing – homes, kids, parents – everything but for those people, the tsunami was a blessing. Now they have the chance to take all the stuff they never had before." Jared Schwalger, Malaela.

"To the people of Coconuts Beach Resort - My wife and I were guests at your resort three years ago. It was the most wonderful vacation I have ever had. It broke my heart and brought tears to my eyes when I heard the news. The owners are wonderful people and the employees are second to none…we are planning to come back next year. Please let me know how I can help, it would be an honour to help such wonderful people. Our prayers and thoughts are with you." Randy Keading, USA. '*Coconuts Resort Website*'.

"No one is ever going to build anything strong enough to withstand a tsunami, there's nothing you can do to build to withstand that. People rebuilding have to realise that anything they're rebuilding – its going to get wiped out if there's another one. The best thing they can do is have a clear cut track straight up the hill. Because with that cliff, people can't go anywhere, they can't go sideways. There's only one way to go and if you've got to climb over trees and bloody run through peoples houses and having dogs bite you – I don't know how the hell people that ran away, actually got away. You've got to run past toilets, pushbikes and whatever else they've got at the back of their houses, over pig pens, through chicken pens, around fales and then you've got to run through the bush. About three hundred meters of dense, tropical bush. Yes, you'll get up that hill if you had a nine meter wave chasing you – but the hard part is getting over all the shit first before you can get to the hill! If you're going build out there again, they're going to have to put in clean, clear tracks where people can just run to get up the hill." Vaughan Simpson, Ca'bella Construction, Volunteer Search and Rescue Worker, Samoa.

"To this day, I have a small carry-bag that I take with me everywhere. Everyone is wondering what is in this bag and it's the clothes my children died in…I have washed them clean and folded them and put them in this bag, so wherever I go, whether it's a family thing, or an errand, even to government offices, I enter that office with this bag which has my dead children's clothes in it. That's the only thing that's in this bag, because it's very painful." Mamea Mikaele 'Mika', Lalomanu.

"I want to help Rita and Romeo and their family at Lalomanu who lost everything. I spent five days in 2003 at Romeo's Beach Fales. I stayed in a rustic fale and loved it. I love the family, they took me to their mom's house. They live a very simple life on a beautiful island in open fales near the beach and rent fales to visitors. They have lost everything except their lives. I am so happy they were not killed. They are not well known to the tourists like the fancier and larger tourist operations, this made them very special." Beth Fridinger, Canada.

"We are so relieved to see and hear that the passion and resolve of the Taufua family and the people of Lalomanu remains strong. We stayed with you for ten days in August 2009 and the news of the tsunami gutted us. We lost dear friends in nearby Satitoa as well as Lalomanu. We haven't heard any news of our guide and dear friend, Mika, who took us for a bushwalk at his home and later guided us on a day tour of Upolu. Please send us any news of him. You have been and always will remain in our thoughts and prayers." Shane Muldrew and Norma Flach, Victoria, B.C, Canada. *'Taufua Resort Facebook Page.'*

"I would like to send my thanks and a big hug to Mika the tour guide. We were lucky to share the most amazing day with him and his family the day before the tsunami. We could never have imagined that it would be the last day for so many members of his family and also for our friend Mary Ann. With his help we fitted so much into that day, including preparing the *umu*. Mary Ann picked up the basket making quite quickly and her *palusami*-making skills put mine to shame! It really just reminds you that just like squeezing the cream out of the coconut, you have to squeeze the most out of every day." Diane Cederman, Raglan, New Zealand. Guest at Taufua Beach Resort.

"There's no other way for you to be strong and there's no other way but for us to work together to get things going back again. My children are all alive, thank God. For me, it's okay, tsunami take my business, take my house, take my cars but please, please don't take my kids because if any of my kids was taken that day, I wouldn't know what to do…I don't care if I have to go and sweep the trash in the street and work another job to make money to get our stuff together, but my kids are alive. I say – take the cars, take the house, take the business, but please don't take my family." Eta Wyberski, Pago, American Samoa.

"When we finish our evening prayers, the first thing I explain to the children is that if they have long lives, God willing – that they should take what the elders say to them and use it and keep it safe in their hearts for their future. When their children grow up, they must explain to them as well. All these things like tsunami and other disaster, come down to our faith in God. People say its due to natural causes but I don't believe it. All these things come down to God. It doesn't matter if you are clever at school and go study overseas – that will mean nothing in these disasters. These things are works of God." Lesa Foti Tuvale, Lalomanu.

"I see now that there were signs that something bad was going to happen…especially when my daughter who died in the tsunami woke me up at two in the morning so we could pray together. Then the next day the earthquake hit and she was saying goodbye. I am so thankful that our other child was saved. It was the will of God. I thank God for his love that he gave me back our other child. This is the child who will have a future." Ave Mikaele, Lalomanu

"During the tsunami I hugged my son to my chest and prayed to God to help us in our struggle. The only thing I was thinking was hoping my son would live, that was what I was hoping for, if anything happened to me but that one of my children would live. And now, me and my children were spared because of God's love." Maua Toa, Lalomanu.

"Funerals are meant to provide relief (financially and emotionally) and do justice, dignify the memory and legacy of the deceased. Instead Samoan funerals have become very expensive and stressful, with some families getting into grave debt by the end of it…the seeming ordinariness of the tsunami funerals, with the minimum fuss and bother that surrounded them, did not, however, lose any face by their simplicity. Instead they gained, in that they reminded us of what really mattered. In this instance, rather than raging menace the tsunami chastened and cleansed. We might say that it forced us to front up to our vanities…the tsunami forced us to ask – are our families suffering because of our own misplaced and inflated expectations? If the answer is yes, then we must take pause to sort out why this is so." Tui Atua Tupua Tamasese Ta'isi Efi, Head of State, Samoa.

"I have realized that there is a very special feeling amongst the people that were staying here with us during the tsunami. We have been brought together like family now, so we have many *palagi* families amongst us! That's one good thing I felt when we were in hospital. We were very concerned about all the guests – not knowing that they were also very concerned about us. That brings a very comforting feeling and it's become like an obligation for our family to treat those that were here like relatives as well." Sili Apelu, Taufua Beach Fales, Lalomanu.

"I think the teaching of Samoan history is very important. In the Phillipines, students complain *why do we always have to do this same stuff over and over again?* But when they get older they realize, how important it is. Learning about past disasters and events is important because I can apply it now to the present and even the future. We should teach more about history, about what happened to Samoa years ago, in terms of weather conditions and war and everything like that so people will be informed and think *oh yeah, its happened before, it can happen again...other people suffered and they did this and that to recover and progress.* Many people didn't know that Samoa had tsunamis before, they think this is something very new, that it can only happen overseas. But now we know that's not true." Melody Lesa, Nurse, Lalomanu Hospital.

"Late that afternoon we got a ride to Apia on a ute. We drove down the hill and past the Lalomanu Hospital. A truck was driving in the gate with a whole lot of men in medical masks and what looked like either wounded or dead people in the tray.We drove back through what was left of our resort which was nothing but scattered rubble. We were all very quiet. Sticking out of the sand was an arm and a leg, the rest of the body was buried under the sand." Christopher Wilson, New Zealand. Guest at Litia Sini Resort, Lalomanu.

"The Samoans were the kindest people I have ever met. On that day they looked after us before they looked after their own families. They lost everything and most had lost family and friends. We are collecting money to help them rebuild their lives, so please give generously. We are returning back to Litia Sini once it's rebuilt. We want to go back for opening night." Max Wilson, Age 12, Auckland, New Zealand. Guest at Litia Sini Resort, Lalomanu.

"It's hard to explain how I felt when our home was destroyed. You want to get mad, but who are you going to get mad at? You work so hard, for so long to build something then it's all taken away in five minutes. You feel like you've lost everything, lost hope, that's how I felt. It was good that nobody in my family got hurt, that would have been worse. But it's just like watching everything collapse in front of you." Francis Keil, Leone, American Samoa.

"I will go back to Samoa, because it's paradise, but any beach I visit now, I will be looking for the escape route." Wendy Pearce, Christchurch, NZ. Guest at Litia Sini Resort. *'NZ Women's Weekly' Magazine.*

"Now, I know how important it is to be prepared. Every night, if I hear something, I get up and go check it out. Now, every night before I sleep I think what I will do if anything happens. I already have a plan ready for my family, a way to save them if anything happens, how to get them out fast. Yes, be prepared. Everyone should have a plan for their family and prepare for anything because anything can happen at anytime." Iuliano Falaniko, Leone, American Samoa.

"Some lessons learned from the tsunami in Pago Pago: It may sound trite but it is *so* true that you never know when the last time will be for most things. Living in the moment, maximizing every opportunity are attitude and behaviour to live by rather than cute phrases and platitudes. When it is all said and done, people, friendships, and relationships are all that really matter... The best place to be when trouble strikes is ON your boat and out in open water. Get there and stay there at almost any cost...In times of great stress and disaster, human nature is on full spectrum display and is the same in all places and cultures...Put your faith and optimism in people because there is so much more good in the world than evil, many more good people than bad...Mother Nature is a majestic and powerful force on a scale that is truly humbling, its likely a good thing to be reminded from time to time just how small and puny we are." Wayne Hodgins and Ruby the Wonderdog Blog, aboard the *Learnativity*, Pago Pago Harbor, American Samoa.

"Each scientist who goes to an area devastated by a tsunami has a unique experience, but all are moved by the evidence of human loss. There is no better way to learn about a tsunami than to go quickly to a stricken area and collect ephemeral data before it is lost, but there is a price to pay for gathering this knowledge. While in American Samoa, I came across a small shrine on the foundation of the destroyed house of three girls who had died in the tsunami. It contained pictures of the girls smiling, their sparkly shoes and their teddy bears. Being the father of a young girl, I found it difficult to continue working, but I did. Like other scientists working in the aftermath, I suppressed my reaction and concentrated on collecting data needed to improve out understanding of tsunamis. The thought that the work we do helps to save lives focused our teams on the task at hand. At the opposite end of the emotional spectrum we also experienced the joy of helping. Many people wanted to know more about this tsunami and to understand the risk from future events. We tried to answer their questions and felt glad to know that our research will help them and other communities, plan for future tsunamis." Bruce Jaffe, USG Scientist, USA.

"This week in October has been set aside for national fasting and prayer. At school I asked people why and they answered, *to avoid another tsunami.* With the tsunami only three weeks behind us, other volunteers also reported being encouraged to pray very hard in order to avoid a tsunami...I have no problem with this national mood and I think its probably for the best that people show more concern about tsunami preparedness. During the national drill last year most people in our village didn't participate. The explanation from one family? *God will save us*. Even the morning of the Sept earthquake, most students acted as if nothing had happened. So it's good that there's a stronger cause-and-effect relationship between earthquakes and tsunamis now. That said, it's a little discomforting that the focus is more on pray and fast than duck-and-cover." Matthew Leal, Volunteer teacher, Samoa.

"Building a house for rich people back home in New Zealand – they're only showing off to their friends, they don't need it. You build homes here in Samoa for the people of the tsunami – you're not only building homes, but lives as well." Jared Suisted, 'Habitat for Humanity' volunteer, Cambridge, New Zealand. '*Waikato Times*' *New Zealand.*

"I cannot begin to express my heartfelt thanks, the thanks of the Government of Samoa, the CEO of Health and all the staff within the health sector, for everything that you have offered to the people affected by the tsunami and for the people of Samoa. You brought with you expert skills, love, the commitment to help those in need and the sacrifice your families and work colleagues have taken, that has made your mark in making the rescue and recovery process of this tragedy, a reality. We have all worked well together...you have represented the goodness that God has blessed each and every one of us with. It will be our challenge to return this goodness some day to those also in need. We want you to know that your contribution has made a difference to so many lives and has lessened the suffering and pain for so many people." Gatoloai Amataga Gidlow, Minister of Health, Samoa.

"My daughter ten year old Losivale had a passion for acting and every Christmas she would participate in the church skits. Last year she was the angel Gabriel, the one who brought the good news to Mary. And two year old Lutia loved to dance, last Christmas she was one of the sheep of the flock...so every time we see the other village kids rehearsing for the Christmas program this year, we cannot help but think what it would be like if Losi and Lutia were still here. Seeing the other children participate in the Christmas skit practice brought back memories. I miss my girls. I couldn't stop the tears from flooding down." Mekala Tainano, Lepa. '*Samoa Observer*' *Newspaper*

"From the time I was asked to go to Samoa to when we flew out, was less than three hours. I only had time to call my very supportive wife and pack... It was my first time in a disaster zone, it was pretty devastating, everything the tsunami touched, it destroyed. We started working early in the morning and finished at about 7:30 every night. One of the girls we operated on- she'd lost her baby and two other children as well. The wound care nurses would spend a lot time talking to families. Being a doctor, we're very privileged that we can help people. I'd definitely do it again." Robert Rowan, Orthopaedic Surgeon, Wellington Hospital. *'The Dominion Post' New Zealand.*

"It's been really good. It's very different; hot, sweaty. Man, I've never sweated so much in my life. The work is slow. We're working with wet timber, limited gear. We've got two hammers for four guys, one bench saw. We're doing our best with what we've got." Glen Cane, Habitat for Humanity Volunteer builder, Matamata, New Zealand. *'Waikato Times' New Zealand.*

"I didn't know that Mother Nature could be like this. I had trained for it but to have it happen in our own back yard, I just couldn't believe it. Cyclones Ofa and Val were nothing like this. Out of the blue like this. The amount of lives that were lost …If it had hit Apia, we would have lost so much more. We were lucky it was early hours of the morning. People were awake. People were evacuating. We were very lucky that it was only a very small percentage of Samoa that got hit. It's a wake up call. Its telling us – hey, you've got to be prepared. Out of all this, we've been very fortunate, we've got a lot of equipment given to us by the Australians and New Zealanders when they were here helping us out… What it's saying to me now, is that you've got the equipment and you've got to put it into place and be ready." Commissioner Tony Hill, Fire and Emergency Services Authority, Samoa.

"I promised the children of Manono that I would help build a school for them…if we all get together then we can make it, it's all about the difference we can make in someone else's life. We are blessed in New Zealand…to have peace here in our lives. We have clean drinking water which they do not have. So far we have had people that have been donating kindly. I was surprised when a single mother of three children donated $160. Single drops will fill the bucket." Richard Khouzame, Waikanae, New Zealand. *'Samoa Observer' Newspaper.*

"Families were so welcoming of us. They were living under tarpaulin but they'd invite you in to offer you drink and food, listen to you and respect you. I was so impressed that after all they'd been through, they were still so positive. But people were deeply affected. They're telling you how children were washed out of their arms. You listen to them through the interpreters and then you have to say, *Well, that's sad, but you really need to dig a toilet. ….*" Matt Molloy, District Health Protection Officer, Nelson, New Zealand.

"When we re-open one of the changes we will be putting in place is curfews for evening prayers, even for the tourists. This is the experience that many *palagis* come to Samoa to get, often we think that we should ignore some of the cultural and religious traditions because of the visitors – but then the tourists don't get that real Samoan experience...whether they are religious or not, curfew will be implemented and if they don't pray then they will be asked to acknowledge that time for the Samoans on the island to hold their evening prayers and dinner will be served afterwards." Sara Ulberg, Namu'a Island Beach Fales, Samoa. '*Woman's Time' Magazine, Samoa.*

"The thing that lessened our burden was the help from our Samoan people. The people from other villages would drop by and see how we were – even though they don't get anything out of it, but just to talk and to help – there were a lot of people like that. That's why I admire our Samoan people, because they help out." Maiava Leota, Manager, Tafuna Airport. Leone.

"Last year we stayed at Taufua Beach Fales with our four daughters. We made new friends and were treated like royalty by the Taufua family. Please accept these photos of the better times and the memories we have as a gift from us to sooth your hearts while you go through the hardest of times. Our love to those that have been lost and our love to those who remain. Your memories and hospitality will never be forgotten." The Grey Family, Nelson, New Zealand. *'YouTube'.*

"There were so many things to learn from this, it was amazing the unity we all felt during that time. Everyone wanted to help – we had all sorts of people just getting in their cars, making their own way out to Lalomanu Hospital and coming to help." Dr Teriki Puni, National Hospital, Samoa.

"This morning when my first student arrived in my room where I teach English one-to-one I could just feel the sadness before he sat down. I asked if he was ok, no answer. When I asked again, with as much English as he could muster, he said, *I don't know what has happened.* At that point he pretty much broke down crying and although I'd been practically holding my breath until that point, I, too, started to tear up. His aunt, sister and five year old niece are missing, but since his village has been wiped out he's already decided they are dead. ..he wanted to talk about life in Samoa. His reflections seemed to be asking why is he here in New Zealand when he should be back in Samoa where life is simple and you don't have to pay for food, housing and water? You just build your house, grow your food and spend time with your family...I felt ridiculously helpless and sad." Threespoons Blog, English teacher, New Zealand.

"God is telling Samoa to go back to the old ways. He holds all in His hands, he has the authority over all of us. There were many warning signs before the tsunami – it was not an isolated accident. There was the swine flu scare, then the road switch, the fishing boat stuck on the reef in the harbor and of course the Virgin Mary appearing on the wall of the John Williams building; these were all signs of the building tsunami. Mary was warning the people of Samoa but we did not heed her warning." Anon. Samoa

"The reality is, like everywhere else in the world, we are a not a country of angels. Some people, with the emphasis on the some, will take advantage of any situation for their own selfish benefit, tell untruths to aid deliverers to get more aid…half-cooked, unsubstantiated media reports [about the misuse of tsunami aid] belittle the efforts of the hundreds of those who've been working on the emergency and recovery frontline for many weeks…The real victims are those affected by the tsunami. The real heroes are those on the emergency and recovery frontline, day and night away from their families committed for the benefit of others." Tupuola Terry Tavita, Editor, *Savali Newspaper,* Samoa.

"Having lived through a hurricane, and covered cyclones in Samoa, it's hard to be a detached journalist. I cannot separate my feelings from being a former resident. Now I and other media have left, but I would hope that the media won't forget about Samoa, that they go back from time to time to check the status…to find out what the special needs for children are, and the special needs for people who had babies ripped from their hands. There are so many social needs that will continue, yet there will not be as much media attention as there has been during the first week after the tsunami. I fear they'll be forgotten and left to themselves with all that destruction." Barry Markowitz, Photographer, American Samoa/Hawaii.

"I learned that we can't have that stubborn attitude that everything's going to be okay when an earthquake hits or any kind of flooding or anything. You have to listen to what the warnings are telling you. If an earthquake hits, first thing you do is run uphill. This disaster has brought a lot of people closer and helped them build a stronger relationship with the Lord. It has tested our faith." Kathleen Pearson, American Samoa.

"This showed me that if there's something you want to do in life – something bad like a disaster could happen at any time and take it away, so you might as well live. If you want to do something, you make sure you do it; do it well and do it when you can, as soon as you can. Don't wait. Don't put it off." Matt Ansell, Age 16, New Zealand. Guest at Faofao Beach Fales, Saleapaga.

"I want everyone out there to know how grateful we are for their love and support. There were so many people – even strangers – sending us messages of comfort on the internet and praying for us. I miss my daughter Vaijoresa so much, we have fixed up our house and everything, but it still doesn't feel right because she's not here to share it with us. It helps me to know that so many people out there are caring and hurting for us. There is no way I can thank all these people. They've been so supportive to me and my family. Please tell them in your book – thank you." Taitasi Fitiao, American Samoa.

"We need to get our kids to be more able in water. We live on an island surrounded by water and yet most people can't swim. When I first moved back here, there were at least six drownings in the first three years and that's way too high for a place that's surrounded by water. During the tsunami, I saw people that were too scared to go and help others – because they couldn't swim. If more of our people had been confident in the water, then maybe there wouldn't have been so many deaths." Robert Toelupe, US Navy Veteran, Leone, American Samoa.

"This event has reaffirmed and heightened my love and affection for our family. I have a stronger sense of belief in what we can achieve by working together, each individuals unique actions contributing to the whole. There is no doubt that this was a tragedy but through it we have become a more unified family. One small moment in our lives has taught me a huge amount about myself and my family. We are an awesome bunch of people." Ngatau Hemopo, Wellington, New Zealand. Guest at Faofao Beach Fales.

"It was like being chased by a monster but ten times worse. You get this feeling right in the pit of your stomach – like you could lose everyone you loved. But I feel so thankful and incredibly lucky that we weren't one of the unfortunates who lost everything…seeing everyone of my family on the hill gave me a feeling of happiness. I don't care about the new phone I lost, or my new clothes, family is all that matters. We can replace my wardrobe, but you can't replace family." Tia Ansell, Age 14, Auckland, New Zealand. Guest at Faofao Beach Fales.

"One of the biggest impacts that this experience has had on me is how life is so precious and how we need to live every moment the best that you can. Take nothing for granted and be grateful for what you have. People who worry about what they don't have - often need to recognise all the wonderful things they *do* have. Life is short, make the most of it!" Rebecca Hamilton, Gisborne, New Zealand. Guests at Litia Sini Beach Resort.

"We want to say how fantastic the fire service was, and the medical teams. They were there so quickly. Anyone who ever says that there was not enough fast action after the tsunami – they're wrong. It was unbelievably good, especially for a country like Samoa where you wouldn't think they would be that well organized. If we had that sort of reaction time in New Zealand, you'd be thrilled. It was sensational. I take my hat off to whoever helps organize Civil Defence in Samoa." Graham Ansell, New Zealand. Guest at FaoFao Beach Fales, Saleapaga.

“For the first three years of my posting here, I told the Samoan Government and people that they mattered to Australia. For the past seven months [since 29/09], I hope I’ve been able to prove it.” Matt Anderson, Australian High Commissioner to Samoa, Apia.

“My job is to encourage people to still visit our islands – is it not still the treasured island of the Pacific? How do you do that with a smile and an encouraging word when the heart feels something else? …but the mind is strong and will go on and the heart will follow.” Gardenia Elisaia, Samoa Touristm Authority, Samoa.

“To me, there’s a lot that I’ve learned from this. About surviving. I’ve learnt how to survive without a home and lots of technology. My family used to have everything and anything. We watched TV, we slept on beds, we had computers, nice cars and nice things. Now we’re living in a tent and living in a shack. I realize now how other people live who don’t have lots of things. Now I appreciate life more and I’m thankful for it – even if I don’t have anything at all.” Simao Masoa, Poloa, American Samoa.

“There has been so much kindness. Everything our family needed has been provided. After the [tsunami] we had nothing but we have been given so many things – a fridge, pots, plates and so much clothing and other things all the kindness and love of people. I just thank the goodness of people from our own country and all those from outside too.” Tofilau Afatasi, Poutasi, Samoa.

“The psalms we recite, both from memory and from anguish,
O Ieova lou leoleo mamoe o ia, e leai se mea ou te mativa ai
The wails get louder; people huddle and grieve openly together, proud people broken
Though I walk through the valley of the shadow of death, I shall fear no evil.
Even the children are not spared the knowledge, they have seen it, smelt death as its aroma lingers over stagnant water left by the killer wave,
Yet still we sing out loud in church, *Pe solo mai mauga, lulu ina mea uma* and *When sorrows like sea billows roll, whatever my lot, it is well with my soul.*
The tears run until there is no more to shed, and people gather together as one
E faavae Samoa i le Atua, Samoa is founded on God, resonates through church halls.
But we will never forget that day as long as we live.”
Aurora Elisaia, Poet, Samoa

"I don't consider myself a victim, I consider myself a survivor. A lucky person to have gone through that. But the real victims are the ones back in our homeland. Who have to live there, be reminded of that day, everyday, who have to rebuild their lives again, move on and slowly start life again...no matter what happens to Samoa, all our people around the world will be affected because we are all joined, we are all one. I am proud to say I am a Samoan." Steph Fa'amatala, Australia. Visitor at Sa'anapu. *'Pacific Eyewitness Web'.*

"Sometimes people are ungrateful, very ungrateful... Whatever donations we get, we give out. If we get boxes of eleni (mackerel) then we give out boxes of eleni. And some families say, *we don't eat eleni, our family don't want this.* And I just get disgusted by their ungratefulness, some of them don't value the donations from people. I advise my staff, if the people don't want it, just bring it back. But you know what? They don't want to return it, the people complain but they still keep the eleni, they'll eat it!" Filomena Nelson, Chief Disaster Management Officer, Samoa.

"Moving forward obviously education is the key and you'd think this generation would have learnt from [29/09] but some haven't. When we had the Chile earthquake and subsequent tsunami warning in February 2010, I heard how some young village men were running out into the sea bed and picking up fish when we were having the sea surges. Some might see that as funny, but I'd call that completely stupid." Vaughan Simpson, Volunteer Emergency services, Ca'bella Construction, Apia

"As we all contemplate the utter horror of the event we share a common sense of the beauty and fragility of life. No one can walk away from this event without being reminded of how grateful we all are to be breathing fresh air, not salt water. As I write this, I am grateful to reach over and touch the warm body of my husband. We sit in a hospital room in Samoa for the second time in a week. I am grateful for the opportunity to sit with him. He has a black eye, scrapes all over his body, two broken ribs, a puncture wound in his knee, a broken toe and an empty spot where pliers just removed his toenail. But he is here. We moved to Samoa to be together every day. We were hoping to simplify our lives. I am grateful for the opportunity to joke with him about how the tsunami simplified it real quick. I am grateful for the opportunity to tell him that I love him a hundred times a day, and I am utterly grateful to have him here to respond that he loves me back." Rachel Lynn, Coconuts Resort, Siumu

"When I think about what happened, it's like it all comes flooding back, the pain and the hurt. I wish I could forget my injuries and the death of my child, but I can't. Some days I will just be sitting here then I remember my child and I weep." Leua Vi'iga, Saleapaga.

“Samoans will remember for some time the power of this tsunami. We will remember this one not just for the paradoxes of its rage, but more poignantly for its rallying, reaffirming and cleansing of the ideals of family. The essence of family I believe is its ability to come together in times of need. The Pasifika family extends beyond the shores of the Pacific. The love and support to Tonga and Samoa that poured in from countries all over the world demonstrates our shared humanity… The response of New Zealanders to the September 29th tragedy can only be described in terms of what would be the response of loving kin. The same must be said of Australia… *O le e lave i tiga, o le ivi, le toto ma le aano.* He who rallies in my hour of need is my kin.” Tui Atua Tupua Tamasese Ta’isi Efi, Head of State of the Independent State of Samoa. Keynote Address, NZ Families Commission, Pasifika Families Fono, 3 November 2009.

“The most important lesson for me from this tsunami – is that I never took it seriously. When they were going through the tsunami drills and educating us about what to do, it wasn’t a big thing to me. It was only now that its happened, that we’ve learnt our lesson. We must learn, listen well to the warnings and the education that is provided by those groups who teach us about *galu afi* and other natural disasters…but for now, we just have to thank God for being alive.” Vaelupe Olosepu, Malaela.

“I am thankful that my children and grandchildren have witnessed a disaster of this magnitude and survived. They are full of fear and don’t want to experience this again. It’s best that they come to higher ground to live but they need to remember what happened in this year so as time goes on they won’t forget. Because as they grow older, they will leave their parents and have their own families, but they need to remember…even though you just want to forget why all these lives were lost.” Peniana Iuni, Lalomanu.

November 2009.

I am in quiet awe.

There are frangipani trees in bloom along the shore of the tsunami zone. Gnarled grey barked trees reach heavenward with delicate white blossoms. New green grass at their feet. Lining a shoreline scattered with the heaps of twisted corrugated iron, jagged wood planks and the fast rusting carcasses of once-beloved new cars.

There are frangipani trees in bloom. In spite of the salt. And the ravage of eight foot waves that sounded like "the crashing roar of a hundred heavy duty machines". The panicked desperation of those who clung to tree branches to escape the leeching swirl of the receding water.

In spite of the desolation. Spindly shelters where families guard their meagre belongings against the ongoing thievery as both young and old pick through the remains of everyone else's lives. Even the hub caps on your truck are not safe.

In spite of the grief. A woman sits by the road crying on a freshly poured cement block. Her mother's grave. I miss my mother she sobs to us, complete strangers driving by on a sun filled afternoon.

There are frangipani trees in bloom along the shore of the tsunami zone.

I am in quiet awe.

Of the father who surveys the nothingness that once was his store and no less than 'three fale palagi'. From the shade of his tarpaulin hut. And tells me. 'It is like I am at the beginning all over again. When I was just starting out and we had nothing. All our hard work is gone. But none of it matters. My wife and children are alive. We are still here together. Possessions come and go. They mean nothing. I thank God for our lives.' He smiles to tell us about the turtles, the fish, the sharks that came in with the wave. 'We shared the water. Together we all searched for safety. For escape. We had no fear of each other.'

Smiling laughing teenagers out from school,making their way back to the makeshift camps. In a few weeks they will take their School C and PSSC final exams.

Houses still standing are newly painted. Glorious loud colours. Pumpkin orange walls with hot pink trim. Sunny yellow and fuchsia. Electric green and purple. Curlicues and twirls. Lacy curtains blowing in the breeze. Family photos with cracked glass carefully hang on the wall.

Children scuffle by the dusty roadside. Smiling.

There are frangipani trees in bloom along the tsunami shore.

They are beautiful.

Epilogue: The Future

Rudyard Kipling said – "If history were taught in the form of stories, it would never be forgotten." Throughout the course of my work gathering survivor stories, I have been amazed at our lack of historical knowledge of previous tsunami in Samoa. Even the very old were shaking their heads, "No, my grandparents never told me anything about such things…No, this has never happened before in Samoa…No I have never seen or heard of any such thing before – this is the first time this has happened to our country."

The science history tells us this is simply not true. In 1980, the International Tsunami Information center released a report compiled by George Carayannis – 'The Catalog of Tsunamis in the Samoan Islands.' For this report, they researched old newspapers and other publications stored in the Hawaii, New Zealand and American Samoa archives, as well as reports from the Apia Observatory. Beginning in 1837 and through to 1964, a total of *sixty tsunami events* were located.

I am not a scientist. But it doesn't take much more than common sense to make several observations. First - tsunami/*galu lolo*/*galu afi* are NOT rare or uncommon events for Samoa. And scanning through the Carayannis report shows, tsunami usually hit the same areas of our islands every time – Aleipata is a recurring favourite. Because of our location, because of the wild nature of Mafui'e, because of that deep chasm called the Tongan Trench – tsunamis have happened to Samoa with enough regularity that we should not be completely taken off guard when they do. Most tsunamis in the past 150 years have been small, barely noticeable. Some were mildly unsettling. And once in a rare while, they were immense and cataclysmic. But what a Samoa tsunami should *never* be, is a complete shock.

Second - previous tsunami, even the big ones, didn't have a deadly impact on us. Again and again the report read, *No lives were lost.* Yes, houses were wiped out. And vegetation was left brown and the land scarred. But there are no reports of deaths on the same scale as Sept 29/09. Why? I echo many others when I state the obvious…because back then we didn't live on the shore like we do now. And it was because back then, we knew enough about the natural warning signs to move. Run. Climb. As soon as the earth shook. Long and hard. A hundred years ago, we didn't need the Pacific Tsunami Warning Center to tell us a tsunami was coming. We didn't wait for a cell phone text to make us evacuate to higher ground. A hundred years ago, grandparents told their children stories about '*galu afi*'. About what they should do when there was a *mafui'e*. They told them about what they themselves had experienced. What they had seen. They told them what the ocean sounded like… "a hungry beast that roars…an animal that jumps out of the sea and runs like fire across the land…" And children would listen. Wide eyed. And remember.

What of now? The last major tsunami to hit Samoa was only fifty years ago. Some of us were not born then, but many others were. They walk among us. Where are their voices? What of their stories? I have interviewed over one hundred survivors living in the tsunami affected areas. Only eight people spoke of previous tsunami events. Many more survivors spoke of the media awareness campaign carried out by the DMO. Because of the ads on television, they knew about the signs of tsunami and what they should do in case of an earthquake. However, knowing and acting are two harshly different things. Too many times, survivors said – we remembered what the television told us but **we '*never thought such things could happen in Samoa. Tsunami only happen in other countries. Not here.***' Some even ran towards the beach to see what the wave looked like "because I never knew what this thing tsunami was.."

People tell me they won't need a book to remember the tsunami of 29/09. They say their children will never forget what happened that day either. And as I look into their eyes where the ocean still haunts, I believe them. Indeed, now that we have seen one, it is difficult to understand how Samoans ever 'forgot' what *galu afi* were in the first place. For those in particular who were there, who fought with the waves for survival, who saw children taken from them, the memory will always remain.

But what of our grandchildren? And their children? Who will make sure they remember? And why would you want them to remember? There is safety in the remembering and valuable life-saving messages in the stories of survivors. There can also be healing in the sharing of such stories. While there are those who do not wish to speak of their trauma, there are so many more who want to be heard.

This book can only scratch the surface of preserving community cultural memory. What is needed is a concerted effort to share, record and preserve the stories of the 29/09 tsunami survivors. Teachers can encourage children to write stories and draw pictures of what they saw and felt that day. Local media can seek out survivors and offer them a forum to have their stories video taped or transcribed. Individual village councils could organize to gather the stories of their area. Thanks to modern technology, recording stories is not difficult or expensive. Many cell phones can record, as do MP3 and iPod players. Utilize the techno-wisdom of teenagers and their gadgets. Then take the time to sit, ask, listen and record.

What can we do with the stories? In Hawaii there is a Pacific Tsunami Museum that has an extensive archive of video interviews of tsunami survivors. Shortly after 29/09, a team visited Samoa and did many video interviews with survivors to include in their archive. We too can certainly put our tsunami stories in Samoa's National Museum. And the Nelson Memorial Library here in Apia. But we can also put them into daily living. Curriculum planners can incorporate such stories into school resources. What better way to teach a unit on seismic geology or oceanography than to discuss the subducting fault lines that lie so close to us? How better to illustrate the need to preserve our coastal environment than by using the UNESCO 29/09 report that clearly shows how reefs and mangrove forests offered protection against the tsunami's wrath? And future disaster awareness campaigns could surely be made even more effective if they included the stories, the living words of survivors. Wouldn't you be more likely to listen, believe and act - if a disaster awareness advertisement included a clip of a survivor telling what happened to them? The same could be applied to cyclone safety messages. Where are the stories of those who experienced the hurricanes of Ofa and Val?

There are many clamouring for a 'better' national tsunami warning system to be put into place with sirens everywhere. Faster seismic measuring, quicker wave detection, better satellite hook-up, fancier lightning-like communication devices. And certainly there is merit in all those ideas. But any disaster management expert will tell you – the best way to be safe in any disaster, is to be individually prepared. To be aware, to be educated about the hazards, the risks and the strategies to deal with them. Have that escape route marked out for your family. Know what the signs of oncoming disasters are and what to do. Be ready to act on that knowledge. As we turn to modern technology to get ourselves better prepared for any and all disasters, let us not overlook the power that can be found in the sharing of our stories, the passing on of experiences through the generations. Because as our ancestors well knew – sometimes, the best way to teach anyone anything – is by telling them a story.

Go to Aleipata. Climb to the cliff tops of Saleapaga and Lalomanu that overlook the wild ocean. Stand on the stone remnants where once our people loved, lived, worked and played. Stand and look towards the faraway tectonic meet. Stand and imagine that long ago, when the earth rumbled and the waters swelled, a people could watch from the safety of their skyward home and marvel at the awesome power of 'galu afi'. And warn their children to be ever watchful of the sea.

Stand and remember, that on the 29th of September, 2009, the earth rumbled and the waters swelled. Stand and remember - and warn our children to be ever watchful of the sea.

The End.

Samoa Tsunami 2009 Deceased and Missing/Presumed Dead

List provided by the Samoa Police Department

Name	Sex	Age	Nationality	Location Found	District Found
Pili Poo	M	4	Local	Apolima Tai	
Tautua Silupe	F	78	Local	Haapai	Manono Tai
Laumoe Aleki	F	1	Local	Lalomanu	Aleipata
Gardenia Meredith	F	1	Local	Lalomanu	Aleipata
Tuese Peilua	M	1	Local	Lalomanu	Aleipata
Eliza Taamilo	F	2	Local	Lalomanu	Aleipata
Feagai fatuesi	F	2	Local	Lalomanu	Aleipata
Pritchard Siu	F	2	Local	Lalomanu	Aleipata
Seea Peilua	F	3	Local	Lalomanu	Aleipata
Malo Miriama Mikaele	F	3	Local	Lalomanu	Aleipata
Ardmore Meredith	M	3	Local	Lalomanu	Aleipata
Gali Anesone	F	3	Local	Lalomanu	Aleipata
Tiloni Sio Pati	M	3	Local	Lalomanu	Aleipata
Gwenalyn Taufua	F	4	Local	Lalomanu	Aleipata
William Tau Faaaliga	M	4	Local	Lalomanu	Aleipata
Aliceannie Meredith	F	4	Local	Lalomanu	Aleipata
PelesasaEtimani	M	4	Local	Lalomanu	Aleipata
Malo Vai	M	4	Local	Lalomanu	Aleipata
Maliumai Anetone	F	5	Local	Lalomanu	Aleipata
Ronaldo Aleni	M	5	Local	Lalomanu	Aleipata
Jayson Siu	M	6	Local	Lalomanu	Aleipata
Lauti Sio	M	8	Local	Lalomanu	Aleipata
Moanalei Sarah Long	F	9	Local	Lalomanu	Aleipata
Nifo Siu	F	10	Local	Lalomanu	Aleipata
Leueta Lesa	F	20	Local	Lalomanu	Aleipata
Anita Sulusi Fetu (Nuualiitia)	F	30	Local	Lalomanu	Aleipata
Anesone Tafia	M	33	Local	Lalomanu	Aleipata
Teleise Aiesi	F	39	Local	Lalomanu	Aleipata
Logotaeao Faaaliga	M	40	Local	Lalomanu	Aleipata
Siimoa Taufua Lauina	F	60	Local	Lalomanu	Aleipata
Lua Tavale	F	65	Local	Lalomanu	Aleipata
Mafatua Su'a	F	67	Local	Lalomanu	Aleipata
Malo Uili Fanene	F	70	Local	Lalomanu	Aleipata
Vinesa Toa	F	80	Local	Lalomanu	Aleipata
Taufua Leifi	M	97	Local	Lalomanu	Aleipata
Shanna Faith Lanu	F	1.9 months	Local	Lalomanu	Aleipata

Sili Taufua	M	11 months	Local	Lalomanu	Aleipata
Frazer Faaleaga	M	2 months	Local	Lalomanu	Aleipata
Etimani Junior Taufua	M	9 months	Local	Lalomanu	Aleipata
Lafotua Aiesi	F	N/A	Local	Lalomanu	Aleipata
Sara Amosa	F	N/A	Local	Lalomanu	Aleipata
Mauapi Amosa	M	N/A	Local	Lalomanu	Aleipata
Sue Taufua	F	N/A	Local	Lalomanu	Aleipata
Sieni Ropeti	F	N/A	Local	Lalomanu	Aleipata
Paepae Faia Toafa	F	N/A	Local	Lalomanu	Aleipata
Sapi Pelesia	F	N/A	Local	Lalomanu	Aleipata
Mauapi Mafoe Seuala	M	N/A	Local	Lalomanu	Aleipata
Matala Taufua	F	N/A	Local	Lalomanu	Aleipata
Lote Tuese	F	N/A	Local	Lalomanu	Aleipata
Palenapa Moeaia	M	N/A	Local	Lalomanu	Aleipata
Aneanea Anetone	F	N/A	Local	Lalomanu	Aleipata
Gali Tafia	F	N/A	Local	Lalomanu	Aleipata
Baby Tsunami	F	N/A	Local	Lalomanu	Aleipata
Falevalu Segifili	F	9	Local	Leauvaa	
Lutia Faapoi	F	2	Local	Lepa	Lotofaga
Losivale Faapoi	F	10	Local	Lepa	Lotofaga
Puni Afoa	M	67	Local	Lotofaga	Safata
Abish Schwalger	F	1	Local	Malaela	Aleipata
Teancum Charles Schwalger	M	2	Local	Malaela	Aleipata
Joseph Purcell	M	4	Local	Malaela	Aleipata
Faapopo Toeuli	F	40	Local	Malaela	Aleipata
Metita Sui (Koreti) Schwaiger	F	62	Local	Malaela	Aleipata
Soaai Jerry Schwaiger	M	72	Local	Malaela	Aleipata
Saeni Ioana Purcell	F	78	Local	Malaela	Aleipata
Maupenei Tofilau	F	1	Local	Poutasi	Falealili
Silvia Eteuati	M	1	Local	Poutasi	Falealili
Nonumaifele Tofilau	M	3	Local	Poutasi	Falealili
Amataga Tiotio	M	11	Local	Poutasi	Falealili
Tui Joe Annandale	F	40	Local	Poutasi	Aleipata
Logo Letone Uili	F	47	Local	Poutasi	Falealili
Vaimoana Faasavalu Fiu	F	67	Local	Poutasi	Falealili
Filisi Tavita	M	11 months	Local	Poutasi	Falealili
Anna Schafaussen	F	N/A	Local	Poutasi	Falealili
Siaea Areta	F	1	Local	Saaga	Siumu
Sima Okei Sepelini	F	3	Local	Saaga	Siumu
Pula Okei Sepelini	F	30	Local	Saaga	Siumu
Tauavaga Tupuola	F	84	Local	Saaga	Siumu
Peter Leitu	M	31	Local	Safatoa	Lefaga
Manino Faaaliga	F	2	Local	Saleapaga	Aleipata

Iefata Sa'u	M	2	Local	Saleapaga	Aleipata
Aneti Luafitu	M	2	Local	Saleapaga	Aleipata
Jay Jay Ulugia	M	2	Local	Saleapaga	Aleipata
Togafalea Alesana	F	3	Local	Saleapaga	Aleipata
Kapenata Viiga	M	3	Local	Saleapaga	Aleipata
Alema Tofu	M	3	Local	Saleapaga	Aleipata
Marilyn Ulugia	F	3	Local	Saleapaga	Aleipata
Ana Iulai	F	5	Local	Saleapaga	Aleipata
Precious Malaga	F	5	Local	Saleapaga	Aleipata
Rachel Leuelu	F	5	Local	Saleapaga	Aleipata
Willie Leio Taamu	M	5	Local	Saleapaga	Aleipata
Aloalo Sao	M	6	Local	Saleapaga	Aleipata
Niusilani Mania Agalenuu	F	30	Local	Saleapaga	Aleipata
Naiuli Faaaliga	F	30	Local	Saleapaga	Aleipata
Tini Suafai	M	33	Local	Saleapaga	Aleipata
Alavina Aiotaota	F	35	Local	Saleapaga	Aleipata
Simone Anae	M	51	Local	Saleapaga	Aleipata
Taua Sagale	F	60	Local	Saleapaga	Aleipata
Sui Mata Esera	F	66	Local	Saleapaga	Aleipata
Salamina Talosaga	F	73	Local	Saleapaga	Aleipata
Maka Esera	M	79	Local	Saleapaga	Aleipata
Otto Leuluaiailii	M	31	Local	Saleaumua	Aleipata
Tusipepa Hanipale	F	42	Local	Saleaumua	Aleipata
Ulalia Siatualauvao Lauvai	F	60	Local	Saleaumua	Aleipata
Milo Soo	F	76	Local	Saleaumua	Aleipata
Milo Too Paipa	F	76	Local	Saleaumua	Aleipata
Tauaipolu Purcell	F	N/A	Local	Saleaumua	Aleipata
Selau Farani	F	21	Local	Samusu Tai	Aleipata
Ranui Simanu	M	N/A	Local	Satalo	Falealili
Satelite ?	F	1	Local	Satitoa	Aleipata
Junior Livigisitone	M	2	Local	Satitoa	Aleipata
Quezon Lesa	M	3	Local	Satitoa	Aleipata
Taimane Tapuloa	F	4	Local	Satitoa	Aleipata
Mirimama Faletoi Fatu	F	25	Local	Satitoa	Aleipata
Fili Tialeaiga	F	31	Local	Satitoa	Aleipata
Sivaisua Livigisitone	F	32	Local	Satitoa	Aleipata
Sefulu Taamu	F	37	Local	Satitoa	Aleipata
Asipau Faofua Alataua	F	60	Local	Satitoa	Aleipata
Faanunu Talapa	F	106	Local	Satitoa	Aleipata
David Sootaga	M	7 months	Local	Satitoa	Aleipata
Merina Lesa	F	N/A	Local	Satitoa	Aleipata
Maria Tinei	F	N/A	Local	Satitoa	Aleipata
Lafitaga Suafai	F	2	Local	Unknown	
Siimamao Suani	F	N/A	Local	Unknown	
Sesa Toomalatai	F	N/A	Local	Vailoa	Aleipata
Hatonaina Lauvai	M	2	Local	Vaovai	Falealili

Vaisigano Lauvai	M	3	Local	Vaovai	Falealili
Rosa Lafaua Vili	F	3	Local	Vaovai	Falealili
Masela Lafaua	F	43	Local	Vaovai	Falealili
Isabel Ana Pinheiro Da Silva	F	41	Brazil	Lalomanu	Aleipata
Clea Salabert Wykes	F	6	Australia	Lalomanu	Aleipata
Vivian Hodgins	F	56	Australia	Lalomanu	Aleipata
Maree Ann Blacker	F	50	Australia	Maninoa	Siumu
Dimitry Anlaga Kikhter	M	18 months	Australia	Saleapaga	Aleipata
Lieselore Rock	F	N/A	Germany	Lalomanu	Aleipata
Lolani Ikenasio Mania-Tevaga	M	N/A	Unknown	Lalomanu	Aleipata
Petria Martin	F	N/A	New Zealand	Lalomanu	Aleipata
Rebecca Martin	F	N/A	New Zealand	Lalomanu	Aleipata
Mary Anne White	F	N/A	New Zealand	Lalomanu	Aleipata

Missing/Presumed dead	**Sex**	**Age**	**Nationality**	**Village/Country**
Aleki Vai Mikaele	M	1	Local	Lalomanu
Alfie Cunliffe	M	2	New Zealand	New Zealand
Savelio Taeao	M	3	Local	Saleapaga
Faatamalii Shen Yi	M	3	Local	Saleapaga
Jamie Viliamu	M	3	Local	Saleapaga

American Samoa List of Dead.

1. Aterei Taua'a
2. Atuama-Salavea Fuatino
3. Billie Leota
4. Columbus Sauiluma (missing)
5. Daniel Peter Olszewski
6. Elena Salavea
7. Eneliko Michelle
8. Eveline Charlotte Seumanu
9. Eveline Tuaato
10. Fa'atamali'i Fanene Lea'e So'oto
11. Fofoga Timoteo
12. Godafreda Palma
13. Helen Taumaloto Seui
14. Kapisi Sima
15. Lauofo Folalela
16. Leasiolagi Aienamu
17. Lee In Shengi
18. Li Chen
19. Mija Shin
20. Nguyen Van Tho
21. Palolo Viena
22. Perise Sula
23. Pita Tanielu
24. Sabrina Seevaetasi
25. Siasaga Tufu
26. Sin Ah Kang
27. Sooalo-Mua Faaumaolagi
28. Taua Liutolo
29. Tuaopepe Aiailua Joe
30. Tuluiga Sauiluma
31. Tusitala Matanana
32. Ulisese Vaijoresa Fitiao
33. Wei Hong Min

Tonga List of Dead

1. Losalio Lefai
2. Sefo Lefai
3. Toni Lefai
4. Tina Lefai
5. Kalolo Kivalu
6. Lesina Tupouto'a
7. Tu'ulomia Tavake
8. Heneli Losalu
9. Lupe Pukenga

People

The stories/ reflections/comments of those listed in italics were obtained via email or news media.
Unless otherwise indicated all are tsunami survivors.

NAME	RESIDENCE	
Annandale, Sose	Siumu	Manager, Sinalei Reef Resort
Annandale, Tuatagaloa Joe	Poutasi	Owner, Sinalei Reef Resort
Afatasi, Tofilau	Poutasi	
Aviga, Henrietta	Thursday Is., Australia	Nurse
Apelu, Sili	Lalomanu	Owner, Taufua Beach Fales
Apelu, Faafetai Taufua	Lalomanu	Owner, Taufua Beach Fales
Ale, Sina	Lalomanu	
Afei, Maka	Vaipoa, Niuatoputapu	Tonga
Anderson, Matt	Australia	High Commissioner to Samoa
Adye, Mike	New Zealand	Holiday, Lalomanu
Adye, Sera	New Zealand	Holiday, Lalomanu
Ansell, Graham	New Zealand	Holiday, Saleapaga
Ansell, Diann	New Zealand	Holiday, Saleapaga
Ansell, Matt	New Zealand	Holiday, Saleapaga
Ansell, Tia	New Zealand	Holiday, Saleapaga
Ansell, Michaela	New Zealand	Holiday, Saleapaga
Ansell, Conor	New Zealand	Holiday, Saleapaga
Bourassa, Stephen	American Samoa	Visitor, Pago Harbor
Bell, Lina	Siumu	Manager, Coconuts Resort
Blacker, John	Australia	Holiday, Coconuts Resort
Barron, Jill	New Zealand	Holiday, Lalomanu
Bilkey, Caroline	New Zealand	High Commissioner to Samoa
Belford, James	Poutasi	Nurse, Poutasi Hospital
Booth, Chris	Vavau	Owner, Seabreeze Resort
Booth, Wendy	Vavau	Owner, Seabreeze Resort
Browning, Diana	New Zealand	Volunteer, Habitat for Humanity.
Bebrouth, Rob	Australia	Holiday, Lalomanu
Buckingham, Peter	Tafitoala	
Brown, Gillian	Australia	Holiday, Lalomanu
Borrero, Dr.Jose	New Zealand/USA	Tsunami scientist
Chague-Goff, Dr. Catherine	Australia	Geologist/geochemist, UNESCO-IOC International Tsunami Survey Team.
Cederman, Brent	New Zealand	Holiday, Lalomanu

Cederman, Dianne	New Zealand	Holiday, Lalomanu
Chapman, Wayne	New Zealand	Volunteer, HFH
De Souza, Tina	New Zealand	Aid organizer for Lalomanu
Davidson, Sarah	New Zealand	Holiday, Lalomanu
Dean, Dr. Stanley	Samoa	General manager, National Health Services
Dolphin, David	New Zealand	Deputy High Commissioner to Samoa
Elisaia, Aurora	Apia	Poet
Elisaia, Gardenia	Apia	Samoa Tourism Authority
Emau, Solonaima	Saleapaga	Principal, Saleapaga Primary School
Epa, Vaosa	Samoa	Secretary to Cabinet.
Fatuesiafioga, Tufaiga	Lalomanu	
Fa'apale, Avega To'oto'o	Poutasi	
Falaniko, Iuliano	American Samoa	Customs Officer, K9 Division, Tafuna Airport
Fatialofa, Sarai	Lepa	
Faleone, Sinapo	Satalo	
Fua'ava, Tuilaepa	Manono	
Fua'ava, Afuafu	Manono	
Fa'amatala, Steph	Australia	Holiday, Sa'anapu
Fonoti, Luaoa Taipisa	Vaovai	
Fruean, Pepe Christian	Apia	CEO, Digicel Samoa
Fitiao, Taitasi	American Samoa	
Fakaosi, Leone	Falehau, Niuatoputapu	Tonga
Fakaosi, Sefilino	Falehau, Niuatoputapu	Tonga
Glew, Becky	UK/New Zealand	Holiday, Lalomanu
Gogola, Thomas	United Kingdom	Holiday, Siumu
Greenwell, Darren	New Zealand	Visitor, Siumu
Hunt, Beth	Apia	Principal, AhMu Academy
Hamilton, Rebecca	New Zealand	Holiday, Lalomanu
Hui, Vatau	Falehau, Niuatoputapu	Tonga
Hamilton, Jana	New Zealand	Oxfam
Hellesoe, Lucia	Apia	LDS Institute Office
Hemopo, Ngatau	New Zealand	Holiday, Saleapaga
Hemopo, Suzanne	New Zealand	Holiday, Saleapaga
Hemopo, Teone	New Zealand	Holiday, Saleapaga
Hemopo, Maia	New Zealand	Holiday, Saleapaga
Hemopo, Margaret	New Zealand	Holiday, Saleapaga
Hemopo, Jordan	New Zealand	Holiday, Saleapaga
Hodgins- May, Carla	Australia	Aid relief organizer
Hodgins-May, Stephenie	Australia	Aid relief organizer
Hill,Tony	Apia	Commissioner, FESA, Samoa
Iuni, Peniana	Lalomanu	

Jones, Larry	American Samoa	Visitor, Pago Harbor
Jaffe, Bruce	USA	USGS Science team
Kitiona, Seuala Taua	Lalomanu	Minister of Agriculture and Fisheries, Samoa
Keil, Hans Joachim	Apia	Member of Parliament
Khouzame, Richard	New Zealand	Aid organizer for Manono
Kini, Vaomalo	Apia	Laboratory, National Hospital
Keil, Francis	American Samoa	
Lumsden, Neil	USA	Visitor, Siumu
Lelea, Heimuli	Tonga	Malaela, Missionary, LDS Church
Leota, Maiava Dallas	American Samoa	Manager, Tafuna Airport
Leota, Tasi	American Samoa	
Lynn, Rachel	USA	Staff, Coconuts Resort
Lewis, Madeline	New Zealand	Holiday, Vavau
Lewis, Tony	New Zealand	Holiday, Vavau
Laufatu, Lualei	Vaovai	
Lefai, Kalala	Hihifo, Niutoputapu	Tonga
Losalu, Sulifa	Hihifo, Niuatoputapu	Tonga
Losalu, Lomano	Hihifo, Niutaoputapu	Tonga
Losalu, Uluaki	Hihifo, Niuatoputapu	Tonga
Langi, Uikelotu	Falehau, Niuatoputapu	Tonga
Leasiolagi, Emmanuel	USA	Aid relief organizer
Leasiolagi, Shammah	USA	Aid relief organizer
Laufatu, Fa'aumu	Vaovai	
Laupu'e, Tipaula	Apia	Deputy Commissioner, FESA
Laufiso, Tua	Lalomanu	
Lui, Faletaulupe	Saleaumua	
Lata, Suliane	Saleaumua	
Lilio, Evelyn	American Samoa	Principal, South Pacific Academy
Lefai, Fokikovi	Hihifo, Niuatoputapu	Tonga
Letau, Nofo	Lalomanu	
Lautusi, Avaia Tuilaepa	Lalomanu	Nurse, Lalomanu Hospital
Lesa, Melody	Satitoa	Nurse, Lalomanu Hospital
Logotaeao, Fatima	Saleapaga	
Lal, Matt	Apia	Teacher, Maluafou College
Mautu, Falima	Satitoa	
Matalavea, Dr. Ben	Apia	Lalomanu Hospital
Malaki, Leu	Saleapaga	
Morton, Allan E.	Arizona, USA	Tonga tsunami researcher
Mikaele, Ave	Lalomanu	
Meredith, Daniel	Apia	Volunteer. LDS tsunami house project.

Mikaele, Mamea 'Mika'	Lalomanu	
Mini, Malakai	Falehau, Niuatoputapu	Tonga
Masoa, Simao	American Samoa	Teacher, Poloa School
Manu, Maulolo	Falehau, Niuatoputapu	Tonga
Maia, Otalose	Falehau, Niuatoputapu	Tonga
Meauma, Yolanda	American Samoa	
Mortenson, Apulu Veronika	American Samoa	Volunteer, Samoa Red Cross
McGeorge, Kirk	USA	Visitor, Pago Harbor
Manuleleua, Malia	New Zealand	Writer & Counselor, Psycho-Social Response team.
Molloy, Matt	New Zealand	District Health Protection Officer, NZ Disaster Response team.
Malua, Lavasa	Apia	CEO, Ministry of Natural Resources and Environment, Samoa
Miller, Wade	USA	Malaela, Missionary, LDS Church
Moors, Mark	Apia	President, US Veteran's Association.
Mauala, Tautala	Apia	Secretary-General, Samoa Red Cross
Matu'u Debra	American Samoa	
Meauma, Jonika	American Samoa	
Markowitz, Barry	USA	Photographer/journalist
Mauga, Petone	Apia	CO Nafanua Patrol Boat
Niumatapele, Lotolua	Lepa	
Niusila, Tina	Saleapaga	
Niuwenhiuzein, Vanessa	Apia	Volunteer, Samoa Red Cross
North, Rosemarie	New Zealand	Volunteer, International Red Cross
Nelson, Filomena	Apia	Principle Disaster Management Officer, Samoa
Onesi, Maleko	Hihifo, Niuatoputapu	Tonga
Olosepu, Vaelupe	Malaela	
Olosepu, Maele	Malaela	
Ofa, Toleafoa Ropati	American Samoa	
Petrini, Kevin	Apia	Aid relief organizer, 'Raw Shakti.'
Palmer, Terisa So'oto	American Samoa	
Potoi, Oselani	Saleaumua	
Porter, Liam	New Zealand	Aid relief organizer
Pol, Emily	New Zealand	Aid relief organizer
Palupe, Letiu Lee	Saleaumua	Owner, retreat.
Penn, Michelle	New Zealand	Aid relief organizer
Pearson, Kathleen	American Samoa	

Pearce, Wendy	New Zealand	Holiday, Lalomanu
Percival, Steve	Apia	Filmographer, 29/09 Tsunami documentary.
Puni, Dr. Teriki	Apia	Lalomanu Hospital
Peteru, Elena	Apia	Volunteer counselor, Pscho-social Response team, Samoa
Peteru, Andrew	Apia	Assistant CEO, Health Dept, Samoa
Popese, Tilasi	Matautu, Falealili	
Pritchard, Molesi	Saleapaga	
Pritchard, Silaumea	Saleapaga	
Purcell, Amy Upu	Malaela	
Petelo, Junior	Saleaumua	
Petelo, Aiga	Saleaumua	
Romeo, Rita	Lalomanu	
Ropati, Leaitutulia	Australia	Aid relief organizer
Reeves, Sara	Apia	Teacher, Wesley College
Reeves, Cale	Apia	Teacher, Wesley College
Rowlands, Claire	Australia	Holiday, Lalomanu
Rowan, Dr. Robert	New Zealand	Surgeon, NZ Emergency Response team.
Rankin, Daniel	Australia	Holiday, Lalomanu
Sang Yum, Tafi	Saleapaga	
Schreiber, Arne	Germany	Honorary Consul to Samoa
Semo, Notuse	Saleapaga	
Sa'aga, Christine	Apia	NZ High Commission
Simpson, Vaughan	Apia	Volunteer, Emergency Services
Sagale, Fuailuma	Saleapaga	
Sagale, Fineaso	Saleapaga	
Susuga, Europa	Saleapaga	
Susuga, Peleiupu	Saleapaga	
Schwalger, Jared	Malaela	
Schwalger, Antonetta	Malaela	
Sua, Leota	Malaela	
Sasse, Nynette	Apia	CEO, Samoa Hotels Association
Solialofi, Sia	Vaovai	
Shen Yi, Iota	Saleapaga	
Salapo, Peka	Apia	Secretary, FESA
Siaosi, Saunoa	Lepa	
Steffany, Tereza	American Samoa	
Seager, Sydnee	USA	Aid relief organizer.
Tago, Lagi Samuelu	Vavau	
Tainano, Mekala	Lepa	

Tuatagaloa, Pauta	Poutasi	
Tauinaola, Henry	Poutasi	
Toala, Tasi	USA	Cargo worker, Salt Lake International Airport.
Taufua, Otele	Lalomanu	
Tavita, Tupuola Terry	Apia	Editor, 'Savali' Newspaper
Tuiolemotu, Val	USA	Aid relief organizer
Tapelu, Saleapaga	Saleapaga	
Talalotu, Rosa	American Samoa	
Tauiliili, Toetu	American Samoa	
Toelupe, Robert	American Samoa	
Tapaleao, Vaimoana	New Zealand	Reporter, NZ Herald
Toa, Maua	Lalomanu	
Tupou, Mine Epenesa	American Samoa	
Toelupe, Palantina	Apia	CEO, National Health Services, Samoa
Tuala, Lomitusi	Saleapaga	
Togafau, Pelesala	Saleaumua	Aleipata Secondary School
Tagaleoa, Iosefa	Saleaumua	
Toilolo, Sapua	Malaela	
Taiva'asele, Sally	Malaela	
Tuvale, Lesa Foti	Lalomanu	
Tumanuvao, Moimoi	Poutasi	
Tafiti, Marge Sunema	American Samoa	
Trafford, Sarah	New Zealand	Aid organizer for Lalomanu
Tominiko, Lesa	Manono	
To'omalatai, Ian	Vaovai	
To'omalatai, Melissa	Vaovai	
To'omalatai, Lauvai	Vaovai	
To'omalatai, Talavou	Vaovai	
Turton, Louise	New Zealand	Holiday, Litia Sini, Lalomanu
Utaulu, Tuluiga	Lepa	
Unaite, Leiataua	Manono	
Ulberg, Salaevalu	Namua Island	Resort Owner
Villamin, Jennifer	USA	Volunteer, Samoa Red Cross
Williams, Dr. Joe	New Zealand	Volunteer, 'Tsunami Samoa 2009' medical response team.
Vaikata, Ipeni	Niuatoputapu	Tonga
Vaea	American Samoa	
Vi'iga, Fuimaono	Saleapaga	
Vi'iga, Leua	Saleapaga	
Vivi, Savaia	American Samoa	
Vivi, Larisa	American Samoa	
Wooten, Sean	USA	Volunteer, Samoa Red Cross

Williams, Steve	Australia	Volunteer Paramedic, FESA
Wutzler, Abby	New Zealand	Holiday, Saleapaga
Whittaker, Mike	New Zealand	Holiday, Lalomanu
Wyberski, Ray	American Samoa	
Wyberski, Eta	American Samoa	
White, Andrew	New Zealand	Holiday, Lalomanu
Wendt, Dr. Felix	Apia	Director, LDS Humanitarian and Welfare Services, Samoa/Tonga
Woodward, Ngaire	New Zealand	Holiday, Lalomanu
Wilson, Christopher	New Zealand	Holiday, Lalomanu
Wilson, Maxwell	New Zealand	Holiday, Lalomanu
Wilson, Poppy	New Zealand	Holiday, Lalomanu
Youngmeyer, David	New Zealand	UNICEF Response team

Glossary

aiga - Nuclear or extended family.

auala galue – roads or tracks going inland to access plantations and settlements

fale – House.

faleo'o – Small dwelling.

faletele – Round house suitable for entertaining visitors and holding meetings.

fao – Small tree often found by the seashore.

fau – Hibiscus tree.

fue – General name given to creepers, vines and lianes.

fati – Of waves, to break or beat down. The surge and breaking of waves.

galu – Wave

galu lolo – Big or deep wave.

galu afi – Fire wave

lavalava – Clothing. A piece of material that can be wrapped around the body. A garment worn by both men and women.

matai – Titled head of a Samoan extended family, formally elected and honored as such.

mafui'e – Earthquake.

pulenu'u – Official representing central government in a village, mayor.

puletasi – Two piece garment worn by women with a floor-length wraparound skirt and co-ordinating top. Formal wear.

tamaligi – Name given to the tamarind tree and to other introduced trees with similar foliage.

taule'ale'a – Name given to a young man or youth. Or to those without chiefly titles.

ulu – Breadfruit tree

umu – Stone oven, consisting of a shallow cavity lined with stones on which a fire is lit and cleared away before the food is laid on the hot rocks.

Bibliography

Books

McCullough, David — *The Johnstown Flood.* Simon and Schuster Paperbacks, New York, USA. 1968

Meleisea, Malama — *Lagaga, A Short History of Western Samoa.* The University of the South Pacific, Fiji. 1987

Milner, G.B — *Samoan Dictionary.* Pasifika Press, Auckland, New Zealand.

Reports, Papers and Addresses

An Ode to 'People of the Sea'. Aurora Elisaia, Samoa. 2009.

Carayannis, George and Dong, Bonnie *The Catalogue of Tsunamis in the Samoan Islands.* UNESCO. 1980

Christmas Message, Address to the nation. Given by Tui Atua Tupua Tamasese Ta'isi, Head of State of Samoa, Apia, Samoa. December 2009.

Dudley, Chagues-Goff and Johnston *Capturing the Next Generation of Cultural Memories - The process of video-interviewing tsunami survivors.* Science of Tsunami Hazards, Vol.28. 2009

Eulogy given for Tuifeamalo Tuatagaloa Annandale. Given by Tui Atua Tupua Tamasese Ta'isi, Head of State of Samoa, Tanumapua Plantation. 29 September, 2009.

Eulogy given for Mary Ann White. Given by Brent and Diane Cederman, Raglan, NZ. October, 2009.

Exploring the Status of Tsunami Early Warning Systems in Samoa. Shaun Williams and Aliimalemanu F.M. Leavasa. Apia, Samoa.

Field Survey of the Samoa Tsunami of 29 September, 2009. Seisomological Research Letters, in Preparation, 13 January, 2010. Okal, Fritz, Synolakis, Borrero, Weiss, Lynett, Titov, Foteinis, Jaffe, Liu and Chan.

Habitat for Humanity Update Report, 10 November, 2009. New Zealand.

Jaffe, Bruce — *Surprises from the Deadly September 29, 2009 Samoa Tsunami.* USGS. December, 2009.

Joint Ministry Report regarding bodies of the deceased from the tsunami held at all morgues. Ministry of Health and Ministry of Police. 3 October, 2009.

Keynote Address to the New Zealand Families Commission Pasifika Families' Fono. Given by Tui Atua Tupua Tamasese Ta'isi, Head of State of Samoa. Auckland, New Zealand. 3 November, 2009.

Preliminary Analysis of the Sept 29, 2009 Samoa Tsunami Southwest Pacific Ocean computer animations of wave propagation. USGS.

Report on the FESA Response to the Samoa Tsunami. Commissioner Seve Tony Hill, Fire and Emergency Services Authority, Samoa. 10 October, 2009.

Report on Visit to Niuatoputapu, 11-17 November, 2009. Galumalemana Steven Percival, Paradigm Documentaries, Samoa.

Report of the EERI Learning from Earthquakes Project, Samoa Earthquake and Tsunami 29/09 USA. January, 2010.

Report of the Population and Housing Census 2006. Samoa Bureau of Statistics, July 2008.

Report of the Lalomanu Village Trust on the Lalomanu Reconstruction. Brent Cederman, Samoa. April 2010.

Report of the Psycho Social Response Team, Samoa. Ministry of Health and National Health Services. 29 October, 2009

Report to the National Disaster Council regarding arrangements for Samoan tsunami victims who died in American Samoa. Samoa Red Cross, Apia. 15 October, 2009.

Report on Samoa Relief Wave Tsunami Donation Program distributed January 1, 2010. Afoafouvale Mark Moors, US Veterans of Samoa.

Samoa's National Disaster Management Plan, 2006-2009. Approved by the National Disaster Council under Part III Section 9 of the *Disaster and Emergency Management Act 2006* on 2nd November, 2006.

Samoa Tourism Authority Annual Report 2007 – 2008. Apia, Samoa.

Tsunami Event Timeline, Preliminary Draft Report issued by USGS Authoritative Hypocenter. Compiled by UNESCO/IOC –NOAA ITIC. Issued 2 October, 2009.

Tsunami Recovery Plan, Niuatoputapu, Kingdom of Tonga October 2009. Compiled by the Government of Tonga with the assistance of the UNDP.

Trust Deed evidencing the Lalomanu Village Samoa Trust. Prepared by Morrison Kent Lawyers. Trustees Andrew White, Brent Cederman, John Baddeley of Raglan, NZ. 2009

UNESCO-IOC International Tsunami Survey Team Samoa (ITST) Interim Report of Field Survey, 14-21 October, 2009. Presented to the Government of Samoa, 26 October, 2009.

Magazines and Newspapers

Anchorage Daily News, USA
Blenheim Express, New Zealand
Fiji Times, Fiji
Gisborne Herald, New Zealand
Herald Sun, Australia.
Islands Business Magazine
Manukau Courier, New Zealand
Otago Daily Times, New Zealand.
Raglan Chronicle, New Zealand
Samoa Observer Newspaper, Samoa
Savali Newspaper, Samoa
Sydney Morning Herald, Australia
Standard-Examiner, USA
Sunday Star Times, New Zealand
Sunday Mirror, United Kingdom
The New Zealand Herald, New Zealand.
The Press, New Zealand
The Courier, Australia
Woman's Time Magazine, Samoa
Waikato Times, New Zealand

Websites for the following:

Coconuts Beach Resort
Event Polynesia
Habitat for Humanity, New Zealand
Islamic Relief USA
Jennifer Villamin Travel Blog
Matthew Leal, Samoa Blog
Matangitonga News Website
Mahina Tiare, Amanda Neal
Mormon Times, Deseret News, USA
New Zealand Civil Defence and Emergency Management
Oxfam, New Zealand
Pacific Eyewitness.Org
Raw Shakti Samoa
Schwalger Family Tsunami Fund
See Reeves Blog
Seabreeze Resort Vavau
Sinalei Reef Resort
Samoa Red Cross
Scientology Volunteer Ministers Program
The Vivien Hodgins Foundation
Taufua Beach Fales Facebook Fan Page
Tsunami Samoa 2009 Appeal
United States Embassy in Samoa
UNICEF New Zealand

Audiovisual Resources

Australia's Tsunami Response, Photo and video montage. Compiled by Australian High Commissioner to Samoa, His Excellency Matt Anderson. 2009.

'Ko Niuatoputapu Hili a e Peau Kula he 2009.' Documentary of Tonga tsunami survivor stories, Government of the Kingdom of Tonga.

Lalomanu Tsunami Memorial Church Service. TV3 Samoa. Filmed and reported, 3 October, 2009.

Post-Tsunami Footage, TV3 Samoa. Filmed and reported, 29 October, 2009.

Saleapaga Memorial Service for Missing Children, TV3 Samoa. Filmed and reported, 22 October, 2009

Tsunami i Samoa, Aso Lua Pogisa, Black Tuesday 29 September, 2009. Film documentary, ETV Productions, Samoa Visual Magazine, Samoa.

Tsunami National Memorial Service, Apia Park., TV3 Samoa. Filmed and transmitted live, 8 October, 2009.

Tsunami National Burial Service, Tafaigata Cemetery. TV3 Samoa. Filmed and transmitted live, 8 October, 2009.

Photographs Contributed by the following:

Australian Defence Force
Bharat Chovan
Christopher Wilson
Diann Ansell
Filomena Nelson
Kathleen Pearson
Lou Anderson
Matthew Leal
Mike Adye
New Zealand Defence Force
Petone Mauga
Pele Wendt
Rob Bebrouth
Rudy Bartley (WT Media)
Sarah Trafford
Tina DeSuza
Vaughan Simpson
Wendy Booth
Robert Toelupe

Appendix

1. The Gathering of Stories -

<u>Samoa and American Samoa interviews:</u>

Interviews of survivors in the Samoa disaster zone were conducted in people's homes, shelters and plantations. We had no set list of interviewee criteria. We drove through the zone and stopped to speak with people wherever they were. We would first introduce ourselves, explain the purpose for our visit, the purpose of the book and then ask if the person was willing to be recorded and share their story for inclusion. Only one person declined. The interviews were carried out in Samoan. They were unstructured conversations, recorded using an MP3 device. A list of focus topics was prepared beforehand as a guide but at all times, the interviewee was asked open-ended questions and invited to respond in any manner that they chose. Many spoke at length for 30min to an hour and a half about their experiences. Some would then refer us to other family members or friends in the same village, for further interviews. At the close of the interviews, permission was sought to take a photo of the interviewee for possible inclusion in the book.

Interviews of rescue/medical and aid workers were usually more structured. Appointments were made and an explanation of the purpose was given before the meeting was arranged. Written consent was sought and obtained from several government ministries/bodies for permission to interview their employees. One interviewee requested a copy of interview questions ahead of the scheduled appointment to assist with his preparation for the meeting.

I went to American Samoa three times for the purpose of obtaining survivor stories. It was decided that the stories of rescue/medical and aid workers there were beyond the scope of this project. Interviews of survivors in American Samoa were carried out primarily in English. Appointments were scheduled, the purpose of the meeting explained and permission sought for participation ahead of time – to better maximize my limited time during my visits there.

All of the recorded interviews were translated into English and transcribed by Noa and Roberta Tiatia.

Tonga interviews:

I did not travel to Tonga to interview survivors. The Tongan Government assisted with the project by supplying us with a television documentary of filmed survivor interviews. The documentary was in Tongan and the interviews were translated to English for me by Caroline Sopu. Additional material was used from interviews of three survivors carried out in Nukualofa Hospital – conducted by the 'Matangitonga' news website.

New Zealand, Australia, UK:

The assistance of the New Zealand and Australian High Commission consular offices in Apia was sought. They each forwarded to their list of tsunami contacts – my letter explaining the purpose of the book, an outline of the information I needed and an invitation to contact me via email should they wish to participate. Several survivors who had been on holiday here chose to respond and shared their stories and photographs. The Ansell family of Auckland, NZ was interviewed in person by Carmel Peteru. She also recorded and transcribed their interviews. The extended family had already compiled a written record of their experiences, a 'family scrapbook' and a copy of it was sent to me for inclusion.

Survivors were also contacted via the assistance of Diane Cederman of Raglan, NZ. While in Samoa, she had started collecting contact details and once back home, she went on to organize several gatherings for fellow survivors. She forwarded my letter to her contacts and several of them chose to share their stories and photographs via email.

Other survivors read about the project in the media and contacted me directly to offer their stories. Some had already shared their stories with overseas media and attempts were made to obtain their express permission to refer to their stories in the book. The different tsunami survey teams that visited Samoa after the disaster, also carried out survivor interviews and I have referred to several of their transcripts in the book.

As the purpose of this project has always been to ensure a record is kept of people's tsunami experiences for the benefit of both present and future generations, the transcripts of all interviews will eventually be collated in archival form and copies given to the Nelson Memorial Library in Apia and the Samoa Museum. Also, a selection of transcript excerpts with corresponding photographs is immediately accessible online at the 'Galu Afi' Samoa tsunami website.

2. The Writing of the Book –

I was asked to pattern the structure and format of this book along similar lines as that of '*The Johnstown Flood.*' written by David McCullough, which tells the story of the 1889 tragedy that killed more than 2,000 people. Using the 'documentary method' McCullough weaves together the personal accounts of survivors, rescuers, medical workers and others as well as drawing on various other written reports of the disaster. It was my aim to write a book that would 'read like a movie', that would give the reader the sense of 'really being there' and NOT like an academic essay. For the sake of better flow and ease of reading, it was decided to follow the example of McCullough's book and NOT use a detailed cross referencing system for source material. Rather, all quotes taken from interview transcripts or other written/audio material are written in quotation marks and all sources are listed in the 'People' section and the Bibliography. After the tsunami, I had written several journal/blog entries reflecting on the event, one of which was published in the 'Samoa Observer.' Throughout the project, I continued to keep a journal and some of these entries feature throughout the book. Hans Joachim Keil helped guide the direction of the writing with suggestions and by arranging for several key interviews.

Several people read initial drafts and excerpts of the book and gave valuable feedback. These included Dr. Felix Wendt, Prof. Albert Wendt, James Winegar, Celine Keil, and Sade Young. All editing and proofreading was done in co-ordination with Hans Joachim Keil.

About the author.

Lani Wendt was born and raised in Samoa. She received her tertiary education in the USA and New Zealand. She has a degree in English Literature and Womens Studies from Victoria University, Wellington and a Diploma in Education. Upon her return home to Samoa she worked as a secondary school English teacher and writer. Her short fiction has been published in Samoa, NZ, Australia and the United Kingdom. She won the National University of Samoa writing competition in 2002 with *'A Sister's Story'*. Her short story '*The Beast that came from the Sea.*' was a winner in the 2010 Commonwealth Short Story Competition. *'Pacific Tsunami- Galu Afi'* is her first book. Lani is married to Darren Young and they have five children.

